Optimal Control with Aerospace Applications

SPACE TECHNOLOGY LIBRARY

Published jointly by Microcosm Press and Springer

For further volumes:
http://www.springer.com/series/6575

James M. Longuski • José J. Guzmán
John E. Prussing

Optimal Control with Aerospace Applications

Published jointly by
Microcosm Press
El Segundo, California

Springer

James M. Longuski
Purdue University
Lafayette, IN, USA

José J. Guzmán
Orbital Sciences Corporation
Chantilly, VA, USA

John E. Prussing
University of Illinois
at Urbana-Champaign
Urbana, IL, USA

ISBN 978-1-4939-4917-5 ISBN 978-1-4614-8945-0 (eBook)
DOI 10.1007/978-1-4614-8945-0
Springer New York Heidelberg Dordrecht London

Softcover reprint of the hardcover 1st edition 2014

Printed on acid-free paper

Springer is part of Springer Science+Business Media (www.springer.com)

Preface

Optimal control theory has become such an important field in aerospace engineering that no graduate student or practicing engineer can afford to be without a working knowledge of it. Unfortunately, there is no modern text which begins from scratch to teach the reader the basic principles of the calculus of variations, to develop the necessary conditions step-by-step, and to introduce the elementary computational techniques of optimal control.

Our book assumes that the reader has only the usual background of an undergraduate engineering, science, or mathematics program, namely, calculus, differential equations, and numerical integration.

We assume no other knowledge. We do not require the reader to know what calculus of variations is, what necessary conditions mean, nor what a two-point boundary-value problem entails. It does not matter if the reader has never heard of the Euler-Lagrange theorem, the Weierstrass condition, Pontryagin's Minimum Principle, or Lawden's primer vector.

Our goal is to provide the reader with sufficient knowledge so that he or she cannot only read the literature and study the next-level textbook (such as Bryson and Ho's *Applied Optimal Control*) but also apply the theory to find optimal solutions in practice.

To accomplish the goals of this introductory text, we have incorporated a number of features as follows. Several theorems are presented along with "proof outlines" that favor a heuristic understanding over mathematical rigor. Numerous rigorous treatments are cited in the references and the book bibliography to support the reader's advanced studies.

In presenting the Euler-Lagrange theorem, we treat two different versions which appear in the literature. In the first method (followed by Bryson and Ho [1975]), we adjoin terminal constraints to the cost functional through the use of additional Lagrange multipliers. We refer to this approach as the "adjoined method" and note that it has become a sort of gold standard in the literature since the revised printing of Bryson and Ho's *Applied Optimal Control* in 1975. This approach leads to a form of the transversality condition which we refer to as the "algebraic form." In publications prior to 1975, a number of authors use an approach which we refer to as the “un-adjoined method” which does not adjoin the terminal constraints to the cost functional and hence does not introduce any additional multipliers to be solved. The un-adjoined method leads to a “differential form” of the transversality condition as given by Citron [1969], Hestenes [1966], Kenneth and McGill [1966], and Pierre [1969]. The un-adjoined method is particularly amenable to simple problems in which the terminal constraints are algebraically eliminated from the transversality condition. Each method has its strengths and weaknesses as observed by Citron [1969], who is one of the few authors who discuss both methods. Introducing the reader to both methods, with applications to current aerospace problems, is an important feature of the present text.

Throughout the book, we make use of the time-optimal launch of a satellite into orbit as an important case study, and we provide a detailed analysis of two examples: launch from the Moon and launch from the Earth. In the Moon-launch case, we assume constant acceleration (from thrusters), no drag, and uniform flat-Moon gravity. For Earth launch we include time-varying acceleration, drag from an exponential atmosphere, and uniform flat-Earth gravity. Appendices A and B provide MATLAB code to solve the resulting two-point boundary-value problems. In Appendix C, we also set up and provide MATLAB code to solve a geocentric low-thrust transfer problem.

A modern approach to Lawden’s primer vector theory is presented for optimal rocket trajectories. The important special cases of constant-specific impulse and variable-specific impulse are treated in detail.

An extensive annotated book bibliography lists the references we found most useful in the preparation of this text. These sources range from highly pragmatic application approaches (for engineers) to rigorous, theoretical treatments (for mathematicians). The second bibliography lists numerous papers and reports that demonstrate the vast range of related aerospace applications.

Finally, for the weary and the worried, we provide a few “Curious Quotations” (in Appendix D) to let the reader know that many great minds and renowned authors have expressed their own concerns, often in humble and humorous ways, about the vast challenges that the calculus of variations and optimal control present to all of us.

Lafayette, IN, USA — James M. Longuski
Chantilly, VA, USA — José J. Guzmán
Urbana, IL, USA — John E. Prussing

Acknowledgments

First and foremost I thank Dr. William F. Powers, who taught me his graduate course on optimization of space trajectories in the winter of 1974 when he was a professor of aerospace engineering at the University of Michigan. Dr. Powers has had an illustrious career as a leader in the application of control theory, as a professor, as a NASA consultant for the Space Shuttle, and as the vice president of research for the Ford Motor Company. He is a member of the National Academy of Engineering, the Royal Swedish Academy of Engineering Sciences, and the University of Florida Foundation Board of Directors.

This book is largely based on the lecture notes I obtained from Dr. Powers, and I have received his permission and encouragement to flesh out those notes to form the present text. I have extended the material over my years of teaching optimization of space trajectories. In my effort to teach and learn more about this fascinating and important subject, I have often marveled at what an excellent course Dr. Powers taught me. I hasten to add that any errors in the text are entirely my own—and should not reflect in any way on Dr. Powers.

I thank my colleagues, my doctoral students, the graduate and undergraduate students who took my course on Optimization in Aerospace Engineering (AAE 508), and the staff at Purdue University who helped in many significant ways to make the final manuscript as clear, accurate, and useful as possible.

In particular I thank Prof. Mohammad Ayoubi, Dr. Julia L. Bell, Ms. Nicole Bryan, Ms. Erin Calderwood, Ms. Jasmine Cashbaugh, Mr. Alan Castillejo Robles, Dr. K. Joseph Chen, Dr. Diane Craig, Mr. Michael Croon, Ms. Ashwati Das, Ms. Meredith Evans, Lt. Jennifer Fuka, Mr. Giacinto Genco, Mr. Filippo Genco, Mr. Seung Yeob Han, Mr. Adam Harden, Mr. Patrick Hayes, Mr. Evan Helmeid, Mr. Gregory Henning, Mr. Kyle Hughes, Mr. Junichi Kanehara, Ms. Aizhan Kinzhebayeva, Mr. Michael Kean, Dr. Kevin Kloster, Mr. Richard Lang, Mr. Frank Laipert, Mr. Karl Madon, Mr. Nicholas Makarowski, Dr. Belinda Marchand, Ms. Kaela Martin (nee Rasmussen), Dr. T. Troy McConaghy, Mr. Wesley McDonald, Prof. Ken Mease, Mr. James Moore, Ms. Bhuvi Nirudhoddi, Dr. Jeffrey Onken, Mr. José F. Paz Soldán Guerra, Dr. Anastassios Petropoulos, Ms. Lucie Poulet, Mr. Blake Rogers, Mr. Saverio Rotella, Mr. Sarag J. Saikia, Dr. Oleg Sindiy, Ms. Nissa Smith, Ms. Tracey Smith,

Mr. Christopher Spreen, Mr. Nathan Strange, Prof. Dengfeng Sun, Prof. Steven G. Tragesser, Prof. Brad Wall, Dr. Geoff Wawrzyniak, Ms. Rozaine Wijekularatne, Mr. Andy Wiratama, and Dr. Chit Hong "Hippo" Yam.

Special thanks go to Mr. Peter Edelman, Mr. Rob Falck, Dr. Joseph Gangestad, Mr. Kshitij Mall, Dr. George E. Pollock IV, and Prof. Marc Williams.

Thanks to the Purdue physics librarians, Ms. Lil Conarroe and Ms. Donna Slone, for their professionalism and their patience in helping me find important references. I also thank Ms. Karen Johnson, Ms. Jennifer LaGuire, and Ms. Vickie Schlene for their secretarial assistance.

I thank the editorial staff at Springer for their professional and unswerving support, in particular Dr. Harry (J.J.) Blom, Ms. Maury Solomon, and Ms. Nora Rawn. Thanks also to my friend and colleague, Dr. James R. Wertz of Microcosm Press for co-publishing our book.

Finally my most grateful thanks to my wife and best friend, Holly, for her unwavering support, her enthusiastic encouragement, and especially for her love.

I apologize to any contributor I should have acknowledged—please let me know who you are so I can thank you in a future edition.

J.M. Longuski

It is often difficult for the working engineer to use calculus of variations (COV) in the normal course of business. We hope that this book makes that process easier and faster for problems amenable to the application of COV. For example, for continuous thrust-steering systems encountered during powered ascent (launch), powered descent (landing), and low-thrust trajectories, COV can provide a great set of tools for the mission analyst. The material presented in this book and in the extensive list of references will guide the engineer in using the theory for practical problems. Perhaps more importantly, the practicing engineer could use this book to understand the software tools available for trajectory optimization and to provide these software tools with initial approximations that will speed up the convergence process. I would like to thank the professors that inspired me to pursue trajectory optimization: Dr. Kathleen C. Howell, Dr. James M. Longuski, Dr. Stephen J. Citron, and Dr. Martin J. Corless. I would like to thank a.i. solutions (Mr. Daryl Carrington, Mr. Jeff Dibble, Dr. Conrad Schiff, Dr. Ariel Edery, Dr. Peter Demarest, and Mrs. Laurie Mann (nee Mailhe)) and NASA Goddard Space Flight Center (Mr. David Folta, Mr. Mark Beckmnan, Mr. Steven Cooley, and Mr. Steven Hughes) all with whom I worked and discussed many challenging optimization problems. I would like to thank the Johns Hopkins Applied Physics Laboratory (APL) for providing funding under the Stuart S. Janney program (spring 2005). This funding allowed the development of some of the numerical work and illustrations in the book. Also at APL, I would like to thank Dr. Robert W. Farquhar, Dr. David W. Dunham, Mr. Peter J. Sharer, Mr. James T. Kaidy, Dr. J. Courtney Ray, Dr. Uday Shankar, and Dr. Thomas E. Strikwerda for invaluable discussions on mission design, trajectory optimization, and attitude control. While at APL, I also had a chance to work with Mr. Jerry L. Horsewood from SpaceFlightSolutions. Working with Jerry allowed me to understand some of the finer points of low-thrust trajectory optimization. Also, many thanks to Dr. Gregory Chirikjian and Dr. Joseph Katz from the Johns Hopkins University and to Dr. Chris D. Hall from Virginia Tech for part-time teaching opportunities at their respective institutions. I would like to thank Orbital Sciences and Dr. James W. Gearhart for providing a challenging work environment and great applied problems to work on. Finally, I'd like to thank my wife, Natalia, and my daughter, Sofia, for their love and patience.

J.J. Guzmán

First, my thanks to three giants in the field of spacecraft trajectory optimization: John V. Breakwell, Theodore N. Edelbaum, and Derek F. Lawden. They had profound influences on my education and my career.

Prof. Breakwell's pioneering 1959 article *The Optimization of Trajectories* launched the sustained exploration of that field in the USA. I was fortunate to know John and interact with him at conferences. Ted Edelbaum was my doctoral thesis research advisor, even though he did not have a PhD and was not at that time affiliated with MIT. He started me on an interesting and productive lifelong journey. Prof. Lawden's 1963 book *Optimal Trajectories for Space Navigation* was for many years the only book in the field and it inspired many to apply and extend his results through the 1960s, continuing to this day. I was fortunate to correspond with Derek after his retirement in England.

I also thank my colleagues in the field, my doctoral and master's thesis students, and all the students who took optimal control theory from me and those who took my second-level graduate course *Optimal Space Trajectories*, offered every other year since 1988 and continuing in my retirement. (What a great way to "flunk retirement!") Their observations and comments have been invaluable, and Chap. 10 in this book is based on that course and on my research over the years.

Finally, my heartfelt thanks to my wife, Laurel, currently Mayor of Urbana, Illinois, for her steadfast support over our 48 years together.

J.E. Prussing

About the Authors

James M. Longuski, PhD

After completing his doctoral dissertation, Analytic Theory of Orbit Contraction and Ballistic Entry into Planetary Atmospheres, at The University of Michigan in 1979, Dr. Longuski (lŏng-gŭs'-skē.) worked at the Jet Propulsion Laboratory as a Maneuver Analyst and as a Mission Designer.

In 1988 he joined the faculty of the School of Aeronautics and Astronautics at Purdue University where he teaches courses in astrodynamics, trajectory optimization, and spacecraft and mission design. Dr. Longuski is coinventor of a "Method of Velocity Precision Pointing in Spin-Stabilized Spacecraft or Rockets" and is an Associate Fellow of the American Institute of Aeronautics and Astronautics (AIAA).

Dr. Longuski has published over 200 conference and journal papers in the area of astrodynamics on topics that involve designing spacecraft trajectories that explore the Solar System and a new idea to test Einstein's General Theory of Relativity. He also coauthored several papers with Dr. Buzz Aldrin on a human Earth-to-Mars transportation system, known as the "Aldrin Cycler".

He has published two other books, *Advice to Rocket Scientists* (AIAA, 2004) and *The Seven Secrets on How to Think Like a Rocket Scientist* (Springer, 2007). In 2008 Dr. Longuski was inducted into Purdue University's Book of Great Teachers.

José J. Guzmán, PhD

Dr. Guzmán obtained his Aeronautical and Astronautical Engineering BS, MS and PhD degrees from Purdue University. He joined a.i. solutions in 2001 and was a member of the NASA's Wilkinson Microwave Anisotropy Probe (WMAP) trajectory design and maneuver team. WMAP is the first spacecraft to be stationed in the vicinity of the Sun-Earth L_2 point for its complete science mission duration. The trajectory included Earth phasing loops (highly eccentric orbits) with maneuvers that boosted the spacecraft to lunar orbit distance. A lunar flyby was then performed to insert the spacecraft into a Lissajous orbit.

From 2004 to 2009, Dr. Guzmán was a senior member of the technical staff at the Johns Hopkins University Applied Physics Laboratory (APL). At APL, he enjoyed working on low-thrust trajectories to comets and on lunar mission studies. He was also in the trajectory design and analysis team for the STEREO (Solar TErrestrial RElations Observatory) mission. STEREO is the first mission to use Earth phasing loops and lunar swingbys for two spacecraft simultaneously. The paper "STEREO Trajectory and Maneuver Design," written by Dr. David W. Dunham, Dr. Guzmán, and Mr. Peter Sharer in the Johns Hopkins APL Technical Digest, 28(2):104–125, won the 2009 Walter G. Berl Award for Outstanding Paper in the *APL Technical Digest*.

Dr. Guzmán is currently a principal senior engineer at Orbital Sciences Corporation. At Orbital he has worked on the mission design and planning for cargo missions to the International Space Station. Dr. Guzmán has also helped with several orbit transfers to the Geosynchronous belt and has provided expertise for proposals and new business opportunities.

Dr. Guzmán is a member of the American Astronautical Society and a senior member of the American Institute of Aeronautics and Astronautics. He has been a lecturer at The Johns Hopkins University and the Virginia Polytechnic Institute and State University and has served as associate editor for *The Journal of the Astronautical Sciences*.

John E. Prussing, ScD

Dr. Prussing received his SB, SM, and ScD degrees in aerospace engineering from MIT, culminating in his 1967 doctoral thesis, Optimal Multiple-Impulse Orbital Rendezvous. He accepted a postdoctoral position at the University of California at San Diego and in 1969 joined the faculty of aerospace engineering at the University of Illinois at Urbana-Champaign. His primary teaching and research areas are astrodynamics, optimal control theory, and optimal spacecraft trajectories.

Prussing is a Fellow of the American Institute of Aeronautics and Astronautics (AIAA), a Fellow of the American Astronautical Society (AAS), and has received the AIAA Mechanics and Control of Flight Award and the AAS Dirk Brouwer Award for his research contributions. His research has been referenced in 62 archival journals in English, and also in Russian, Chinese, French, and Portuguese journals.

In 1993 Prussing and Bruce A. Conway published their textbook *Orbital Mechanics* (Oxford University Press), which is available in 300 public and university libraries worldwide and is in its second edition.

Advance Praise for
OPTIMAL CONTROL WITH AEROSPACE APPLICATIONS

Optimal Control with Aerospace Applications fills a huge void between Derek Lawden's dated but classic 1963 text, *Optimal Trajectories for Space Navigation* and the exhaustive, and exhausting, treatment in *Applied Optimal Control* by Bryson and Ho. Modern in its approach and in its treatment of applications, this thorough but very accessible text is destined to become an instant classic.

Michael D. Griffin, Ph.D.
NASA Administrator 2005–2009
Author (with James R. French) of *Space Vehicle Design*

This book represents a new approach to the teaching of optimal control theory; it is much more accessible than extant texts but does not omit any of the important subjects. The critical theorems are accompanied with proofs but these are clearly presented and explained. The classic, historical examples are included but many new solved problems, especially relevant to space trajectory optimization, are also included. As a "bonus" it also has the most complete list of relevant books and journal articles that I've ever seen collected in one source.

Bruce A. Conway. Ph.D.
Professor of Aerospace Engineering
University of Illinois at Urbana-Champaign
Co-author of *Orbital Mechanics*
Editor of *Spacecraft Trajectory Optimization*

Optimal Control with Aerospace Applications is a clear and concise treatment of optimal control theory and its application to spacecraft trajectories. The authors use a conversational tone to successfully explain the challenging mathematics behind optimal control and therefore demystify the associated theorems and necessary conditions. This textbook updates the classic optimal-control examples (such as launch trajectories and orbit transfers) by presenting the practical issues associated with their numerical solution and by providing the MATLAB code to obtain their numerical solution. I found this textbook to be an excellent introduction to optimal control theory which I believe will serve the student as well as the practicing engineer.

Craig A. Kluever, Ph.D.
C.W. LaPierre Professor
Mechanical and Aerospace Engineering
University of Missouri-Columbia

An instant neo-classic on the calculus of variations for the rocket scientist! Scratch that. The optimization of space trajectories starts here!

I. Michael Ross, Ph.D.
Author of *A Primer on Pontryagin's Principle in Optimal Control*

This is a long-overdue text that comprehensively covers optimal control as applied to aerospace vehicle (i.e., aircraft and spacecraft) trajectories, replacing the combined classical treatises by Lawden, Vinh and Marec. The authors cover the essential topics and applications in this field with a unique combination of rigor and readability that is generally lacking in optimal control texts. I intend to use this book in my graduate-level optimal trajectories class.

Daniel J. Scheeres, Ph.D.
A. Richard Seebass Endowed Chair Professor
Department of Aerospace Engineering Sciences
The University of Colorado
Author of *Orbital Motion in Strongly Perturbed Environments: Applications to Asteroid, Comet and Planetary Satellite Orbiters*

This book is an introductory treatment of optimal control (particularly with regard to aerospace vehicles), presenting the essence of a highly mathematical subject in a simplified and easily understood manner. The book is clearly written, focuses on practical applications, and includes numerous examples. It can serve quite effectively either as a first textbook on aerospace optimal control for those who will later explore this field in greater depth or as the only textbook for those interested just in gaining some exposure to this area.

Lincoln J. Wood, Ph.D.
Dr. Wood is currently a Principal Engineer at the Jet Propulsion Laboratory, California Institute of Technology

Contents

Chapter 1

Parameter Optimization

1.1. Introduction

Two major branches of optimization are: parameter optimization and optimal control theory. In parameter optimization (a problem of finite dimensions, that is, where the parameters are not functions of time) we minimize a function of a finite number of parameters. We only provide an overview of parameter optimization in this chapter, since the main topic of this book is optimal control theory. Optimal control (a problem of infinite dimensions where the parameters are functions of time) seeks $\boldsymbol{x}(t)$, an n-vector, that minimizes something called a functional, which will be defined in Chap. 2.

Parameter optimization, the theory of ordinary maxima and minima, is based on calculus. In general, the (unconstrained) problem could be stated as:
Find:

$$\boldsymbol{x}$$

to minimize:

$$J = f(\boldsymbol{x}) \tag{1.1}$$

where J is the scalar cost function or index of performance and $\boldsymbol{x}$ is a constant n-vector.

If the x_i are independent and all the partial derivatives of f are continuous, then a stationary solution, $\boldsymbol{x}^*$, is determined by

$$\frac{\partial f}{\partial x_i} = 0, \quad i = 1, 2, 3, \ldots, n \tag{1.2}$$

Equation (1.2) is a necessary condition for an extremum (a maximum or a minimum). We note that f can be maximized by minimizing $-f$.

The stationary point $\boldsymbol{x}^*$ is a local minimum if the matrix formed by the components, $\frac{\partial^2 f}{\partial x_i \partial x_j}$ (evaluated at $\boldsymbol{x}^*$), is a positive-definite matrix, which provides a sufficient condition for a local minimum. To ensure that the matrix is well defined, all the second partials of f must be continuous.

J.M. Longuski et al., *Optimal Control with Aerospace Applications*,
Space Technology Library 32, DOI 10.1007/978-1-4614-8945-0_1,

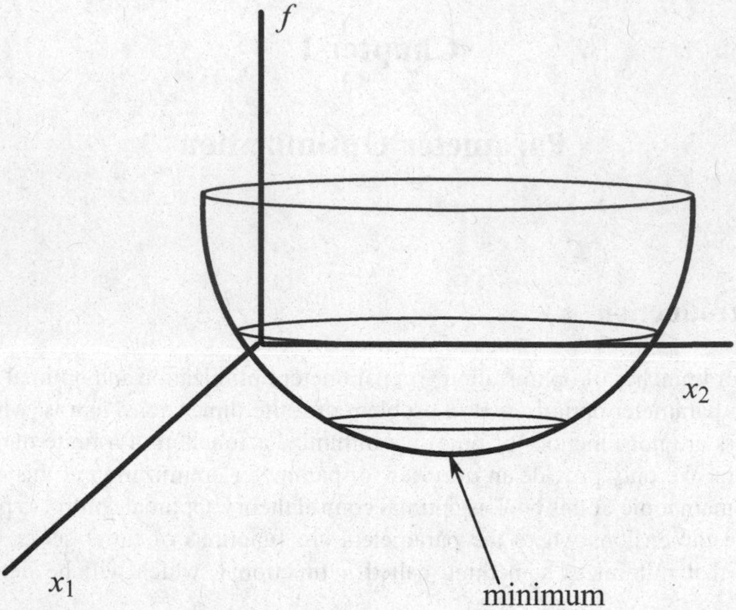

Figure 1.1. The minimum of a function where x is a 2-vector.

In Fig. 1.1 we illustrate an example where $\boldsymbol{x}$ is a 2-vector. In this case, for $\min f(x_1, x_2)$, the necessary and sufficient conditions are:

$$\frac{\partial f}{\partial x_1} = 0 \tag{1.3a}$$

$$\frac{\partial f}{\partial x_2} = 0 \tag{1.3b}$$

and

$$\begin{bmatrix} \delta x_1 & \delta x_2 \end{bmatrix} \begin{bmatrix} \frac{\partial^2 f}{\partial x_1 \partial x_1} & \frac{\partial^2 f}{\partial x_1 \partial x_2} \\ \frac{\partial^2 f}{\partial x_2 \partial x_1} & \frac{\partial^2 f}{\partial x_2 \partial x_2} \end{bmatrix}\Bigg|_{\boldsymbol{x}^*} \begin{bmatrix} \delta x_1 \\ \delta x_2 \end{bmatrix} > 0 \tag{1.4}$$

where δx_1 and δx_2 are infinitesimal arbitrary displacements (variations) from x_1^* and x_2^*. The inequality must hold for all $\delta x_1, \delta x_2 \neq 0$ (then by definition the matrix is *positive definite*). For the 2×2 matrix, Eq. (1.4) is satisfied if both of the leading principal minors are positive:

$$\frac{\partial^2 f}{\partial x_1 \partial x_1} > 0 \tag{1.5a}$$

$$\frac{\partial^2 f}{\partial x_1 \partial x_1} \frac{\partial^2 f}{\partial x_2 \partial x_2} - \left[\frac{\partial^2 f}{\partial x_1 \partial x_2} \right]^2 > 0 \tag{1.5b}$$

where the left-hand side of Eq. (1.5b) is the determinant of the matrix in Eq. (1.4).

We note that we can write the "variation":

$$\delta J = \frac{\partial f}{\partial x_1} \delta x_1 + \cdots + \frac{\partial f}{\partial x_n} \delta x_n \tag{1.6}$$

for an infinitesimal, arbitrary displacement $\delta \boldsymbol{x}$ and conclude that at a stationary point (i.e., Eq. (1.2)): $\delta J = 0$. In the general case when $\boldsymbol{x}$ is an n-vector, the matrix (formed by $\frac{\partial^2 f}{\partial x_i \partial x_j}$) is positive definite if all n of the leading principal minors are positive.

1.2. Parameter Optimization with Constraints

To include constraints in the optimization problem, it is necessary to describe them algebraically. For example, to describe the condition that a point (x_1, x_2) is constrained to lie on a circle of radius R and centered at the origin, we write

$$x_1^2 + x_2^2 - R^2 = 0 \tag{1.7}$$

This statement can be generalized as follows. The n variables $x_1, x_2, \ldots, x_n$ may be subject to certain relations, called constraints, of the form:

$$\begin{aligned} \phi_1(x_1, \ldots, x_n) &= 0 \\ &\vdots \\ \phi_m(x_1, \ldots, x_n) &= 0 \end{aligned} \tag{1.8}$$

with $m < n$ so there are $n - m$ independent variables. Of course if $m = n$ then the problem is so constrained that there is only one possible solution (or there may be no solution) and we have no optimization problem.

1.2.1. Lagrange Multipliers

The n variables $x_1, x_2, \ldots, x_n$ can be treated as if they are "independent" by introducing a set of m constants $\lambda_1, \lambda_2, \ldots, \lambda_m$, called Lagrange multipliers. With the Lagrange multipliers we can couple the constraints to the cost function. Explicitly, we consider the augmented function, F:

$$F(x_1, \ldots, x_n, \lambda_1, \ldots, \lambda_m) = J + \lambda_1 \phi_1 + \ldots + \lambda_m \phi_m \tag{1.9}$$

As long as the x_i satisfy the constraints ($\phi_j = 0$), a stationary value of J corresponds to a stationary value of F. We are led to a problem of unconstrained optimization, $\delta F = 0$, where:

$$
\begin{aligned}
\frac{\partial F}{\partial x_1} &= 0 \\
&\vdots \\
\frac{\partial F}{\partial x_n} &= 0
\end{aligned}
\tag{1.10}
$$

$$
\begin{aligned}
\frac{\partial F}{\partial \lambda_1} &= \phi_1 = 0 \\
&\vdots \\
\frac{\partial F}{\partial \lambda_m} &= \phi_m = 0
\end{aligned}
\tag{1.11}
$$

Thus, there are $n+m$ equations and $n+m$ unknowns (n x's and m λ's). Introducing m more unknowns (the λ_j) seems counterproductive. In principle we could solve the constraints $\phi_j = 0$ for m of the x_i, leaving only $n-m$ of the x_i to be solved for. However, the resulting equations—while fewer—are in practice usually more complicated than Eqs. (1.10) and (1.11), so it is easier to solve the $n+m$ equations.

Example 1.1 Application of Lagrange multipliers.

Determine the rectangle of maximum perimeter that can be inscribed in a unit circle. Let the center of the circle lie at $(x, y) = (0, 0)$ and locate the corners of the rectangle at $(\pm x, \pm y)$. The problem is to minimize

$$J = f(x, y) = -4(x+y) \tag{1.12}$$

subject to the constraint

$$\phi(x, y) = x^2 + y^2 - 1 = 0 \tag{1.13}$$

To solve the problem using the Lagrange multiplier method we construct the augmented function F:

$$
\begin{aligned}
F(x, y, \lambda) &= f(x, y) + \lambda\ \phi(x, y) \\
&= -4(x+y) + \lambda\ (x^2 + y^2 - 1)
\end{aligned}
\tag{1.14}
$$

The first-order necessary conditions are

$$F_x = \frac{\partial F}{\partial x} = -4 + 2\lambda x = 0 \tag{1.15}$$

$$F_y = \frac{\partial F}{\partial y} = -4 + 2\lambda y = 0 \tag{1.16}$$

and the constraint

$$F_\lambda = \frac{\partial F}{\partial \lambda} = x^2 + y^2 - 1 = 0 \tag{1.17}$$

where we introduce the subscript notation for partial derivatives. Equations (1.15)–(1.17) are easily solved to yield

$$x^* = y^* = \frac{\sqrt{2}}{2} \tag{1.18}$$

$$\lambda = 2\sqrt{2} \tag{1.19}$$

where Eq. (1.18) indicates that the rectangle of maximum perimeter is a square.

To check that our stationary solution $(x^*, y^*) = (\frac{\sqrt{2}}{2}, \frac{\sqrt{2}}{2})$ is a minimum we note that there are two variables and one constraint, leaving one free variable. Arbitrarily choosing y as the free variable, we calculate the constrained second derivative of our function (see Sect. 1.3 of Bryson and Ho [1975]):

$$\begin{aligned}\left(\frac{\partial^2 f}{\partial y^2}\right)_{\phi=0} &= F_{yy} - F_{yx}\phi_x^{-1}\phi_y - \phi_y\phi_x^{-1}F_{xy} + \phi_y\phi_x^{-1}F_{xx}\phi_x^{-1}\phi_y \\ &= 4\sqrt{2} - 0 - 0 + 4\sqrt{2} = 8\sqrt{2} > 0\end{aligned} \tag{1.20}$$

This constrained second derivative being positive (positive definite in the multi-dimensional case) satisfies the sufficient condition for a minimum of J, resulting in a maximum perimeter of $4\sqrt{2}$.

1.2.2. Parameter Optimization: The Hohmann Transfer (1925)

As an important example in parameter optimization, we present the Hohmann transfer. In 1925, Walter Hohmann introduced a two-impulse scheme to transfer a spacecraft from a lower circular orbit to a higher circular orbit. (In 1960, NASA published a Technical Translation, NASA TT F-44 of Hohmann's original work titled, "The Attainability of Heavenly Bodies." See Hohmann [1960].) In Soviet literature the transfer is sometimes referred to as the Hohmann-Vetchinkin transfer orbit to credit the Russian mathematician who presented lectures on the transfer in the early 1920s. (See Ulivi and Harland [2007].)

The assumptions Hohmann made were:

1. There is a central gravitational field that obeys Newton's law, $F = G(Mm)/r^2$, where F is the force of gravity, G is Newton's universal constant, M is the mass of the central body, m is the mass of the spacecraft, and r is the distance from the center of the central body to the spacecraft.

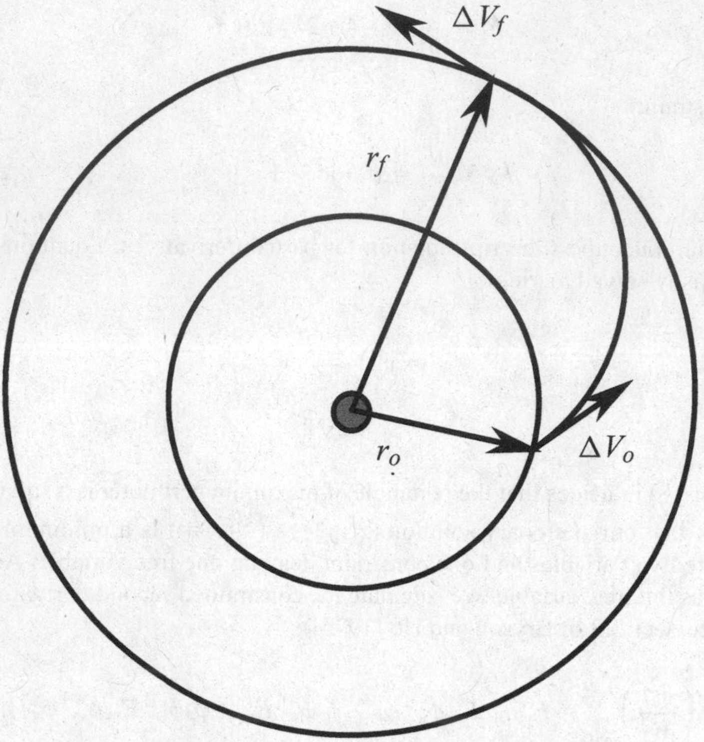

Figure 1.2. General two-impulse transfer geometry from one coplanar circular orbit to another. In this case ΔV_o and ΔV_f are not necessarily tangent to the orbits.

2. The thrust level is unlimited, therefore the change of velocity can be instantaneous. This is known as the impulsive ΔV assumption. Only two impulsive ΔV maneuvers are allowed; together they must minimize the propellant consumed during the orbit transfer.
3. The initial and final orbits are circular and coplanar.

Figure 1.2 depicts the two impulsive $\mathbf{\Delta V}$'s: the initial $\mathbf{\Delta V_o}$ at radial distance r_o and the final $\mathbf{\Delta V_f}$ at radial distance r_f. We note that Fig. 1.2 illustrates the general case, not the Hohmann transfer.

The $\mathbf{\Delta V}$'s could be oriented arbitrarily in space. A more general problem allows for multiple $\mathbf{\Delta V}$'s, which if extended to large numbers could approximate a continuous thrusting problem.

By fixing the number of impulses, the problem is reduced to one of parameter optimization. The Hohmann transfer is a two-impulse, time-free (i.e., unconstrained duration) transfer between circular, coplanar orbits. Therefore, the problem can be stated as:

Minimize:

$$J = |\mathbf{\Delta V_o}| + |\mathbf{\Delta V_f}| \tag{1.21}$$

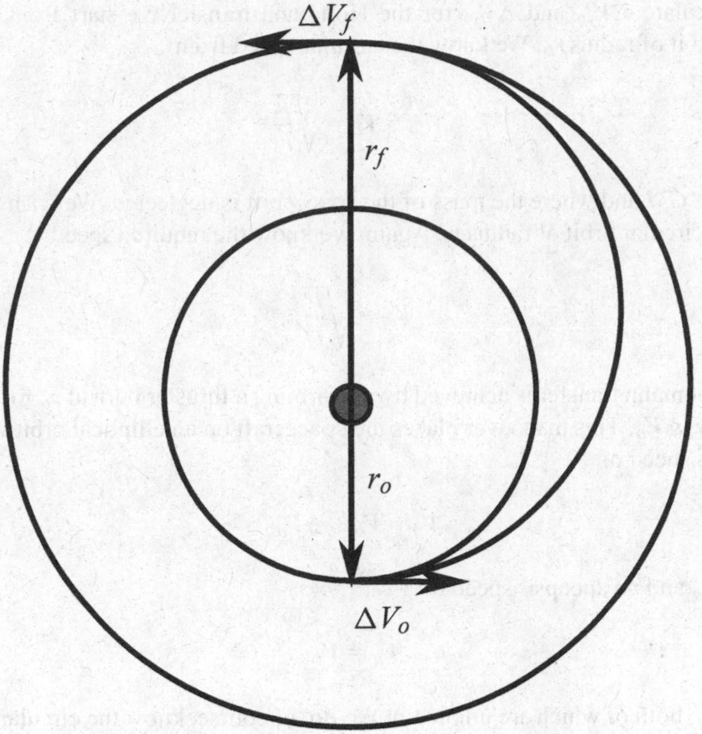

Figure 1.3. Hohmann transfer, in which ΔV_o and ΔV_f are tangential to the circular orbits at r_o and r_f, respectively, and are 180° apart.

subject to:

$$\phi_1(\Delta V_{xo}, \Delta V_{yo}, \Delta V_{xf}, \Delta V_{yf}) = V_{xf} = 0 \tag{1.22a}$$

$$\phi_2(\Delta V_{xo}, \Delta V_{yo}, \Delta V_{xf}, \Delta V_{yf}) = V_{yf} - V_{cf} = 0 \tag{1.22b}$$

where ΔV_{xo} and ΔV_{yo} are the components of the initial impulsive $\boldsymbol{\Delta V}$ along the radial and tangential directions, respectively. Likewise, the subscript f corresponds to the final impulsive $\boldsymbol{\Delta V}$. The velocity components V_{xo}, V_{yo}, V_{xf}, and V_{yf} are defined similarly; V_{co} and V_{cf} are circular speeds at r_o and r_f, respectively. Equations (1.22a) and (1.22b) merely state that the final orbit (of Fig. 1.2) must be circular. We note that minimizing the sum of the $\boldsymbol{\Delta V}$ magnitudes provides the minimum-propellant cost (as discussed in Prussing and Conway [2013]).

By parameter optimization, it can be shown (see Prussing and Conway [2013]) that J is minimized when

$$\Delta V_{xo} = \Delta V_{xf} = 0 \tag{1.23}$$

i.e., all impulses are tangential (as shown in Fig. 1.3).

To calculate ΔV_{yo} and ΔV_{yf} for the Hohmann transfer we start from an initial circular orbit of radius r_o. We know the circular speed from

$$V_{co} = V_o = \sqrt{\frac{\mu}{r_o}} \tag{1.24}$$

where $\mu = GM$ and where the mass of the spacecraft is neglected. We wish to end up on a final circular orbit of radius r_f. Again, we know the required speed

$$V_{cf} = \sqrt{\frac{\mu}{r_f}} \tag{1.25}$$

The Hohmann transfer is achieved by performing a thruster burn at r_o for a change of velocity, ΔV_o. This maneuver places the spacecraft on an elliptical orbit which has a periapsis speed of

$$V_p = V_o + \Delta V_o \tag{1.26}$$

at radius r_o and an apoapsis speed of

$$V_a = V_f \tag{1.27}$$

at radius r_f, both of which are unknown. We do, of course, know the circular speed of the final orbit from Eq. (1.25):

$$V_{cf} = \sqrt{\frac{\mu}{r_f}} = V_f + \Delta V_f \tag{1.28}$$

but V_f and ΔV_f are, so far, unknown. We note that when apoapsis is reached, a ΔV_f must be performed to circularize the orbit (at r_f). Our problem is to find ΔV_o and ΔV_f. We have two unknowns so we need two equations to solve for them.

The problem can be solved by determining the transfer orbit. To do this, we use two equations corresponding to the conservation of angular momentum and the conservation of total mechanical energy. From conservation of specific angular momentum we have:

$$r_o(V_o + \Delta V_o) = r_f V_f \tag{1.29}$$

so that

$$V_f = (r_o/r_f)(V_o + \Delta V_o) \tag{1.30}$$

From conservation of specific energy we have:

$$\frac{1}{2}(V_o + \Delta V_o)^2 - \frac{\mu}{r_o} = \frac{1}{2}V_f^2 - \frac{\mu}{r_f} \tag{1.31}$$

where the left hand side of Eq. (1.31) gives the energy at periapsis, while the right hand side gives the energy at apoapsis. Next, we substitute the expression for V_f, Eq. (1.30), into Eq. (1.31):

$$\frac{1}{2}(V_o + \Delta V_o)^2 - \frac{\mu}{r_o} = \frac{1}{2}\left(\frac{r_o}{r_f}\right)^2 (V_o + \Delta V_o)^2 - \frac{\mu}{r_f} \tag{1.32}$$

By solving for the unknown, $V_o + \Delta V_o$, in Eq. (1.32):

$$\frac{1}{2}(V_o + \Delta V_o)^2 \left(1 - \frac{r_o^2}{r_f^2}\right) = \frac{\mu}{r_o} - \frac{\mu}{r_f} \tag{1.33}$$

we obtain

$$V_o + \Delta V_o = \sqrt{2\left(\frac{\mu}{r_o} - \frac{\mu}{r_f}\right)\left(\frac{r_f^2}{r_f^2 - r_o^2}\right)} \tag{1.34}$$

Noting that

$$\left(\frac{1}{r_o} - \frac{1}{r_f}\right)\left(\frac{r_f^2}{r_f^2 - r_o^2}\right) = \frac{r_f}{r_o}\frac{1}{(r_f + r_o)} \tag{1.35}$$

we have, from Eq. (1.34):

$$V_o + \Delta V_o = \sqrt{2\mu\frac{r_f}{r_o}\frac{1}{(r_f + r_o)}} \tag{1.36}$$

Substituting for V_o from Eq. (1.24) into Eq. (1.36), we obtain the solution for ΔV_o:

$$\Delta V_o = \sqrt{\frac{2\mu r_f}{r_o(r_f + r_o)}} - \sqrt{\frac{\mu}{r_o}}$$

$$= \sqrt{\frac{\mu}{r_o}}\left(\sqrt{\frac{2r_f}{r_f + r_o}} - 1\right) \tag{1.37}$$

We can find ΔV_f from Eq. (1.28), but first we must find V_f from Eqs. (1.30) and (1.36):

$$V_f = \frac{r_o}{r_f}(V_o + \Delta V_o) = \frac{r_o}{r_f}\sqrt{2\mu\frac{r_f}{r_o}\frac{1}{(r_f + r_o)}} \tag{1.38}$$

so that

$$V_f = \sqrt{2\mu \frac{r_o}{r_f} \frac{1}{(r_f + r_o)}} \tag{1.39}$$

From Eqs. (1.28) and (1.39):

$$\begin{aligned} \Delta V_f &= \sqrt{\frac{\mu}{r_f}} - V_f \\ &= \sqrt{\frac{\mu}{r_f}} - \sqrt{2\mu \frac{r_o}{r_f} \frac{1}{(r_f + r_o)}} \end{aligned} \tag{1.40}$$

So we have the solution for ΔV_f:

$$\Delta V_f = \sqrt{\frac{\mu}{r_f}} \left(1 - \sqrt{\frac{2r_o}{r_f + r_o}} \right) \tag{1.41}$$

1.2.3. Extensions of the Hohmann Transfer (1959)

In the Hohmann transfer, only two impulses are applied. Can the total cost (ΔV) be reduced with more impulses? Edelbaum [1967] discusses extensions of the Hohmann transfer such as the bi-parabolic and bi-elliptic transfers and other minimum-propellant, impulsive transfers. In 1959 three articles appeared independently by Edelbaum, Hoelker and Silber, and Shternfeld which showed that if the radius ratio r_f/r_o is sufficiently large, other transfers exist which require less propellant than the Hohmann transfer.

These other transfers are based on the three-impulse, bi-elliptic transfer shown in Fig. 1.4. Now, in addition to the two ΔVs employed by the Hohmann transfer, we have an intermediate ΔV performed at a radial distance, r_i. The value of r_i is typically much larger than the outer circular orbit. We can imagine the limiting case where $r_i \to \infty$ so that $\Delta V \to 0$ to achieve any desired periapsis value r_f. This case is referred to as a bi-parabolic transfer.

Usually we assume that $r_f > r_o$ so that we have the following interpretation of Fig. 1.4:

1. ΔV_o is applied tangentially and forward (to increase the velocity).
2. ΔV is applied tangentially and forward (to increase the velocity).
3. ΔV_f is applied tangentially and backwards (to decrease the velocity).

When we have the problem of transferring from an initial larger orbit to a final smaller orbit, then $r_f < r_o$ and Fig. 1.4 can be interpreted by "running time backwards" so that the orbits run clockwise. In this case, the three ΔVs have the same values as before, but we are now interpreting r_f to be the initial radius and r_o to be the final

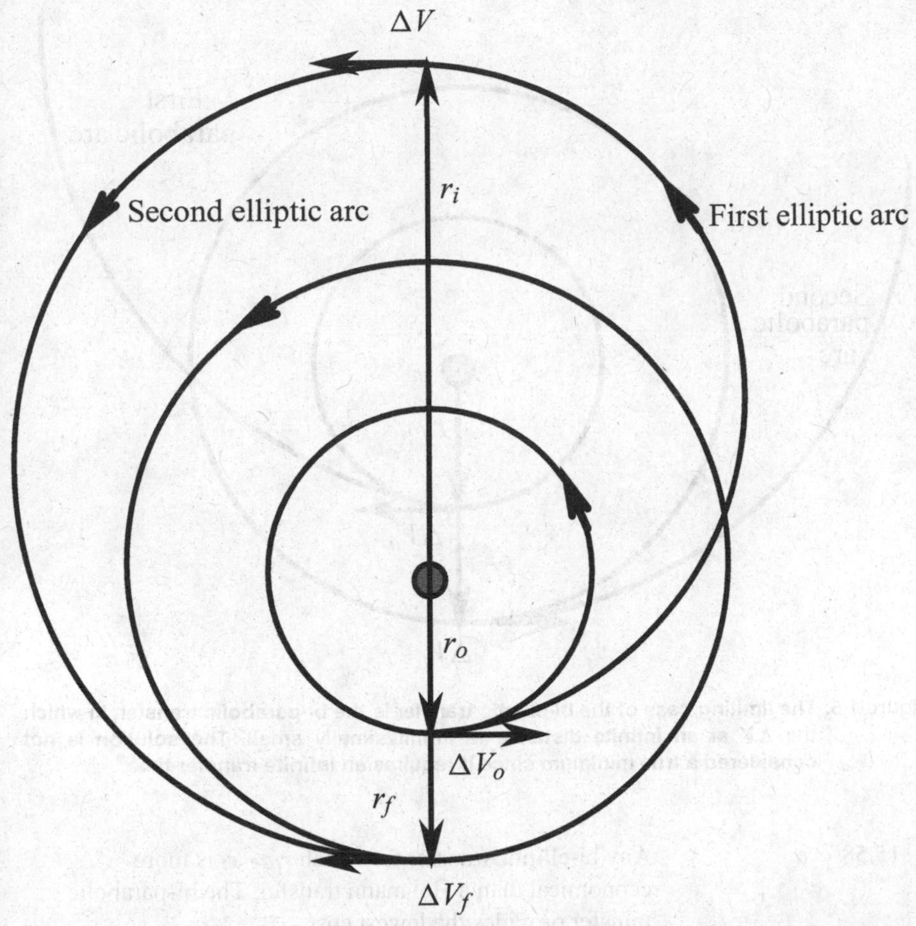

Figure 1.4. Three-impulse, bi-elliptic transfer from circular orbit of radius r_o to circular orbit of radius r_f. If r_i is large enough, the total ΔV cost can be less than that of the Hohmann transfer depending on the ratio r_f/r_o.

radius of the circular orbits. A similar interpretation can be made for the Hohmann transfer when the initial orbit is the larger orbit.

Now, let us summarize what is known about the Hohmann and the bi-elliptic transfers, i.e., which is more economical depending on the ratio $\alpha = r_f/r_o$. Here we assume that $r_f > r_o$ and note that both the bi-elliptic and bi-parabolic are three-impulse transfers.

1	$< \alpha < 11.94$	The Hohmann transfer is the most economical.
11.94	$< \alpha < 15.58$	A bi-elliptic transfer is more economical if r_i (the radius of the intermediate ΔV) is sufficiently large. The bi-parabolic transfer provides the lowest cost, but is impractical. (See Sect. 1.2.4.)

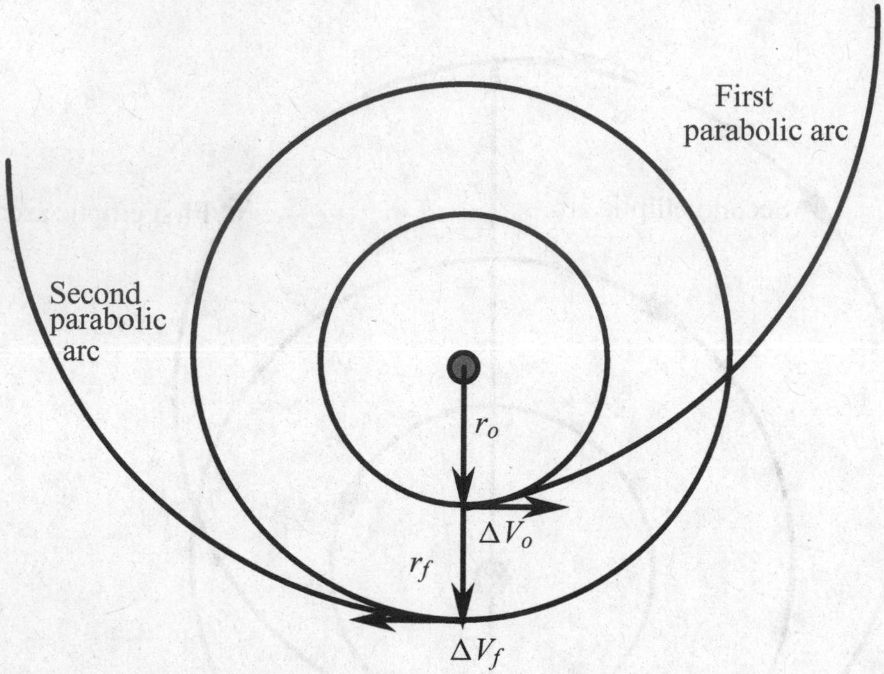

Figure 1.5. **The limiting case of the bi-elliptic transfer is the bi-parabolic transfer, in which the ΔV at an infinite distance is infinitesimally small. The solution is not considered a true minimum since it requires an infinite transfer time.**

$15.58 < \alpha$	Any bi-elliptic transfer for which $r_i > r_f$ is more economical than a Hohmann transfer. The bi-parabolic transfer provides the lowest cost.

The bi-elliptic transfer is significantly more advantageous for a non-coplanar transfer and is an interesting example of a minimum-propellant transfer that requires more impulses than are necessary to satisfy the boundary conditions. (See Prussing and Conway [2013].)

1.2.4. The Bi-parabolic Transfer

The bi-parabolic transfer (Fig. 1.5) is the limiting case of a bi-elliptic transfer as $r_i \to \infty$. It consists of two finite impulses, ΔV_o and ΔV_f, and a third infinitesimal ΔV at "infinity" which facilitates the transfer from the first parabola to the second parabola.

This transfer is of no practical use since the transfer duration is infinite. However, for $11.94 < \alpha$ and r_i large enough, a bi-elliptic transfer can be found that is better than a Hohmann but worse than a bi-parabolic transfer.

1.3. Exercises

1. Find the minimum of $f(x) = x^2$.
2. Let $f(x) = x^4$. Is $f(0) = 0$ the minimum value? Is the second-order sufficient condition for a minimum satisfied? What can you conclude from this example?
3. The rocket equation is

$$\Delta V = c \ln \tilde{m}$$

where ΔV is the change in velocity, c is the effective exhaust velocity, and

$$\tilde{m} = m_o / m_{bo}$$

is the ratio of initial mass to burnout mass. We note that c is related to specific impulse, I_{sp}, by

$$c = I_{sp} g$$

where g is standard free fall:

$$g \equiv 9.80665 \text{ m/s}^2$$

Assume that the burnout mass consists of the structural mass of the stage, m_s, and the payload mass, P:

$$m_{bo} = m_s + P$$

Also, the initial mass is the burnout mass plus the propellant mass, m_p:

$$m_o = m_{bo} + m_p = P + m_s + m_p$$

The structural factor, β, is defined as

$$\beta = \frac{m_s}{m_p + m_s}$$

which is the mass ratio of the empty stage (structure alone) to the loaded stage (propellant plus structure). Note that this implies that

$$m_s = \beta(m_o - P)$$

3a. Derive an equation which provides the ratio of the initial mass, m_o, to the payload mass, P:

$$m_o / P = f(\Delta V, c, \beta)$$

Make sure your final solution contains *only* ΔV, c, and β on the right hand side.

3b. In order to launch a payload from the surface of the Earth to Mars, the ΔV is approximately 11.6 km/s. Assuming the I_{sp} is 300 s and β is 0.100, will the rocket be able to achieve the required velocity? Explain your answer.

3c. If $\beta = 0.0100$ and all other values are the same, what is the mass ratio m_o/P?

3d. Is $\beta = 0.0100$ a practical value for the structural factor? (That is, does it correspond to launch vehicles being used today? Give an example.)

3e. If the answer to 3 is no, how can the payload be launched to Mars?

4. The rocket equation for a multistage rocket of N stages is

$$\Delta V_{\text{TOT}} = \sum_{i=1}^{N} c_i \ln \tilde{m}_i \tag{1.42}$$

where ΔV_{TOT} is the total velocity change, c_i is the effective exhaust velocity of the ith stage, and $\tilde{m}_i = (m_o/m_{bo})_i$ is the mass ratio of the ith stage.

Our goal is to find the mass ratios, $\tilde{m}_i$, which minimize the overall mass ratio, m_{o1}/P:

$$\text{Min}\, J = \frac{m_{o1}}{P} \tag{1.43}$$

where m_{o1} is the takeoff mass, P is the payload mass, and where ΔV_{TOT} is given (as the final constraint). We will assume that the c_i and β_i are given parameters of the problem.

We note that

$$\frac{m_{o1}}{P} = \left(\frac{m_{o1}}{m_{o1} - m_{p1} - m_{s1}}\right)\left(\frac{m_{o2}}{m_{o2} - m_{p2} - m_{s2}}\right)\cdots\frac{m_{oN}}{P} \tag{1.44}$$

where the mass of the second stage is

$$m_{o2} = m_{o1} - m_{p1} - m_{s1} \tag{1.45}$$

where m_{p1} is the propellant mass of the first stage and m_{s1} is the structural mass of the first stage. An example of the nomenclature is illustrated for a three-stage rocket in Fig. 1.6.

4a. Show that if

$$\tilde{m}_i = \frac{m_{oi}}{m_{oi} - m_{pi}} \tag{1.46}$$

and

$$\beta_i = \frac{m_{si}}{m_{pi} + m_{si}} \tag{1.47}$$

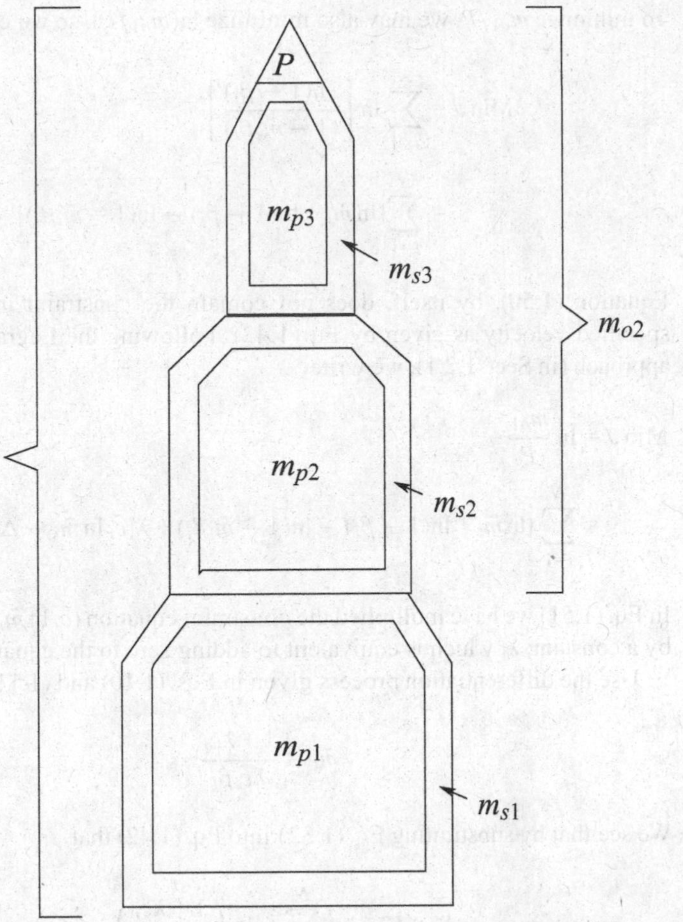

Figure 1.6. Nomenclature applied to 3-stage rocket: *P* is the payload mass, m_{o1} is the total mass at takeoff, m_{p1} is the mass of propellant in first stage, m_{s1} is the mass of structure in first stage, m_{o2} is the total mass at ignition of second stage after dropping first stage, and *N* is number of stages, three.

where β_i are the structural factors, then the ratios in Eq. (1.44) can be written as

$$\frac{m_{oi}}{m_{oi} - m_{pi} - m_{si}} = \frac{\tilde{m}_i(1 - \beta_i)}{1 - \tilde{m}_i \beta_i} \tag{1.48}$$

Thus, Eq. (1.45) can be put in the form

$$\frac{m_{o1}}{P} = \frac{\tilde{m}_1(1 - \beta_1)}{(1 - \tilde{m}_1 \beta_1)} \frac{\tilde{m}_2(1 - \beta_2)}{(1 - \tilde{m}_2 \beta_2)} \cdots \frac{\tilde{m}_N(1 - \beta_N)}{(1 - \tilde{m}_N \beta_N)} = \prod_{i=1}^{N} \frac{\tilde{m}_i(1 - \beta_i)}{1 - \tilde{m}_i \beta_i} \tag{1.49}$$

4b. To minimize m_{o1}/P, we may also minimize $\ln(m_{o1}/P)$, so we can write

$$\begin{aligned}\text{Min } J &= \sum_{i=1}^{N} \ln\left[\frac{\tilde{m}_i(1-\beta_i)}{(1-\tilde{m}_i\beta_i)}\right] \\ &= \sum_{i=1}^{N}[\ln \tilde{m}_i + \ln(1-\beta_i) - \ln(1-\tilde{m}_i\beta_i)] \qquad (1.50)\end{aligned}$$

Equation (1.50), by itself, does not contain the constraint imposed by the specified velocity as given by Eq. (1.42). Following the Lagrange multiplier approach (in Sect. 1.2.1), we write

$$\begin{aligned}\text{Min } J &= \ln\frac{m_{o1}}{P} \\ &= \sum_{i=1}^{N}\{\ln \tilde{m}_i + \ln(1-\beta_i) - \ln(1-\tilde{m}_i\beta_i) + \lambda[c_i \ln \tilde{m}_i - \Delta V_{\text{TOT}}]\} \qquad (1.51)\end{aligned}$$

In Eq. (1.51) we have multiplied the constraint equation ($c_i \ln \tilde{m}_i - \Delta V_{\text{TOT}} = 0$) by a constant, λ, which is equivalent to adding zero to the equation.

Use the differentiation process given in Eqs. (1.10) and (1.11) to obtain

$$\tilde{m}_i = \frac{1+\lambda c_i}{\lambda c_i \beta_i} \qquad (1.52)$$

4c. We see that by substituting Eq. (1.52) into Eq. (1.42) that

$$\Delta V_{\text{TOT}} = \sum_{i=1}^{N} c_i \ln\left(\frac{1+\lambda c_i}{\lambda c_i \beta_i}\right) \qquad (1.53)$$

We can solve for the constant, λ, since ΔV_{TOT}, c_i, and β_i are known. The value of λ determines the mass ratios, $\tilde{m}_i$, of each stage by Eq. (1.52). Show that for the special case where all the c_i (specific impulses) are the same

$$\tilde{m}_i = \frac{1}{\beta_i}\exp\left[\frac{1}{N}\left(\frac{\Delta V_{\text{TOT}}}{c} + \sum_{i=1}^{N}\ln\beta_i\right)\right] \qquad (1.54)$$

(Here we are assuming that $c_1 = c_2 = \cdots = c_N = c$ and thus $\sum_{i=1}^{N} c_i = Nc$. We also note that $\ln\left(\frac{1}{c_i\beta_i}\right) = \ln\left(\frac{1}{c_i}\right) - \ln\beta_i$.)

4d. Show that if c_i and β_i are the same for each stage, the optimal mass ratios, $\tilde{m}_i$, are

$$\tilde{m}_i = \exp[\Delta V_{\text{TOT}}/(Nc)] \qquad (1.55)$$

4e. A two-stage rocket must attain a maximum speed of 7,925 m/s with $I_{sp} = 300$ s for both stages and $\beta_1 = \beta_2$. Determine the mass ratio of each stage ($\tilde{m}_1$, $\tilde{m}_2$) to minimize m_{o1}/P.
4f. For 4e, determine the propellant mass per stage in terms of the initial mass of the stage. Also determine the structural factors β_i, assuming $m_{si} = 0.15 m_{oi}$, and show that the optimal overall mass ratio is $m_{o1}/P = 82.6$.
4g. In designing a two-stage rocket for a maximum speed of 7,925 m/s, assume that $I_{sp} = 250$ s for both stages and $\beta_1 = 0.180$ and $\beta_2 = 0.150$. Show that it is capable of boosting a payload of $0.00169\, m_{o1}$.

5. In solving the following orbit transfer problems, put all of your solutions in terms of the ratios: $\alpha \equiv r_f/r_o$ and $\Delta V_{TOT}/\sqrt{\mu/r_o}$. Simplify your algebraic results to the most compact form.

5a. Describe the Hohmann transfer between circular coplanar orbits and derive a formula for the total velocity change, $\Delta V_{\text{TOT}} = \Delta V_o + \Delta V_f$. Assume that $r_f \geq r_o$.
5b. Derive a formula for ΔV_{TOT} for the bi-parabolic transfer.
5c. Make a plot of $\Delta V_{\text{TOT}}/\sqrt{\mu/r_o}$ versus α showing both types of transfer. Let $1 \leq \alpha \leq 40$.
5d. For what values of α is the bi-parabolic transfer more economical?

6. Determine the rectangle of maximum area with a specified perimeter, p. Use a Lagrange multiplier to incorporate the perimeter constraint and apply the conditions of Sect. 1.2.1.
7. A spacecraft approaches a planet on a hyperbolic trajectory with a given V_∞ (that is, the specific energy is $\frac{1}{2}V_\infty^2$). A ΔV is to be applied at an unspecified periapsis radius r_p to capture the spacecraft into a circular orbit. Assume the ΔV is applied along the velocity vector at periapsis and determine the value of r_p that minimizes ΔV in terms of V_∞ and the gravitational parameter, μ.

References

A.E. Bryson Jr., Y.C. Ho, *Applied Optimal Control* (Hemisphere Publishing Corporation, Washington, DC, 1975)

T.N. Edelbaum, How many impulses? Astronautics and Aeronautics **5**, 64–69 (1967). (now named *Aerospace America*)

W. Hohmann, *Die Erreichbarkeit der Himmelskörper* (1925) (The attainability of heavenly bodies). NASA Technical Translation F-44, National Aeronautics and Space Administration, Washington, DC, Nov 1960

J.E. Prussing, B.A. Conway, *Orbital Mechanics*, 2nd edn. (Oxford University Press, New York, 2013)

P. Ulivi, D.M. Harland, *Robotic Exploration of the Solar System* (Springer, New York, 2007)

Chapter 2

Optimal Control Theory

2.1. Optimal Launch of a Satellite

We start this chapter with what might be called the fundamental trajectory optimization problem: launching a satellite into orbit. Before setting up the mathematical model for this problem, let us first discuss in words what the problem entails.

It is very expensive to launch a satellite into orbit about the Earth because of the tremendously high speed required to achieve a low circular orbit. The lowest energy orbit must be high enough above the sensible atmosphere to avoid immediate dissipation of its energy and subsequent orbit decay and reentry. The lowest "sustainable" orbit is about 160 km (100 miles) above the Earth's surface. At this altitude the satellite must travel at 7.9 km/s (5 miles/s) to stay in circular orbit. In its final flight in 2011, the cost of reaching circular orbit with the United States Space Shuttle was about \$22,000/kg (\$10,000/lb). So if we can save a few kilograms on the way into orbit, we can save a lot of money.

Let's consider what our choices are in launching a satellite into circular orbit. To make things easier, we will cheat a little and ignore atmospheric drag during launch. Let us also assume that the launch vehicle has a predetermined amount of propellant and that it burns the propellant at a constant rate. Then, if the specific impulse is constant, the thrust of the engine, F, is constant. To further simplify, we assume the rocket has a single stage. (There is, of course, a great advantage to staging the engines, but that is a parameter optimization problem as discussed in the exercises in Chap. 1.)

Let us also assume: the Earth is not rotating (so we do not get a boost from the rotation), the circular orbit plane contains the launch site (so there is no out-of-plane dog-leg maneuver), the Earth is flat (we tip our hats to the Flat Earth Society), and the acceleration due to gravity is a constant equal to standard free fall (i.e. $g = 9.80665\,\text{m/s}^2$).

One might well ask, "What is there left to do?" It turns out that we have one control variable. In Fig. 2.1, we illustrate our flat-Earth problem of launching a satellite into circular orbit. The orbital plane is in the xy plane, and the altitude we wish to achieve corresponds to a radial distance from the center of the Earth, r_c. The one control variable we have is the steering angle of the thrust, α.

Consider what happens when a rocket is launched into orbit. First, the rocket lifts off the pad and rises vertically, which corresponds to a steering angle of 90°. Then, as

J.M. Longuski et al., *Optimal Control with Aerospace Applications*, Space Technology Library 32, DOI 10.1007/978-1-4614-8945-0_2,

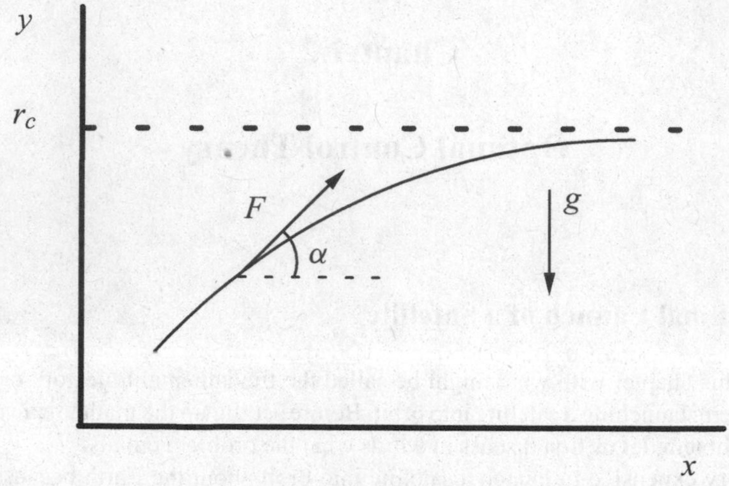

Figure 2.1. **Problem of finding the steering law, $\alpha(t)$, to maximize satellite mass.**

the rocket achieves higher altitudes, it begins to pitch over so that α is less than 90°. Eventually, the rocket is traveling with α near zero (or even negative to cancel upward momentum).

Here we note that actual launches from the Earth have to contend with air drag so that α remains near 90° longer than would be necessary if there were no atmosphere. In launching astronauts from the Moon, the steering angle α approached zero (and below) much more rapidly, because there was no atmosphere to clear (just high lunar mountains).

The time function $\alpha(t)$ could follow an infinite number of profiles that would achieve circular orbit. For every steering profile, there is a corresponding path of the rocket given by $x(t)$ and $y(t)$.

Our problem is to find $\alpha(t)$ such that the maximum payload is delivered into a prescribed circular orbit. We are maximizing final mass, which for a constant burn rate means we are minimizing the time to get into orbit. If we minimize the time for a given burn rate, we minimize the propellant burned. Our infinite-dimensional problem is: find the steering law, $\alpha(t)$, which is a function of time (an infinite number of points) to minimize the final time to orbit, t_f.

Example 2.1 Launch into circular orbit.

We assume that at the initial time, t_o, we know the position and velocity components, x_o, y_o, v_{xo}, and v_{yo}. However, we do not know the final time, t_f, nor the final value, x_f, but we do know the final velocity components $v_{xf} = v_c$ and $v_{yf} = 0$, which are specified for the circular orbit at the known altitude, $y_f = r_c - r_{\text{Earth}}$, where r_{Earth} is the radius of the Earth.

The mass of the rocket is

$$m(t) = m_o + \dot{m}_o(t - t_o) \tag{2.1}$$

where $\dot{m}_o$ is a negative constant.

For our "flat-Earth" problem in Fig. 2.1, the governing differential equations are

$$\dot{x} = v_x \tag{2.2a}$$

$$\dot{y} = v_y \tag{2.2b}$$

$$\dot{v}_x = \frac{F}{m}\cos\alpha \tag{2.2c}$$

$$\dot{v}_y = \frac{F}{m}\sin\alpha - g \tag{2.2d}$$

where Eqs. (2.2c) and (2.2d) are from Newton's second law. In our problem we must find $\alpha(t)$ to minimize the time.

Next, we introduce standard nomenclature from the literature. We write state variables as:

$$x_1 = x \tag{2.3a}$$

$$x_2 = y \tag{2.3b}$$

$$x_3 = v_x \tag{2.3c}$$

$$x_4 = v_y \tag{2.3d}$$

For our control variable we use:

$$u = \alpha \tag{2.4}$$

Using vector notation (indicated by bold) we can write our differential equations, Eqs. (2.2), and initial conditions as follows:

$$\dot{\boldsymbol{x}} = \boldsymbol{f}(t, \boldsymbol{x}, u) \tag{2.5a}$$

$$\boldsymbol{x}(t_o) = \boldsymbol{x}_o \tag{2.5b}$$

where $\boldsymbol{x}$ and $\boldsymbol{f}$ are 4-vectors in this problem (and n-vectors in general).

For our final boundary conditions, we have the components

$$\Psi_1 \equiv x_2(t_f) - r_c + r_{\text{Earth}} = 0 \tag{2.6a}$$

$$\Psi_2 \equiv x_3(t_f) - v_c = 0 \tag{2.6b}$$

$$\Psi_3 \equiv x_4(t_f) = 0 \tag{2.6c}$$

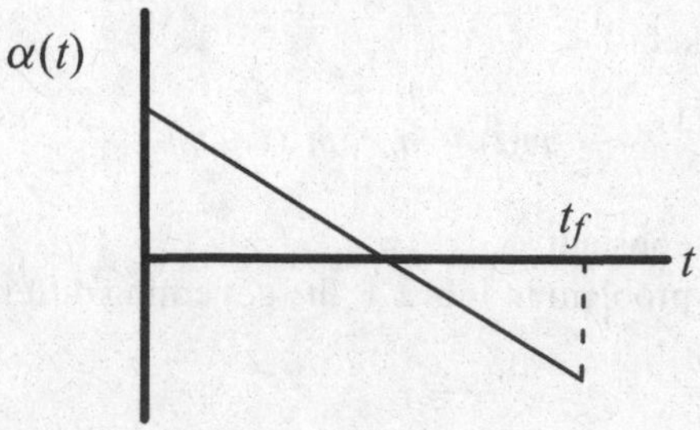

Figure 2.2. A possible steering law for minimum time to orbit.

or

$$\boldsymbol{\Psi}(t_f, \boldsymbol{x}_f) = \boldsymbol{0} \tag{2.7}$$

where $\boldsymbol{\Psi}$ is a 3-vector (q-vector in general).

For the cost functional (or performance index) to be minimized we have:

$$J = t_f \tag{2.8}$$

which minimizes the time (and maximizes the payload). Figure 2.2 shows a representative plot of the optimal steering law. Sometimes the steering law can look that simple. In Chap. 3 we develop the theory and in Chap. 4 we apply the theory to find such control laws.

2.2. General Form of the Problem

The general form for the optimal control problem is expressed as:

Minimize:

$$J = \phi(t_f, \boldsymbol{x}_f) + \int_{t_o}^{t_f} L(t, \boldsymbol{x}, \boldsymbol{u})dt \tag{2.9}$$

subject to:

$$\dot{\boldsymbol{x}} = \boldsymbol{f}(t, \boldsymbol{x}, \boldsymbol{u}) \tag{2.10a}$$

$$\boldsymbol{x}(t_o) = \boldsymbol{x}_o \tag{2.10b}$$

$$\boldsymbol{u} \in U \tag{2.10c}$$

$$\boldsymbol{\Psi}(t_f, \boldsymbol{x}_f) = \boldsymbol{0} \tag{2.10d}$$

with a possible state variable inequality constraint:

$$S(\boldsymbol{x}) \geq 0 \tag{2.11}$$

Here U is the set of admissible controls, which we will discuss later, and $S(\boldsymbol{x}) \geq 0$ is an inequality constraint which, for example, may require a spacecraft or aircraft to never fly below the surface of the Earth.

At this point it is useful to formalize our notation according to standard notation. Lower case letters in bold denote a vector

$$\boldsymbol{x} = \begin{bmatrix} x_1 \\ x_2 \\ \vdots \\ x_n \end{bmatrix} \tag{2.12}$$

A superscript T indicates the transpose of a vector or matrix:

$$\boldsymbol{x}^T\boldsymbol{y} = x_1y_1 + x_2y_2 + \cdots + x_ny_n \tag{2.13}$$

so that $\boldsymbol{x}^T\boldsymbol{y}$ represents the scalar product of vectors $\boldsymbol{x}$ and $\boldsymbol{y}$.

The gradient of a scalar function, J, is defined to be a row vector:

$$\begin{aligned} \nabla J = J_{\boldsymbol{x}} = \frac{\partial J}{\partial \boldsymbol{x}} &= \left[\frac{\partial J}{\partial x_1} \; \frac{\partial J}{\partial x_2} \cdots \frac{\partial J}{\partial x_n} \right] \\ &= \left[J_{x_1} \; J_{x_2} \cdots J_{x_n} \right] \end{aligned} \tag{2.14}$$

The second derivative of a scalar function with respect to vector arguments is:

$$\begin{aligned} J_{\boldsymbol{xy}} &= \frac{\partial}{\partial \boldsymbol{y}}(J_{\boldsymbol{x}}^T) \\ &= \begin{bmatrix} \frac{\partial^2 J}{\partial x_1 \partial y_1} & \frac{\partial^2 J}{\partial x_1 \partial y_2} & \cdots & \frac{\partial^2 J}{\partial x_1 \partial y_m} \\ \vdots & & & \\ \frac{\partial^2 J}{\partial x_n \partial y_1} & & \cdots & \frac{\partial^2 J}{\partial x_n \partial y_m} \end{bmatrix} \end{aligned} \tag{2.15}$$

Example 2.2 The brachistochrone problem.

In 1696, Johann Bernoulli posed and solved the brachistochrone problem (see Bell [1965]). He originally created the problem as a challenge to his brother, Jacob, who also solved it. Figure 2.3 illustrates the problem. The xy plane is vertical, in a uniform gravity field, where the x axis points downward in a uniform gravity field. A particle of mass m is placed (motionless) on a frictionless track that connects the origin to a given point (x_f, y_f). The shape of the track, $y = y(x)$, must be found such that the particle takes the shortest time to travel from the origin to (x_f, y_f). A straight line (i.e., a ramp) does not provide the quickest trajectory.

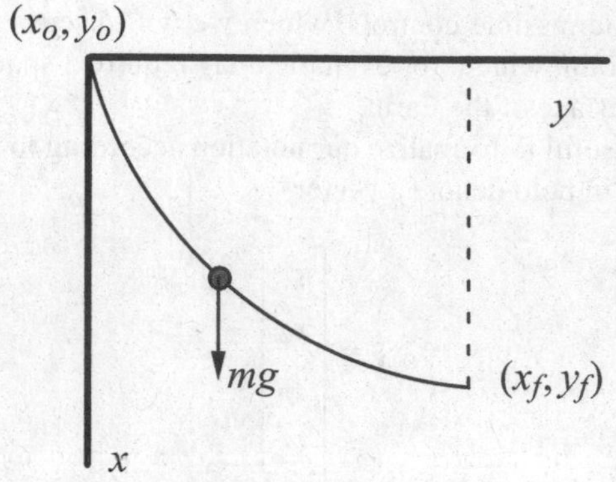

Figure 2.3. **The brachistochrone problem: find $y = y(x)$ to minimize the time for the particle to slide in the vertical plane from $(x_o,\ y_o)$ to $(x_f,\ y_f)$. Note that the straight line solution is not the minimum-time path**

In fact, the fastest trajectory starts with a steeper slope than the ramp to gain speed early and the extra speed more than makes up for the longer path. Since the initial speed of the particle is zero and it falls in a uniform gravity field, its velocity is known from the conservation of total mechanical energy:

$$v = \sqrt{2gx} \tag{2.16}$$

The infinitesimal distance, *ds* along the track can be given as

$$ds = \sqrt{(dx)^2 + (dy)^2} = \sqrt{1 + (dy/dx)^2}\, dx = \sqrt{1 + y'^2}\, dx \tag{2.17}$$

Thus, the time to travel from $x = 0$ to $x = x_f$ is:

$$t = \int_0^{x_f} \frac{\sqrt{1 + y'^2}}{\sqrt{2gx}}\, dx \tag{2.18}$$

The problem is to find the path, $y = y(x)$, to minimize the time given by Eq. (2.18). This problem of quickest descent baffled the mathematicians of Europe for 6 months. Eventually several correct solutions were sent to Johann Bernoulli including an anonymous one from Isaac Newton, who solved the problem in one day. Upon receiving Newton's solution, Bernoulli exclaimed, "tanquam ex ungue leonem," loosely translated by Bell [1965] as "I recognize the lion by his paw print!"

It turns out that the solution is a cycloid, but much more important results ensued. The general problem could be written as

$$J = \int_{x_o}^{x_f} F[y(x), y'(x), x]dx \tag{2.19}$$

where J is a scalar, F is known, but $y(x)$ is unknown. The general problem is to find $y(x)$ that makes J stationary (such that small changes in $y(x)$ make no change in J). This was a new type of problem. How does one go about finding $y(x)$? How can we solve for an entire function?

It was discovered that $y(x)$ must satisfy the Euler-Lagrange equation:

$$\frac{\partial F}{\partial y} - \frac{d}{dx}\left(\frac{\partial F}{\partial y'}\right) = 0 \tag{2.20}$$

Equation (2.20) is a differential equation, and we know how to solve this type of problem. Solving a differential equation is tantamount to finding a function, which can be very difficult if the equation is nonlinear. [The derivation of Eq. (2.20) is given in Sect. 3.2.]

The brachistochrone problem led to a more general optimization problem and to a field of mathematics called the calculus of variations. The problem of launching a satellite into orbit is closely related to the brachistochrone problem. For the launch problem we must find the trajectory $x(t)$, $y(t)$ which gives the shortest time to orbit. The launch problem is an example of an optimal control problem in which we must find the steering law, $\alpha(t)$, to minimize the time.

Johann Bernoulli's discovery also led to the realization that dynamical motion obeys Lagrange's equations. That is, the functional

$$J = \int_{t_0}^{t_f} L\,dt \tag{2.21}$$

has a stationary value. The integrand, L, in Eq. (2.21) is referred to as the *Lagrangian*, which is equal to the difference between the kinetic and potential energies of the system.

This idea is known as Hamilton's principle which indicates that nature obeys an optimization principle. Such concepts occupied the minds of mathematicians and philosophers for several centuries. (For more information see the enthusiastic presentation given by Lanczos [1986]).

2.3. The Problems of Bolza, Lagrange, and Mayer

Throughout this text, we will be mainly concerned with the Problem of Bolza:

Minimize:

$$J = \phi(t_f, \boldsymbol{x}_f) + \int_{t_o}^{t_f} L(t, \boldsymbol{x}, \boldsymbol{u})dt \tag{2.22}$$

subject to:

$$\dot{\boldsymbol{x}} = \boldsymbol{f}(t, \boldsymbol{x}, \boldsymbol{u}) \quad \text{System or Process Equations} \tag{2.23a}$$

$$\boldsymbol{x}(t_o) = \boldsymbol{x}_o \quad \text{Initial Conditions (I.C.s)} \tag{2.23b}$$

$$\boldsymbol{\Psi}(t_f, \boldsymbol{x}_f) = \boldsymbol{0} \quad \text{Terminal Constraints} \tag{2.23c}$$

where,

$\boldsymbol{x}$ is an n-vector

$\boldsymbol{u}$ is an m-vector

$\boldsymbol{\Psi}$ is a $q \leq n$ vector

and J is the scalar cost to be minimized. (See Bolza [1961], Bryson and Ho [1975], and Hull [2003].) The scalar J is called a functional because it maps functions [the path $\boldsymbol{x}(t)$ and control $\boldsymbol{u}(t)$] into a single number. The functional is also called the cost index, the performance index, or sometimes the cost function (though cost functional is more precise). As we have seen, typical examples for J are the propellant used to launch a spacecraft into orbit or the time for a particle to travel between points. Besides the Problem of Bolza, there are two other forms that appear in the literature.

In the Problem of Lagrange we have:

Minimize:

$$J = \int_{t_o}^{t_f} L(t, \boldsymbol{x}, \boldsymbol{u})dt \tag{2.24}$$

subject to the same conditions as the Problem of Bolza.

Equation (2.24) consists of what is sometimes referred to as the "path cost" which is the same form as we saw in the brachistochrone problem [Eq. (2.19)]. We will find this integral form useful when we explore the effect of variations of the path $\boldsymbol{x}$ and the control $\boldsymbol{u}$ from the optimal, which will lead us to the Euler-Lagrange equation, Eq. (2.20). In Eq. (2.24) the integrand, L, is called the Lagrangian. (We note for those familiar with Lagrangian dynamics that the integrand of Eq. (2.24) is more general and may have nothing to do with dynamics.)

In the Problem of Mayer we have:

Minimize:

$$J = \phi(t_f, \boldsymbol{x}_f) \tag{2.25}$$

subject to the same conditions as the Problem of Bolza.

Equation (2.25) is sometimes called the "terminal cost." It has the same form as Eq. (2.8) where we considered the problem of minimizing the final time to launch a satellite into orbit.

Example 2.3 Interchangeability of forms.

The three forms (Bolza, Lagrange, and Mayer) are equally general and interchangeable. (See Bliss [1968], Hull [2003], and Vagners [1983].) As a trivial example, consider:

$$\text{Mayer: Min. } J = t_f \tag{2.26}$$

$$\text{Lagrange: Min. } J = \int_0^{t_f} dt \tag{2.27}$$

$$\text{Bolza: Min. } J = \frac{1}{2}t_f + \frac{1}{2}\int_0^{t_f} dt \tag{2.28}$$

2.3.1. Transformation from Lagrange to Mayer

Let us consider in general how to transform the Problem of Lagrange into the Problem of Mayer. In the Problem of Lagrange we have:

Minimize:

$$J = \int_{t_o}^{t_f} L(t, \boldsymbol{x}, \boldsymbol{u})dt \tag{2.29}$$

Let us define a new variable with zero initial condition:

$$\dot{x}_{n+1} = L(t, \boldsymbol{x}, \boldsymbol{u}) \tag{2.30a}$$

$$x_{n+1}(t_o) = 0 \tag{2.30b}$$

Since $x_{n+1}(t) = \int_{t_o}^{t} L(t, \boldsymbol{x}, \boldsymbol{u})dt + x_{n+1}(t_o)$, the problem becomes:

Minimize:

$$J = x_{n+1}(t_f) \tag{2.31}$$

which is in Mayer form.

2.3.2. Transformation from Mayer to Lagrange

Next we show how to transform a Mayer problem into a Lagrange problem. For the Problem of Mayer we have:
Minimize:

$$J = \phi(t_f, \boldsymbol{x}_f) \tag{2.32}$$

To transform Eq. (2.32) into a Lagrange problem we write:

Minimize:

$$J = \int_{t_o}^{t_f} \frac{d\phi(t, \boldsymbol{x})}{dt} dt \tag{2.33a}$$

subject to:

$$\phi(t_o, \boldsymbol{x}_o) = 0 \tag{2.33b}$$

2.4. A Provocative Example Regarding Admissible Functions

Example 2.4 Admissible functions.

In this example we demonstrate how the class of functions being considered (for the state and the control) can affect the optimal solution we obtain.

Minimize:

$$J = \int_0^3 x^2 dt \tag{2.34}$$

subject to:

$$\dot{x} = u \tag{2.35a}$$

$$x(0) = 1 \tag{2.35b}$$

$$\Psi(t_f, x_f) = x(3) - 1 = 0 \tag{2.35c}$$

where x and u are scalars.

Problem: Determine $u(t) \in$ P.C. [0,3] which causes $x(t) \in C^0$ [0,3].

That is, our problem is to find a scalar control, $u(t)$, that is piecewise continuous over the closed time interval from $t_0 = 0$ to $t_f = 3$ such that $x(t)$ is continuous over the same time interval and such that Eq. (2.34) is minimized. The problem turns out to be a trick question, because there is no such control.

The point of this example is to emphasize the importance of the class of functions we are considering. The class of functions that we allow can have a dramatic effect on the solutions we obtain. For example, in Chap. 1 where we considered two impulsive ΔV maneuvers for orbit transfer between two circular coplanar orbits, we had the Hohmann transfer to minimize the total ΔV. However, when we allowed three impulses, we found it sometimes provided a lower total ΔV. Thus the type of control that we permit (i.e., what we call an admissible control) can affect the optimal solution.

A nice aspect of the problem we are now considering is that we don't need any theorems from optimal control to understand the problem. We can deduce all of our results by inspection. Upon examining the cost functional of Eq. (2.34), we see that

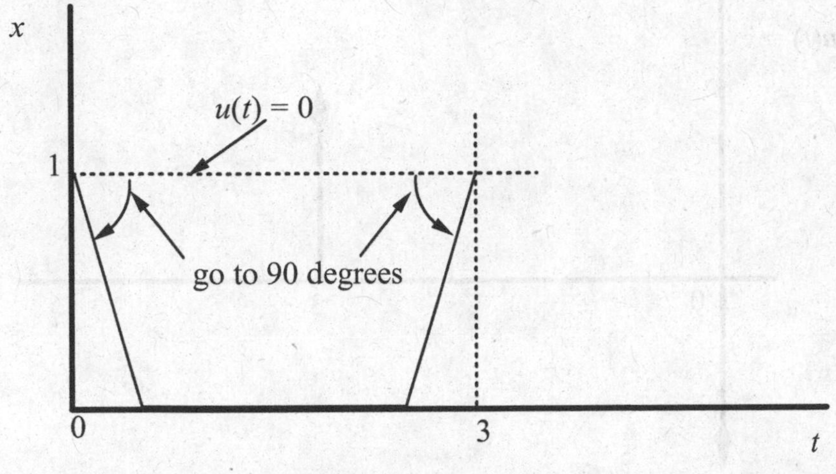

Figure 2.4. A provocative example regarding admissible functions. As the slope of x approaches vertical, the cost, Eq. (2.34), decreases until the slope is vertical when the control consists of Dirac delta functions

the value of x must be as close to zero as possible over the time interval. However, from the initial and final conditions [Eqs. (2.35b) and (2.35c)] the value of x must be unity at $t = 0$ and $t_f = 3$.

Figure 2.4 illustrates the solution for $u(t) \equiv 0$ where $x(t) \equiv 1$ is indicated by a dotted line. Clearly this is not the optimal. We note in the figure that if the value of x slopes downward near $t = 0$ and upward near $t_f = 3$ then the value for J will be small. If we consider the extreme case then x will discontinuously drop to zero at $t_0 = 0$ and will discontinuously jump to unity at $t_f = 3$. This corresponds to the 90° case indicated in Fig. 2.4.

If we allow these discontinuities then $x(t)$ will not be a continuous function as we originally assumed. Furthermore, $u(t)$ will not be a piecewise continuous function, but will instead appear as illustrated in Fig. 2.5, where we represent a Dirac delta function by the vertical arrows. On the other hand, if we allow the slope of $x(t)$ to approach 90° at the initial and final times, but to not actually reach 90°, then $x(t)$ will be continuous and $u(t)$ will be piecewise continuous. However, no optimal solution exists within these classes of functions because for any given slope near 90° there is always a steeper slope which gives a lower cost. When no unique solution exists, the problem has no optimal solution for the given class of functions.

We now provide more formal definitions of some classes of functions.

Piecewise Continuous: $x(t) \in$ P.C. $\left[t_o, t_f\right]$ if $x(t)$ is continuous at each open subinterval and if $\dot{x}$ has finite limits at the ends of the intervals. (See Fig. 2.6.)

m Continuously Differentiable: $x(t) \in C^m \left[t_o, t_f\right]$ if all derivatives of $x(t)$ of order $\leq m$ exist and are continuous. We note that $C^m \in C^{m-1}$ and that $C^1 \in C^0 \in$ P.C.

In Fig. 2.7, we show an example where x is continuous ($x \in C^0$) and $\dot{x}$ is piecewise continuous ($\dot{x} \in$ P.C.). The point on x with the jump is called a *corner*; the class is also referred to as *piecewise smooth*.

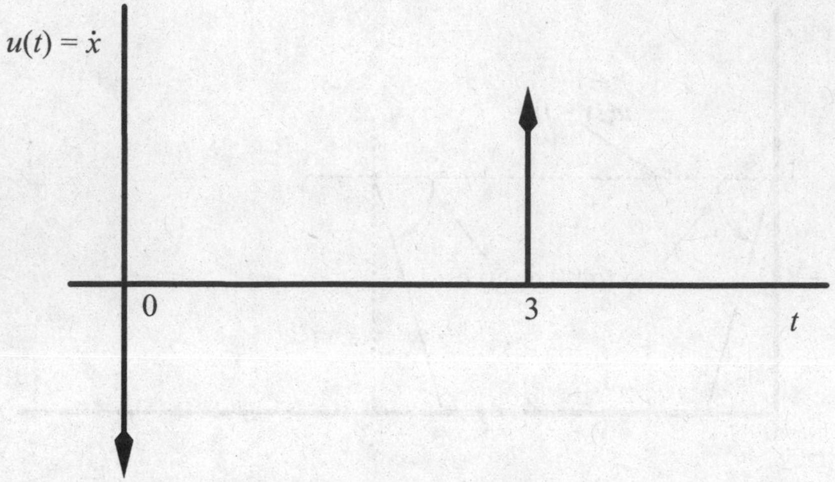

Figure 2.5. **The control for Fig. 2.4, $u(t) = \dot{x}(t)$. The *downward* and *upward arrows* represent Dirac delta functions, $-\delta(t)$ and $\delta(t-3)$, which are considered inadmissible.**

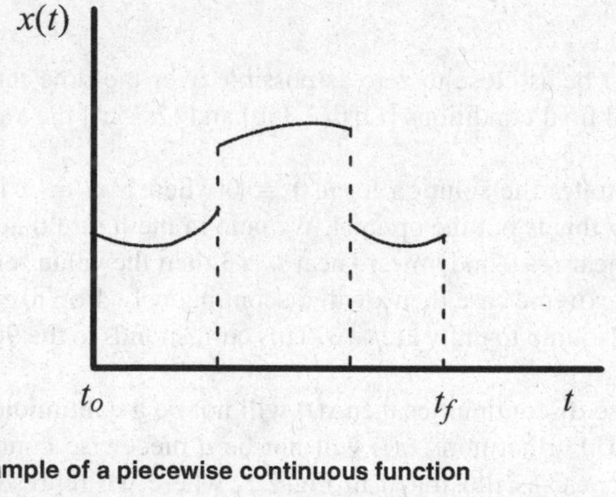

Figure 2.6. **An example of a piecewise continuous function**

The Dirac delta (or unit impulse) *function* has the properties:

$$\delta(t) = 0 \ \ \forall\, t \neq 0 \tag{2.36}$$

that is, the delta function is zero for all time not equal to zero and

$$\int_{-\infty}^{\infty} \delta(t)dt = 1 \tag{2.37}$$

This function has no definitive value at $t = 0$. According to Kaplan [1962], "...no ordinary function can have the properties mentioned. The situation is similar to that encountered in algebra: The equation $x^2 = -1$ can be satisfied by no real number.

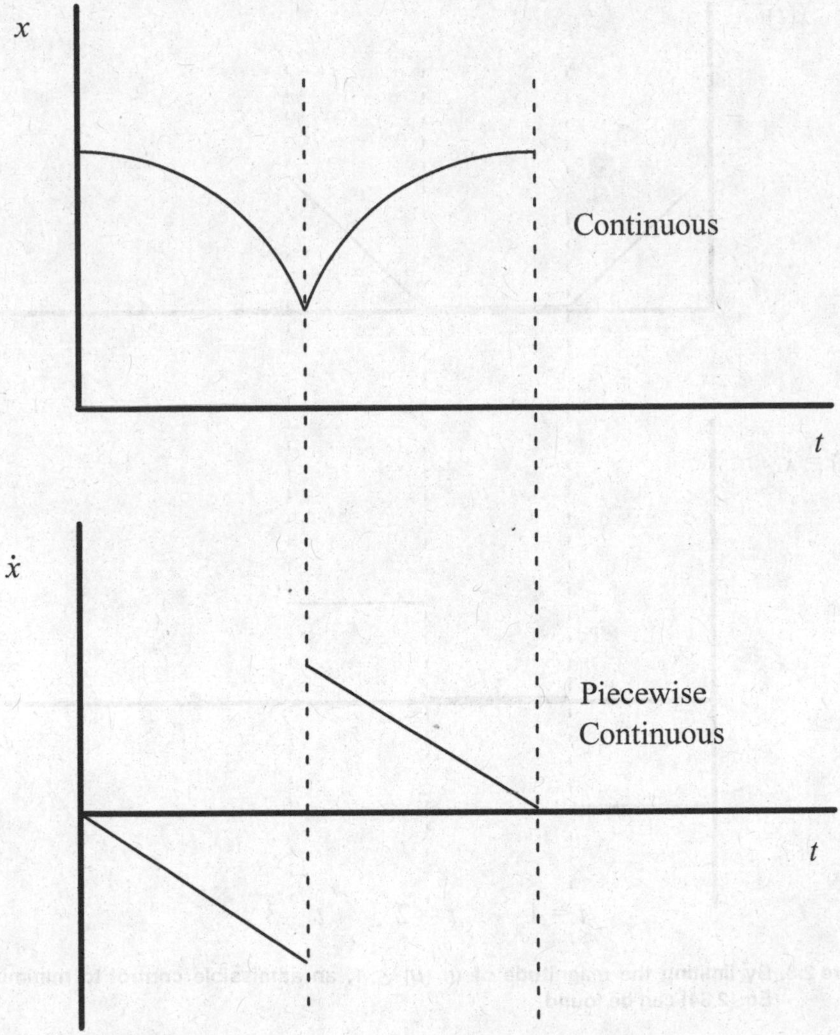

Figure 2.7. An example of a continuous function, $x \in C^0$, and its derivative which is piecewise continuous, i.e. $\dot{x} \in$ P.C. Continuous functions that have corners (discontinuous jumps in slope) are also called *piecewise smooth*

Hence we invent an 'imaginary number i' which has this property: $i^2 = -1$. In the same way we invent an 'imaginary' or 'ideal' function $\delta(t)$ to have the properties above." Kaplan goes on to develop a class of generalized functions that include δ, δ', δ'' etc. There is no mechanical way to generate such generalized functions and so they are not considered admissible.

Now, let us reconsider our problem given in Eqs. (2.34) and (2.35) in which we impose a new condition: $|u| \leq 1$. Then, the solution shown in Fig. 2.8 is obtained where $x(t) \in C^0[0, 3]$ and $u(t) \in P.C.[0, 3]$, which is admissible. We have achieved our goal by simply adding a constraint on the control.

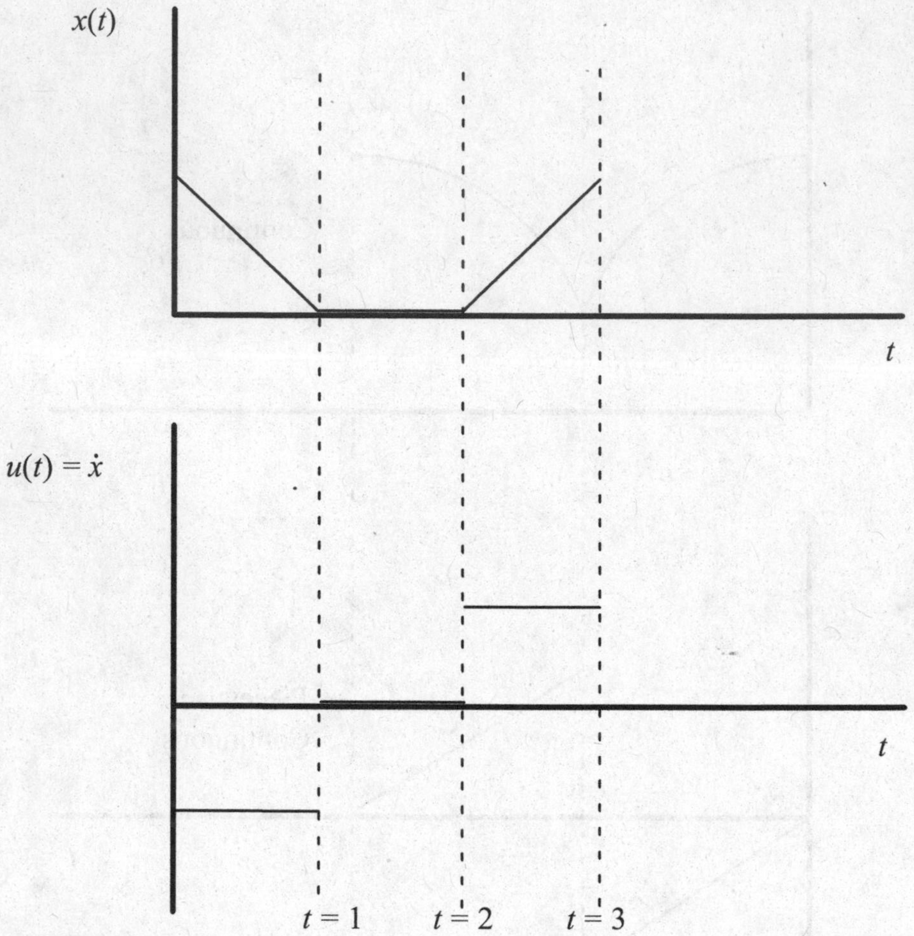

Figure 2.8. By limiting the magnitude of u: $|u| \leq 1$, an admissible control to minimize Eq. (2.34) can be found

Next we consider yet another version of our problem. Suppose for some physical or mechanical reason that P.C. functions are inadmissible for the control. We assume in this case that $u \in C^0$ is admissible.

Let us define the following state and control variables:

$$x_1 = x \tag{2.38a}$$

$$x_2 = u \tag{2.38b}$$

$$\tilde{u} = \dot{u} \tag{2.38c}$$

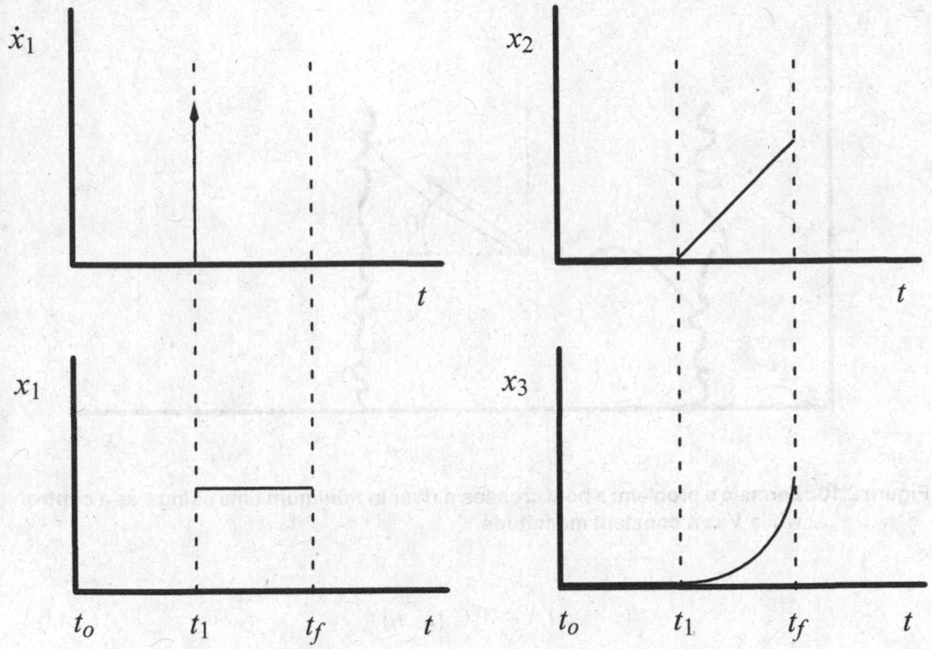

Figure 2.9. Example of classes of functions: $\delta(t - t_1)$, P.C., C^0, and C^1 over $[t_o, t_f]$.

and recalling from Eq. (2.35) that $\dot{x} = u$ we obtain

$$\dot{x}_1 = x_2 \tag{2.39a}$$

$$\dot{x}_2 = \tilde{u} \tag{2.39b}$$

By constraining $|\tilde{u}| \leq k$ (the slope of u), and $|x_2| \leq 1$ (the slope of x_1), we keep $u \in C^0[0, 3]$.

Example 2.5 Classes of functions.

In the following example let

$$\dot{x}_1 = \delta(t - t_1) \quad \text{for } t_o \leq t \leq t_f \tag{2.40a}$$

$$\dot{x}_2 = f(x_1) \tag{2.40b}$$

$$\dot{x}_3 = f(x_2) \tag{2.40c}$$

where $f \in C^\infty[t_o, t_f]$, which means that f is continuous (infinitely continuously differentiable) for all partial derivatives. What class is each of x_1, x_2, and x_3?

We can most easily visualize the solution by letting $f(x) = x$ in Eq. (2.40). Then Fig. 2.9 shows plots of $\dot{x}_1, x_1, x_2$, and x_3 and we have the result that

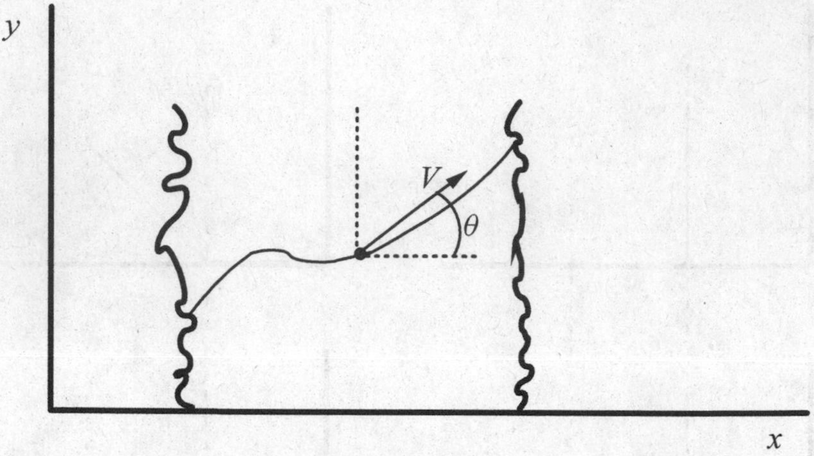

Figure 2.10. Zermelo's problem: a boat crosses a river in minimum time using θ as a control while V is a constant magnitude.

$$x_1(t) \in \text{P.C.}[t_o, t_f] \tag{2.41a}$$

$$x_2(t) \in C^0[t_o, t_f] \tag{2.41b}$$

$$x_3(t) \in C^1[t_o, t_f] \tag{2.41c}$$

Reinstating the function f into Eqs. (2.40) does not alter the conclusions of Eqs. (2.41).

Example 2.6 Zermelo's problem.

This problem involves minimizing the time required for a boat to cross a river, as illustrated in Fig. 2.10. In this example we assume the velocity V, is constant and that the control is the steering angle, θ.

Minimize:

$$J = t_f \tag{2.42}$$

subject to:

$$\dot{x} = V\cos\theta \tag{2.43a}$$

$$\dot{y} = V\sin\theta \tag{2.43b}$$

with B.C.s:

$$x(0) = x_o \tag{2.44a}$$

$$y(0) = y_o \tag{2.44b}$$

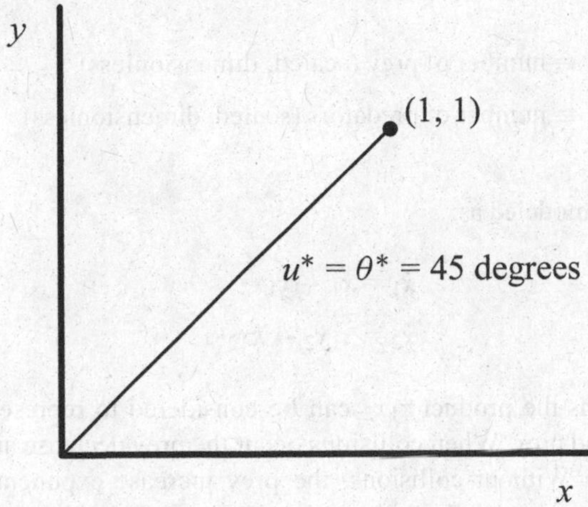

Figure 2.11. Solution to simplified Zermelo's problem: $\theta^* = 45°$.

$$x(t_f) = x_f \tag{2.44c}$$

$$y(t_f) = y_f \tag{2.44d}$$

We can simplify the problem further by assuming

$$x_0 = y_0 = 0 \tag{2.45a}$$

$$x_f = y_f = 1 \tag{2.45b}$$

This version of Zermelo's problem gives the optimal solution $\theta^* = 45°$, as shown in Fig. 2.11.

A more difficult problem considers currents, p and q, in the river, where the state equations are

$$\dot{x} = V\cos\theta + p \tag{2.46a}$$

$$\dot{y} = V\sin\theta + q \tag{2.46b}$$

and where p and q may be functions of time and of the state variables. We will return to examples of Zermelo's problem later in the text. We note in passing that an aerospace application of Zermelo's problem is the problem of an aircraft flying in a crosswind.

Example 2.7 The Lotka-Volterra model for the predator-prey problem.

The Lotka-Volterra model can be used to study population dynamics problems as well as more general problems (such as chemical and nuclear reactions and game theory problems).

Let:

$$x_1 \equiv \text{number of prey (scaled, dimensionless)}$$

$$x_2 \equiv \text{number of predators (scaled, dimensionless)}$$

This system can be modeled as:

$$\dot{x}_1 = x_1 - x_1 x_2 \tag{2.47a}$$

$$\dot{x}_2 = x_1 x_2 - k x_2 \tag{2.47b}$$

In these equations the product $x_1 x_2$ can be considered to represent a "collision" between predator and prey. When collisions occur the prey decrease and the predators increase in number. Without collisions, the prey increase exponentially while the predators decrease exponentially with time constant k. This is actually a dynamics problem but it can be turned into an optimization problem.

Let us imagine the problem of a farmer raising crops (i.e. prey, x_1) which are damaged by insects (i.e. predators, x_2). The farmer wants to maximize profit, which depends on the crops produced at the end of the season, say 100 days. The farmer may elect to introduce insecticide to kill off the insects, but insecticide costs money.

Let the control variable be

$$u \equiv \text{rate of insecticide introduction}$$

Then,

$$\dot{x}_1 = x_1 - x_1 x_2 \tag{2.48a}$$

$$\dot{x}_2 = x_1 x_2 - k x_2 - l x_2 u, \quad (0 \leq l \leq 1) \tag{2.48b}$$

where l is considered the insecticide effectiveness. The farmer's cost function can then be stated as:

Minimize:

$$J = -x_{1f} + \int_0^{t_f} a_1 u \, dt \tag{2.49}$$

where $a_1 > 0$ (a constant), l, and t_f are given. (Here we use the negative sign with x_{1f} because the farmer wants to maximize crop output.)

To complete the setup of this optimization we need to specify I.C.s $x_1(0)$ and $x_2(0)$ as well as the final B.C.s:

$$\Psi_1 = t_f - 100 = 0 \tag{2.50}$$

A similar biological population dynamics problem is discussed in Stengel [1994].

2.5. Summary

The brachistochrone problem presented by Johann Bernoulli at the end of the seventeenth century was a new mathematical problem which required that a path (i.e., a function) be found to minimize a scalar function of that path. In the middle of the twentieth century a closely related problem to the path of quickest descent presented itself: the problem of finding the propellant-optimal trajectory for launching a satellite into orbit. Such problems that map functions (paths) into a scalar are called optimal control problems. They can be expressed generally in three forms: the Mayer problem (a function of the final time or final state), the Lagrange problem (a definite integral over time which includes the path function in the integrand), and the Bolza problem (which is a combination of the Mayer and the Lagrange problems). The three forms are equivalent; the text adopts the Problem of Bolza which most often appears in the literature.

The allowable (or admissible) class of functions for the control or for the trajectory can alter the solution obtained. What is admissible due to engineering or physical constraints may differ from what is mathematically admissible. For example, while the Dirac delta function is amenable to mathematical analysis in the field of generalized functions, it is physically impossible to mechanize as a control. To avoid unacceptable solutions (inadmissible functions) it may be possible to reformulate the problem, for example by putting bounds on the control or on its derivatives.

The Bolza problem is an ideal form for studying space trajectory optimization in which a control (such as the steering law for the thrust vector) is used to direct a launch vehicle into orbit using the least amount of propellant. While this text focuses on space trajectory optimization, it will include other examples of optimal control, such as the classical problem of Zermelo.

2.6. Exercises

1. Find the C space for $f(x_1, x_2) = x_1 + x_2^{3/2}$:

   1a. For $-\infty < x_1 < \infty$, $0 \leq x_2 < \infty$
   1b. For $-\infty < x_1 < \infty$, $0 < x_2 < \infty$

2. Let

$$\dot{x}_1 = \frac{1}{t}$$

$$\dot{x}_2 = x_1$$

   Assume $0 < t < \infty$. What class of functions are x_1 and x_2? (Answer in terms of P.C., C^0, C^1, etc.)

References

E.T. Bell, *Men* [sic] *of Mathematics* (Simon and Schuster, New York, 1965)

G.A. Bliss, *Lectures on the Calculus of Variations*. Phoenix Science Series (The University of Chicago Press, Chicago, 1968)

O. Bolza, *Lectures on the Calculus of Variations* (Dover, New York, 1961)

A.E. Bryson Jr., Y.C. Ho, *Applied Optimal Control* (Hemisphere Publishing, Washington, D.C., 1975)

D.G. Hull, *Optimal Control Theory for Applications* (Springer, New York, 2003)

W. Kaplan, *Operational Methods for Linear Systems* (Addison-Wesley, Reading, 1962)

C. Lanczos, *The Variational Principles of Mechanics*, 4th edn. (Dover, New York, 1986)

R.F. Stengel, *Optimal Control and Estimation* (Dover, New York, 1994)

J. Vagners, Optimization techniques, in *Handbook of Applied Mathematics*, ed. by C.E. Pearson, 2nd edn. (Van Nostrand Reinhold, New York, 1983), pp. 1140–1216

Chapter 3

The Euler-Lagrange Theorem

3.1. The Variation

The brachistochrone problem posed by Johann Bernoulli was a new type of mathematical problem which required a new mathematical approach. Lagrange developed the calculus of variations in which he considered suboptimal paths nearby the optimal one. He then showed that, for arbitrary but infinitesimal variations from the optimal path, the function sought must obey a differential equation now known as the Euler-Lagrange equation.

Let us consider a generalization of Lagrange's original technique in which $x(t, \varepsilon)$ represents a family of curves that are near the optimal path. This one-parameter family is illustrated in Fig. 3.1 for different, infinitesimally small values of ε (e.g., $\varepsilon_1, \varepsilon_2, \varepsilon_3, \ldots$). When ε is set to zero, the curve is the optimal path. Of course we don't know $x(t, 0)$, but we assume it exists and seek conditions which will lead to its solution. Note that varying ε has just as significant an effect as varying time. That is, for a given time, t_1, the value of $x(t_1, \varepsilon)$ changes with ε. In our search for the optimal path we will find that derivatives with respect to parameters (such as ε), as well as derivatives with respect to time, must be taken. Figure 3.2 depicts the optimal solution, $x(t, 0)$, and a nearby non-optimal solution, $x(t, \varepsilon)$.

Now, let us expand $x(t, \varepsilon)$ in a Taylor series about $\varepsilon = 0$ at time t:

$$\begin{aligned} x(t, \varepsilon) &= x(t, 0) + \left.\frac{\partial x}{\partial \varepsilon}\right|_{\varepsilon=0} (\varepsilon - 0) + \mathcal{O}(\varepsilon^2) \\ &\approx x(t, 0) + \left.\frac{\partial x}{\partial \varepsilon}\right|_{\varepsilon=0} \varepsilon \end{aligned} \tag{3.1}$$

where we neglect terms of order ε^2 (and above) as indicated by the "Big $\mathcal{O}$" symbol, $\mathcal{O}(\varepsilon^2)$. The *first variation* of a function $x(t, \varepsilon)$ at time t is defined as:

$$\delta x(t) \equiv \left.\frac{\partial x(t, \varepsilon)}{\partial \varepsilon}\right|_{\varepsilon=0} \varepsilon \tag{3.2}$$

Figure 3.2 illustrates the first variation. Another form of the first variation originally used by Lagrange is:

$$\delta x(t) = \varepsilon \eta(t) \tag{3.3}$$

J.M. Longuski et al., *Optimal Control with Aerospace Applications*,
Space Technology Library 32, DOI 10.1007/978-1-4614-8945-0_3,

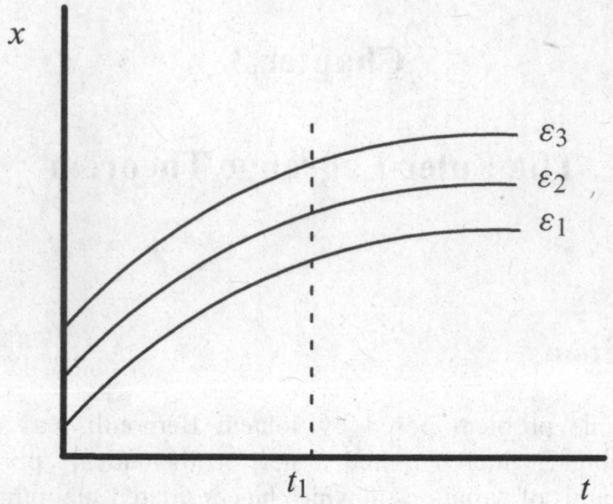

Figure 3.1. **One-parameter family of curves, $x(t, \varepsilon)$. During a variation the value of x changes with ε at a fixed time, t_1.**

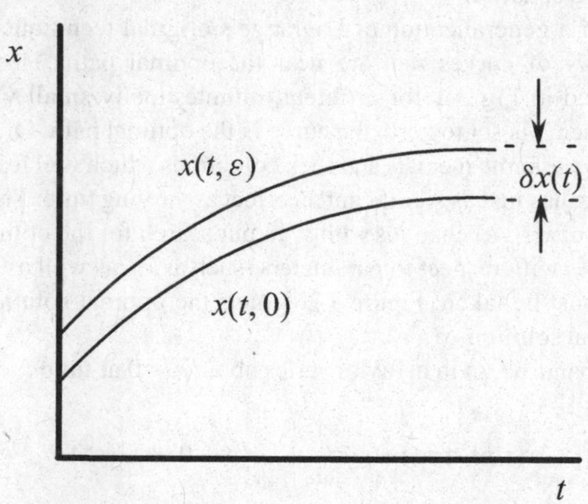

Figure 3.2. **Retaining ε to first-order expansion of Eq. (3.1) about $\varepsilon = 0$ provides the first variation $\delta x \equiv \left.\frac{\partial x}{\partial \varepsilon}\right|_{\varepsilon=0} \varepsilon$.**

We see in Eq. (3.3) the essential features of the variation. First, the amplitude is determined by ε which is always assumed to be infinitesimally small. Second, the function $\eta(t)$ is arbitrary. That is, the function $\eta(t)$ represents an arbitrarily large set of virtual functions; it is not itself a particular function.

Of course, we can always choose specific functions and examine their variations.

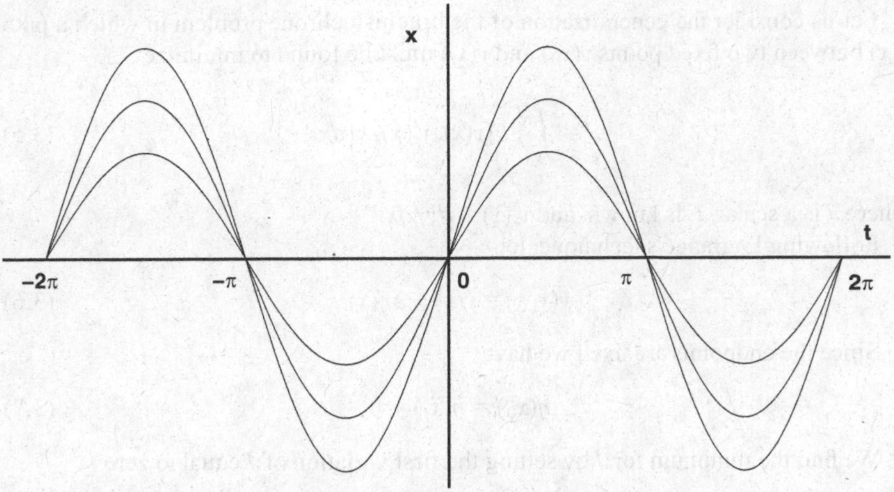

Figure 3.3. An example of expansion about $\varepsilon = 0$ from Eq. (3.4).

Example 3.1 Assumed varied path.

If we let $x(t, \varepsilon) = (\varepsilon + \varepsilon^2)\sin(t)$, then $x(t, 0) = 0$ is the optimal solution and

$$\begin{aligned}\delta x(t) &= \left.\frac{\partial x}{\partial \varepsilon}\right|_{\varepsilon=0} \varepsilon \\ &= [1 + 2\varepsilon]|_{\varepsilon=0}\, \varepsilon \sin t \\ &= \varepsilon \sin t \end{aligned} \tag{3.4}$$

Figure 3.3 illustrates this particular variation. In general the variation merely creates a perturbation on the amplitude.

3.2. The Euler-Lagrange Equation and the Brachistochrone Problem

Before introducing the Euler-Lagrange theorem, which provides the necessary conditions for trajectory optimization (such as the problem of launching a satellite into orbit), we derive the Euler-Lagrange equation and apply it to the brachistochrone problem.

The Euler-Lagrange equation is simpler to derive than the theorem because the Euler-Lagrange equation only solves for a path, such as the shape of $y(x)$ for the brachistochrone problem. The Euler-Lagrange theorem applies to the more difficult case (of optimal control) in which a control input is involved, such as the steering of the thrust vector on a launch vehicle.

Let us consider the generalization of the brachistochrone problem in which a path $y(x)$ between two fixed points $y(x_0)$ and $y(x_f)$ must be found to minimize

$$J = \int_{x_0}^{x_f} F[y(x), y'(x), x]\, dx \tag{3.5}$$

where J is a scalar, F is known, and $y'(x) = dy/dx$.

Following Lagrange's technique, let

$$y(x, \varepsilon) = y(x) + \varepsilon \eta(x) \tag{3.6}$$

Since the endpoints are fixed we have

$$\eta(x_0) = \eta(x_f) = 0 \tag{3.7}$$

We find the minimum for J by setting the first variation of J equal to zero

$$\delta J = \left.\frac{dJ}{d\varepsilon}\right|_{\varepsilon=0} \varepsilon = 0 \tag{3.8}$$

Thus we must have

$$\begin{aligned} \left.\frac{dJ}{d\varepsilon}\right|_{\varepsilon=0} &= \int_{x_0}^{x_f} \left(\frac{\partial F}{\partial y} \left.\frac{\partial y}{\partial \varepsilon}\right|_{\varepsilon=0} + \frac{\partial F}{\partial y'} \left.\frac{\partial y'}{\partial \varepsilon}\right|_{\varepsilon=0} \right) dx \\ &= \int_{x_0}^{x_f} \left[\frac{\partial F}{\partial y} \eta(x) + \frac{\partial F}{\partial y'} \eta'(x) \right] dx = 0 \end{aligned} \tag{3.9}$$

Now, by what Lanczos [1986] called "an ingenious application of the method of integration by parts," we write

$$\int_{x_0}^{x_f} \frac{\partial F}{\partial y'} \eta'(x) dx = \left[\frac{\partial F}{\partial y'} \eta(x) \right]_{x_0}^{x_f} - \int_{x_0}^{x_f} \eta(x) \frac{d}{dx} \left(\frac{\partial F}{\partial y'} \right) dx \tag{3.10}$$

From Eq. (3.7), the first term on the right side drops out. Substituting the remaining term on the right hand side of Eq. (3.10) into Eq. (3.9) we obtain

$$\int_{x_0}^{x_f} \left[\frac{\partial F}{\partial y} - \frac{d}{dx} \left(\frac{\partial F}{\partial y'} \right) \right] \eta(x)\, dx = 0 \tag{3.11}$$

Because $\eta(x)$ is arbitrary (except for restriction upon continuity and the end conditions) it follows that for Eq. (3.11) to hold, a necessary condition is that the integrand vanishes. Thus, we have

$$\frac{\partial F}{\partial y} - \frac{d}{dx} \left(\frac{\partial F}{\partial y'} \right) = 0 \tag{3.12}$$

which is the *Euler-Lagrange equation*.

Next we re-examine the brachistochrone problem of Chap. 2. [See Fig. 2.3 and Eq. (2.18).] The time to be minimized is

$$t = \int_0^{x_f} \frac{\sqrt{1+y'^2}}{\sqrt{2gx}}\,dx \tag{3.13}$$

where we must find the time-optimal path $y = y(x)$. The Euler-Lagrange equation, Eq. (3.12), must be satisfied. Since F, the integrand of Eq. (3.13), is not an explicit function of y the first term of Eq. (3.12) is zero so the term inside the parentheses must be constant:

$$\frac{\partial F}{\partial y'} = \frac{y'}{\sqrt{2gx(1+y'^2)}} = c \tag{3.14}$$

Equation (3.14) can be rearranged as

$$\frac{dy}{dx} = \sqrt{\frac{2gc^2x}{1-2gc^2x}} \tag{3.15}$$

The (not immediately obvious) trigonometric substitution

$$x = a(1-\cos\theta) \tag{3.16}$$

solves the problem where $a = 1/(4gc^2)$. By substituting Eq. (3.16) into Eq. (3.15), we obtain (after simplification):

$$y = \int a(1-\cos\theta)\,d\theta \tag{3.17}$$

which provides

$$y = a(\theta - \sin\theta) \tag{3.18}$$

Equations (3.16) and (3.18) are the parametric equations of a cycloid that begins at the origin. In this analysis, we have shown that the cycloid provides a stationary value for time, but not necessarily the minimum time. Greenwood [1997] points out that a comparison of nearby trajectories confirms that the cycloid does indeed give the minimum-time path.

Example 3.2 A simple problem in calculus of variations.

Assume an optimal $x^*(t)$ exists and take an assumed varied path:

$$x(t,\varepsilon) = x^*(t) + \varepsilon \sin t \tag{3.19}$$

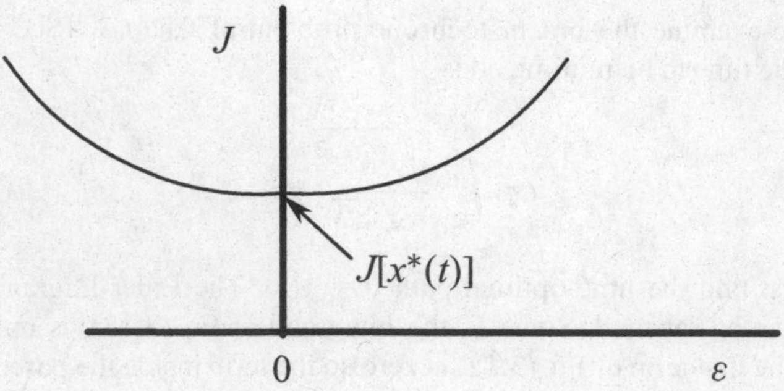

Figure 3.4. **MInimizing the cost functional $J(\varepsilon)$, as in the case of Eq. (3.23), means that $J(0) = J_{min}$.**

In this case the first variation, $\delta x(t) = \varepsilon \sin t$, is the same as given in Eq. (3.4) and illustrated in Fig. 3.3. Find a necessary condition for the problem to minimize:

$$J = \int_0^{\pi} F(t, x, \dot{x}) dt \tag{3.20}$$

with boundary conditions:

$$x(0) = x_o \tag{3.21}$$

$$x(\pi) = x_f \tag{3.22}$$

We can write the cost functional as $J(\varepsilon)$:

$$J(\varepsilon) = \int_0^{\pi} F[t, x^* + \varepsilon \sin t, \dot{x}^* + \varepsilon \cos t] dt \tag{3.23}$$

since $\dot{x}(t, \varepsilon) = \dot{x}^* + \varepsilon \cos t$. Figure 3.4 represents the behavior of $J(\varepsilon)$. Differentiating with respect to ε, we have:

$$\frac{dJ}{d\varepsilon} = \int_0^{\pi} \left(\frac{\partial F}{\partial x} \sin t + \frac{\partial F}{\partial \dot{x}} \cos t \right) dt \tag{3.24}$$

Integrating the last term by parts we obtain:

$$\int_0^{\pi} \frac{\partial F}{\partial \dot{x}} \cos t dt = \left(\frac{\partial F}{\partial \dot{x}} \sin t \right) \Bigg|_0^{\pi} - \int_0^{\pi} \frac{d}{dt} \left(\frac{\partial F}{\partial \dot{x}} \right) \sin t \, dt \tag{3.25}$$

Therefore:

$$\left. \frac{dJ}{d\varepsilon} \right|_{\varepsilon=0} = \int_0^{\pi} \left[\frac{\partial F}{\partial x} - \frac{d}{dt} \left(\frac{\partial F}{\partial \dot{x}} \right) \right] \sin t \, dt = 0 \tag{3.26}$$

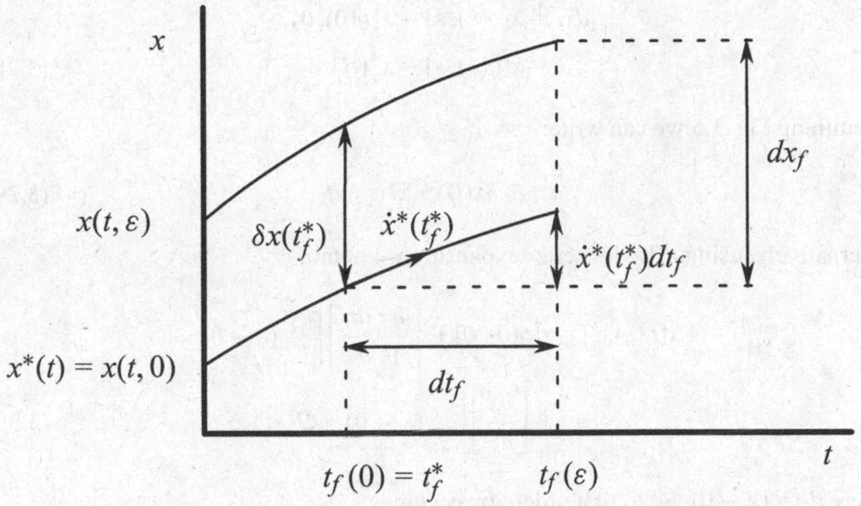

Figure 3.5. The variation at the final time is $\delta x^*(t_f)$; however, the difference in the final state, dx_f, can be greater due to the difference in the final time, dt_f.

In this simple problem, we again deduce that the Euler-Lagrange equation must hold (for this very specific variation):

$$\frac{\partial F}{\partial x} - \frac{d}{dt}\left(\frac{\partial F}{\partial \dot{x}}\right) = 0 \tag{3.27}$$

3.3. The Euler-Lagrange Theorem

To derive the necessary conditions of the Euler-Lagrange theorem, the change in J due to control variations δu (and hence δx) and to the differential change in the terminal time t_f must be found. Here we make an important distinction between the brachistochrone problem, where the endpoints of the trajectory are fixed, and the launch problem, where the final endpoint is not necessarily fixed. In the latter case we must allow nearby suboptimal paths to take more time to achieve orbit than the time-optimal path. In Fig. 3.5, we illustrate the complication that arises from the free final boundary condition. For simplicity, we only show one component of the state variable, where $x(t, \varepsilon)$ represents the varied path and $x^*(t) = x(t, 0)$ is the optimal path. Here we adopt notation in which the optimal path and optimal final time are indicated by an asterisk. The final time varies with nearby suboptimal paths and is represented by $t_f(\varepsilon)$. Thus, the optimal final time is given by $t_f^* = t_f(0)$.

We note in Fig. 3.5 that, at the optimal final time, the variation is $\delta x(t_f^*)$, but because the varied path can take longer (as in the case of a suboptimal launch trajectory requiring more time to achieve orbit than the optimal trajectory), the change in the final state (with respect to the optimal final state) is dx_f. Thus,

$$\begin{aligned} dx_f &\equiv x[t_f(\varepsilon), \varepsilon] - x[t_f(0), 0] \\ &= x[t_f(\varepsilon), \varepsilon] - x^*(t_f^*) \end{aligned} \tag{3.28}$$

Examining Fig. 3.5 we can write:

$$dx_f \approx \delta x(t_f^*) + \dot{x}^*(t_f^*)dt_f \tag{3.29}$$

Alternatively, using a Taylor series expansion we obtain

$$\begin{aligned} x[t_f(\varepsilon), \varepsilon] = x[t_f(0), 0] &+ \left[\frac{\partial x}{\partial t_f}\frac{dt_f}{d\varepsilon}\right]\Bigg|_{\varepsilon=0} (\varepsilon - 0) \\ &+ \left[\frac{\partial x}{\partial \varepsilon}\right]\Bigg|_{\varepsilon=0} (\varepsilon - 0) + \mathcal{O}(\varepsilon^2) \end{aligned} \tag{3.30}$$

where $d\varepsilon = (\varepsilon - 0)$. So to first order, dx_f becomes

$$\begin{aligned} dx_f &\approx \left\{\frac{\partial x[t_f(0), 0]}{\partial t_f}\right\} dt_f + \left\{\frac{\partial x[t_f(0), 0]}{\partial \varepsilon}\right\} \varepsilon \\ &\approx \dot{x}^*(t_f^*)dt_f + \delta x(t_f^*) \end{aligned} \tag{3.31}$$

as before.

3.3.1. Proof Outline of the Euler-Lagrange Theorem

Now we restate the Problem of Bolza (from Chap. 2) for which we will prove (or more precisely provide an outline of the proof of) the Euler-Lagrange theorem:

For a specified t_0 minimize:

$$J = \phi(t_f, \boldsymbol{x}_f) + \int_{t_o}^{t_f} L(t, \boldsymbol{x}, \boldsymbol{u})dt \tag{3.32}$$

subject to:

$$\dot{\boldsymbol{x}} = \boldsymbol{f}(t, \boldsymbol{x}, \boldsymbol{u}) \quad \text{Process Equations} \tag{3.33a}$$

$$\boldsymbol{x}(t_o) = \boldsymbol{x}_o \quad \text{Initial Conditions (I.C.s)} \tag{3.33b}$$

$$\boldsymbol{\Psi}(t_f, \boldsymbol{x}_f) = \boldsymbol{0} \quad \text{Terminal Constraints} \tag{3.33c}$$

where,

$\boldsymbol{x}$ is an n-vector

$\boldsymbol{u}$ is an m-vector

$\boldsymbol{\Psi}$ is a q-vector with $0 \leq q \leq n$

and J is the cost. The upper bound q on the dimension of Ψ is based on the number of independent constraints that determine the final values of all n state variables. [Note: in the present counting scheme, Eq. (3.33b) represents n I.C.s and does not count t_o as an I.C. Later, in Sect. 3.3.3 we will include t_o as an I.C. to obtain $n+1$ I.C.s.]

Assume $\phi, L, f, \Psi \in C^1$ on their respective domains and that the optimal control, $u^*(t)$, is unconstrained. If $u^*(t) \in C^0[t_o, t_f]$ minimizes J, then the Euler-Lagrange theorem states that there exist a time-varying multiplier vector $\boldsymbol{\lambda}^T(t) = (\lambda_1, \lambda_2, \ldots, \lambda_n)$ and a constant multiplier vector $\boldsymbol{\nu}^T = (\nu_1, \nu_2, \ldots, \nu_q)$ such that with the Hamiltonian

$$H(t, \boldsymbol{x}, \boldsymbol{u}, \boldsymbol{\lambda}) \equiv L(t, \boldsymbol{x}, \boldsymbol{u}) + \boldsymbol{\lambda}^T \boldsymbol{f}(t, \boldsymbol{x}, \boldsymbol{u}) \tag{3.34a}$$

and a terminal function

$$\Phi(t_f, \boldsymbol{x}_f) \equiv \phi(t_f, \boldsymbol{x}_f) + \boldsymbol{\nu}^T \boldsymbol{\Psi}(t_f, \boldsymbol{x}_f) \tag{3.34b}$$

the following necessary conditions must hold:

$$\dot{\boldsymbol{\lambda}}^T = -\frac{\partial H^*}{\partial \boldsymbol{x}} = -H_x^* \tag{3.35a}$$

$$\boldsymbol{\lambda}^T(t_f) = \frac{\partial \Phi^*}{\partial \boldsymbol{x}_f} \tag{3.35b}$$

$$H_u^* = \boldsymbol{0}^T \tag{3.35c}$$

and the *transversality condition*:

$$\Omega(t_f, \boldsymbol{x}_f, \boldsymbol{u}_f) \equiv L_f^* + \frac{d\Phi^*}{dt_f} = 0 \tag{3.35d}$$

which applies only if t_f is unspecified (i.e. $dt_f \neq 0$).

The Euler-Lagrange theorem [Eqs. (3.35)] assumes there exists a one-parameter family $\boldsymbol{u}(t, \varepsilon)$ which satisfies the constraints and $\boldsymbol{u}(t, 0) = \boldsymbol{u}^*(t)$. We also note that there exists an associated family $\boldsymbol{x}(t, \varepsilon)$ formed by integrating $\dot{\boldsymbol{x}} = \boldsymbol{f}[t, \boldsymbol{x}, \boldsymbol{u}(t, \varepsilon)]$ with $\boldsymbol{x}(t_o) = \boldsymbol{x}_o$ where $\boldsymbol{x}^*(t) \in C^1$. (The state variable is usually one order higher in continuity class than the control variable due to the integration of $\dot{\boldsymbol{x}} = \boldsymbol{f}$.)

Proof Outline of the Euler-Lagrange theorem. Let $\boldsymbol{u}(t, \varepsilon)$ be a one-parameter family of admissible controls with $\boldsymbol{u}^*(t) = \boldsymbol{u}(t, 0)$. Augment the cost functional with the constraints by some as yet undefined variables $\lambda_1(t), \ldots, \lambda_n(t)$ and $\nu_1, \ldots, \nu_q$:

$$J(\varepsilon) = \phi\{t_f(\varepsilon), \boldsymbol{x}[t_f(\varepsilon), \varepsilon]\} + \boldsymbol{\nu}^T \boldsymbol{\Psi}\{t_f(\varepsilon), \boldsymbol{x}[t_f(\varepsilon), \varepsilon]\}$$
$$+ \int_{t_o}^{t_f(\varepsilon)} \left(L[t, \boldsymbol{x}(t, \varepsilon), \boldsymbol{u}(t, \varepsilon)] + \boldsymbol{\lambda}^T(t) \{\boldsymbol{f}[t, \boldsymbol{x}(t, \varepsilon), \boldsymbol{u}(t, \varepsilon)] - \dot{\boldsymbol{x}}(t, \varepsilon)\} \right) dt \tag{3.36}$$

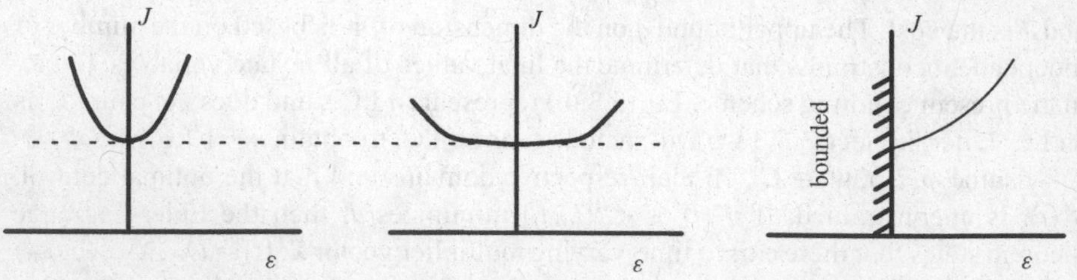

Figure 3.6. Possible graphs of J. The case of $\left.\frac{dJ}{d\varepsilon}\right|_{\varepsilon=0} > 0$ can occur for bounded control problems which are discussed in Chap. 9.

Equation (3.36) provides the cost, $J(\varepsilon)$, for nearby suboptimal solutions. By definition, $J(0)$ is the minimum cost. Following the method of Lagrange, we will show that the problem of finding the control $\boldsymbol{u}^*(t)$ to minimize J can be converted into the problem of solving the algebraic and differential equations, Eqs. (3.35). (The meaning and methods of solving these equations will be discussed later.) We see in the integrand of Eq. (3.36) that the term $\boldsymbol{\lambda}^T\{\boldsymbol{f}-\dot{\boldsymbol{x}}\}$ is zero on the optimal trajectory due to the process equations, Eq. (3.33a). Similarly, the term $\boldsymbol{\nu}^T\boldsymbol{\Psi}$ is zero due to Eq. (3.33c).

Figure 3.6 illustrates how the cost of $J(\varepsilon)$ may vary with ε. For unbounded controls we have:

$$\left.\frac{dJ}{d\varepsilon}\right|_{\varepsilon=0} = 0 \tag{3.37}$$

corresponding to the two diagrams on the left in Fig. 3.6.

For bounded controls:

$$\left.\frac{dJ}{d\varepsilon}\right|_{\varepsilon=0} \geq 0 \tag{3.38}$$

as illustrated in the right-most diagram of Fig. 3.6. In the present proof we assume the control is unconstrained so that only Eq. (3.37) holds. We consider the problem of constrained (or bounded) controls later in the text.

Using the definition of the Hamiltonian, Eq. (3.34a), and the terminal function, Eq. (3.34b), in Eq. (3.36) we obtain

$$J(\varepsilon) = \Phi\{t_f(\varepsilon), \boldsymbol{x}[t_f(\varepsilon), \varepsilon]\} + \int_{t_o}^{t_f(\varepsilon)} \left\{ H[t, \boldsymbol{x}(t,\varepsilon), \boldsymbol{u}(t,\varepsilon), \boldsymbol{\lambda}(t)] - \boldsymbol{\lambda}^T(t)\dot{\boldsymbol{x}}(t,\varepsilon) \right\} dt \tag{3.39}$$

Next we need Leibniz' rule to form $dJ/d\varepsilon$ of the functional $J(\varepsilon)$:

$$\frac{d}{d\varepsilon}\left(\int_{a(\varepsilon)}^{b(\varepsilon)} f(x,\varepsilon)dx\right) = f[b(\varepsilon),\varepsilon]\frac{db(\varepsilon)}{d\varepsilon} - f[a(\varepsilon),\varepsilon]\frac{da(\varepsilon)}{d\varepsilon} + \int_{a(\varepsilon)}^{b(\varepsilon)} \frac{\partial f(x,\varepsilon)}{\partial \varepsilon}dx \quad (3.40)$$

We note that ε appears in the upper bound, $t_f(\varepsilon)$, of the integral in Eq. (3.39) and that ε also appears in the Hamiltonian through $\boldsymbol{x}(t,\varepsilon)$ and $\boldsymbol{u}(t,\varepsilon)$ since $H = H[t,\boldsymbol{x}(t,\varepsilon),\boldsymbol{u}(t,\varepsilon),\boldsymbol{\lambda}(t)]$. Thus, we can write

$$\frac{dJ}{d\varepsilon} = \frac{d\Phi}{d\varepsilon} + L[t_f(\varepsilon)]\frac{dt_f(\varepsilon)}{d\varepsilon} + \int_{t_o}^{t_f(\varepsilon)} \left(\frac{\partial H}{\partial \boldsymbol{x}}\frac{\partial \boldsymbol{x}}{\partial \varepsilon} + \frac{\partial H}{\partial \boldsymbol{u}}\frac{\partial \boldsymbol{u}}{\partial \varepsilon} - \boldsymbol{\lambda}^T(t)\frac{\partial \dot{\boldsymbol{x}}}{\partial \varepsilon}\right) dt \quad (3.41)$$

where $\boldsymbol{\lambda}(t)$ is a function of time only. Setting $\varepsilon = 0$ we obtain

$$\left.\frac{dJ}{d\varepsilon}\right|_{\varepsilon=0} = \frac{d\Phi^*}{d\varepsilon} + L^*(t_f^*)\frac{dt_f}{d\varepsilon} + \int_{t_o}^{t_f^*} \left(H_{\boldsymbol{x}}^* \left.\frac{\partial \boldsymbol{x}}{\partial \varepsilon}\right|_{\varepsilon=0} + H_{\boldsymbol{u}}^* \left.\frac{\partial \boldsymbol{u}}{\partial \varepsilon}\right|_{\varepsilon=0} - \boldsymbol{\lambda}^T(t) \left.\frac{\partial \dot{\boldsymbol{x}}}{\partial \varepsilon}\right|_{\varepsilon=0}\right) dt \quad (3.42)$$

We find $dJ|_{\varepsilon=0}$ by multiplying $\frac{dJ}{d\varepsilon}|_{\varepsilon=0}$ by $d\varepsilon = \varepsilon - 0$ and noting that $d\Phi = \frac{\partial\Phi}{\partial t_f}dt_f + \frac{\partial\Phi}{\partial \boldsymbol{x}_f}d\boldsymbol{x}_f$:

$$dJ|_{\varepsilon=0} = \frac{\partial\Phi^*}{\partial t_f}dt_f + \frac{\partial\Phi^*}{\partial \boldsymbol{x}_f}d\boldsymbol{x}_f + L^*(t_f^*)dt_f + \int_{t_o}^{t_f^*} [H_x^*\delta\boldsymbol{x}(t) + H_u^*\delta\boldsymbol{u}(t) - \boldsymbol{\lambda}^T(t)\delta\dot{\boldsymbol{x}}(t)]dt = 0 \quad (3.43)$$

where we have made use of the definition of the variation, Eq. (3.2), so that $\delta\boldsymbol{u} = \frac{\partial \boldsymbol{u}}{\partial\varepsilon}\big|_{\varepsilon=0}\varepsilon$ and $\delta\dot{\boldsymbol{x}} = \frac{\partial\dot{\boldsymbol{x}}}{\partial\varepsilon}\big|_{\varepsilon=0}\varepsilon$. The third term on the right-hand side of Eq. (3.43) is the contribution to dJ by the integral term in Eq. (3.32), namely the product of the integrand value at t_f times dt_f. Now let us integrate $\int_{t_o}^{t_f^*} -\boldsymbol{\lambda}^T(t)\delta\dot{\boldsymbol{x}}(t)dt$ by parts:

$$\int_{t_o}^{t_f^*} -\boldsymbol{\lambda}^T(t)\delta\dot{\boldsymbol{x}}(t)dt = [-\boldsymbol{\lambda}^T(t)\delta\boldsymbol{x}(t)]\Big|_{t_o}^{t_f^*} + \int_{t_o}^{t_f^*} \dot{\boldsymbol{\lambda}}^T(t)\delta\boldsymbol{x}dt \quad (3.44)$$

We recall that a similar step was made in the derivation of the Euler-Lagrange equation in Eq. (3.10).

Substituting Eq. (3.44) into Eq. (3.43), we obtain:

$$dJ|_{\varepsilon=0} = \frac{\partial\Phi^*}{\partial t_f}dt_f + \frac{\partial\Phi^*}{\partial \boldsymbol{x}_f}d\boldsymbol{x}_f + L^*(t_f^*)dt_f - \boldsymbol{\lambda}^T(t_f^*)\delta\boldsymbol{x}(t_f^*) + \int_{t_o}^{t_f^*} [H_x^*\delta\boldsymbol{x}(t) + H_u^*\delta\boldsymbol{u}(t) + \dot{\boldsymbol{\lambda}}^T(t)\delta\boldsymbol{x}(t)]dt = 0 \quad (3.45)$$

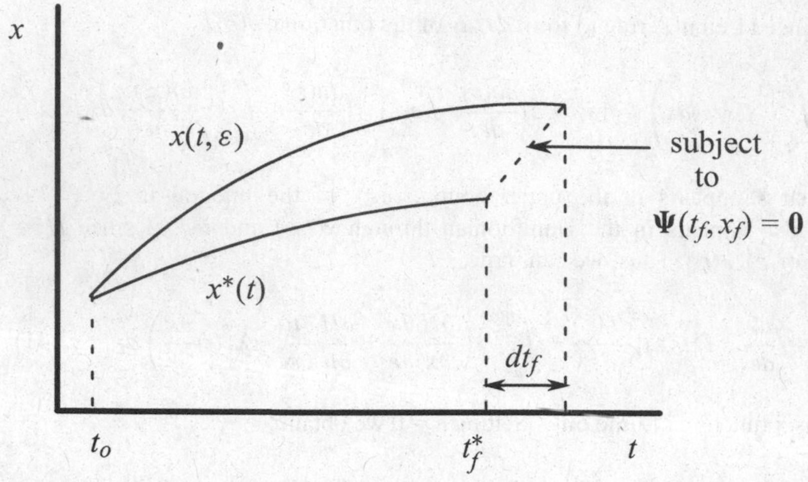

Figure 3.7. **When t_f is free the terminal constraint, $\Psi(t_f, x_f)$ must be satisfied on the varied path.**

where we have used $\delta\boldsymbol{x}(t_o) = 0$. Keeping in mind that the final time, t_f, may be free (as shown in Fig. 3.7), we will replace $\delta\boldsymbol{x}(t_f^*)$ with $\delta\boldsymbol{x}(t_f^*) = d\boldsymbol{x}_f - \dot{\boldsymbol{x}}_f^* dt_f$ [according to Eq. (3.29)], therefore:

$$dJ|_{\varepsilon=0} = \left[\frac{\partial\Phi^*}{\partial t_f} + L^*(t_f^*) + \boldsymbol{\lambda}^T(t_f^*)\dot{\boldsymbol{x}}_f^*\right] dt_f + \left[\frac{\partial\Phi^*}{\partial \boldsymbol{x}_f} - \boldsymbol{\lambda}^T(t_f^*)\right] d\boldsymbol{x}(t_f)$$
$$+ \int_{t_o}^{t_f^*} \{[H_x^* + \dot{\boldsymbol{\lambda}}^T]\delta\boldsymbol{x}(t) + H_u^*\delta\boldsymbol{u}(t)\}dt = 0 \qquad (3.46)$$

We now come to the determination of the Lagrange multipliers, also known as costate variables. For a stationary value of the cost J, the variation dJ in Eq. (3.46) must be zero. For this condition to be true for all $\delta\boldsymbol{x}(t)$ [for arbitrary $\delta\boldsymbol{x}(t)$] it is necessary that

$$\dot{\boldsymbol{\lambda}}^T(t) = -H_{\boldsymbol{x}}[t, \boldsymbol{x}^*(t), \boldsymbol{u}^*(t), \boldsymbol{\lambda}(t)] \qquad (3.47)$$

Similarly, to have $dJ = 0$ for all $d\boldsymbol{x}(t_f)$ it is necessary that

$$\boldsymbol{\lambda}^T(t_f^*) = \frac{\partial\Phi^*}{\partial \boldsymbol{x}_f} \qquad (3.48)$$

Note that Eq. (3.48) provides the required boundary condition to accompany Eq. (3.47). Next, $dJ = 0$ for all dt_f requires its coefficient to be zero in Eq. (3.46). Substituting in Eq. (3.48) this condition is

$$\Omega = \frac{d\Phi^*}{dt_f} + L^*(t_f^*) = 0 \qquad (3.49)$$

This relation is the *transversality condition* that applies only if t_f is unspecified (that is, $dt_f \neq 0$).

Equation (3.49) can be calculated as $\frac{\partial \Phi^*}{\partial t_f} + H^*(t_f^*) = 0$ although, strictly speaking, the arguments of the function Ω in Eq. (3.35d) do not include $\boldsymbol{\lambda}_f^*$ (on which H_f^* explicitly depends). Thus, the only term remaining in Eq. (3.46) is:

$$\begin{aligned} dJ &= \int_{t_o}^{t_f^*} H_{\boldsymbol{u}}^* \delta \boldsymbol{u} dt \\ &= 0 \end{aligned} \tag{3.50}$$

so we conclude that because $\delta \boldsymbol{u}$ is arbitrary, a necessary condition is

$$H_{\boldsymbol{u}}^* = \boldsymbol{0}^T \tag{3.51}$$

which is valid only for unconstrained control. We note that this part of the proof is not rigorous, because $\delta \boldsymbol{u}$ is not arbitrary if there are terminal constraints $\boldsymbol{\Psi}(t_f, \boldsymbol{x}_f) = \boldsymbol{0}$. The admissible $\delta \boldsymbol{u}$ are only those that generate $\delta \boldsymbol{x}$ that satisfy $\boldsymbol{\Psi} = \boldsymbol{0}$ (see Fig. 3.7), but a more rigorous treatment in Bryson and Ho [1975] shows that Eq. (3.51) is a correct necessary condition, even with terminal constraints.

This completes the proof outline of the Euler-Lagrange theorem.

If we retain the term $\boldsymbol{\lambda}^T(t_o)\delta \boldsymbol{x}(t_o)$ from Eq. (3.44) in Eq. (3.45) rather than eliminate it using $\delta \boldsymbol{x}(t_o) = 0$, two useful facts emerge: (i) if any component $x_i(t_o)$ is *not* specified [i.e. $\delta x_i(t_o) \neq 0$], then the condition that will determine its optimal value $x_i^*(t_o)$ is that the corresponding $\lambda_i(t_o) = 0$; and (ii) this provides an *interpretation* for the Lagrange multiplier λ. Each component $\lambda_i(t_o)$ is the first-order sensitivity (gradient) of the cost J due to a differential change in the initial state component $x_i(t_o)$, $\lambda_i(t_o) = \partial J/\partial x_i(t_0)$.

Along an optimal solution, any time t is the "initial time" for the remainder of the solution, so this interpretation of λ also applies at any time t. Because of this interpretation, one can evaluate the small change in cost due to a small change in state at the initial time (or any other time due to a disturbance) by using the value of λ rather than completely re-solving the problem. This observation is especially useful if solutions are determined numerically.

We also note that the function Ω in Eq. (3.35d), which is the coefficient of dt_f in Eq. (3.46), has the interpretation that it is the sensitivity of the change in cost J due to a differential change in the final time t_f, $\Omega = \partial J/\partial t_f$. So when the final time is specified, the algebraic sign of Ω indicates whether a small increase or decrease in the final time will lower the cost and provides a first-order estimate of the cost change.

As shown in Exercise 5, the components of the Lagrange multiplier $\boldsymbol{\nu}$, introduced in Eq. (3.34b), have the interpretation that they relate the change in the cost due to small changes in the constants in the terminal constraints. Again, this is especially useful if solutions are determined numerically, because we do not have to completely re-solve the problem.

3.3.2. Summary of the Euler-Lagrange Theorem

For a minimum of J, the following set of necessary conditions must be met:

$$\dot{\boldsymbol{x}} = \boldsymbol{f}(t, \boldsymbol{x}, \boldsymbol{u}) = H_{\boldsymbol{\lambda}}^T \tag{3.52a}$$

$$\dot{\boldsymbol{\lambda}} = -H_{\boldsymbol{x}}^T \tag{3.52b}$$

$$H_{\boldsymbol{u}}^T = \boldsymbol{0} \tag{3.52c}$$

where Eqs. (3.52a) and (3.52b) each represent n differential equations and Eq. (3.52c) represents m algebraic equations. In addition we have the end conditions:

$$\boldsymbol{x}(t_o) = \boldsymbol{x}_o \tag{3.53a}$$

$$\boldsymbol{\lambda}^T(t_f) = \frac{\partial \Phi}{\partial \boldsymbol{x}_f} \tag{3.53b}$$

$$\boldsymbol{\Psi}(\boldsymbol{x}_f, t_f) = \boldsymbol{0} \tag{3.53c}$$

where Eq. (3.53a) represents n initial conditions (not including the initial time) and Eq. (3.53c) represents q boundary conditions where $q \leq n$. Finally we have the transversality condition if $dt_f \neq 0$:

$$\Omega = L_f + \frac{d\Phi}{dt_f} = 0 \tag{3.54}$$

which can be calculated using the comment after Eq. (3.49) and in Exercise 4. The m algebraic equations obtained from Eq. (3.52c) provide the optimal control, u^*.

There are $2n + m + q + 1$ unknown variables $\boldsymbol{x}, \boldsymbol{\lambda}, \boldsymbol{u}, \boldsymbol{\nu}$, and t_f, and there are a corresponding number of equations to solve for them: Eq. (3.52a) with B.C.s (3.53a), Eq. (3.52b) with B.C.s (3.53b), (3.52c), (3.53c), and (3.54).

Examples of how the Euler-Lagrange theorem leads to a two-point boundary-value problem are given in Chap. 4.

In the next section an alternative formulation of the necessary conditions is discussed which is similar to what has been described above, except that the terminal constraints are satisfied by manually enforcing (directly solving) $d\boldsymbol{\Psi} = \boldsymbol{0}$, rather than by using the Lagrange multiplier vector, $\boldsymbol{\nu}$.

3.3.3. Alternate Form of the Transversality Condition

In Sect. 3.3.1, the terminal constraints, $\boldsymbol{\Psi}(t_f, \boldsymbol{x}_f)$, were adjoined to the cost functional, J, through the use of the additional multiplier vector, $\boldsymbol{\nu}$, as given in Eq. (3.34b). This method has a certain elegance in its structure, and has been adopted by most modern authorities following Bryson and Ho [1975]. It does provide the q components of $\boldsymbol{\nu}$ and these components provide additional information about the optimal solution, namely, the change in the cost due to small changes in the terminal constraints. (See Exercise 5).

There is an alternative approach in which the prescribed terminal constraints are not adjoined to the cost functional, so that the ν_i never appear. In this approach, the transversality condition appears in a differential form which, when combined with the differentials of the terminal constraints, provides the natural boundary conditions.

Citron [1969], Hestenes [1966], Kenneth and McGill [1966], and Pierre [1969] all use the differential form of the transversality condition. Citron demonstrates the equivalence of both methods of employing the transversality condition in an example and states, "It is up to the user to decide which method better suits his [or her] needs in any particular problem."

We can derive the differential form of the transversality condition by a slight modification of the material in Sect. 3.3.1. We begin by dropping $\boldsymbol{\nu}$ from Eq. (3.34b) and, of course, from Eq. (3.36) and note that the terminal B.C.s, $\boldsymbol{\Psi}(\boldsymbol{x}_f, t_f) = \mathbf{0}$, are written as a p-vector where $1 \leq p \leq n+1$. As a result, $\boldsymbol{\Psi}$ must provide at least one stopping condition and may contain all of the final states and the final time. The only change in Eq. (3.39) is that Φ is replaced by ϕ. Equation (3.43) becomes

$$dJ|_{\varepsilon=0} = d\phi^* + (H^* - \boldsymbol{\lambda}^T\dot{\boldsymbol{x}}^*)|_{t_f^*} dt_f + \int_{t_o}^{t_f^*} [H_{\boldsymbol{x}}^*\delta\boldsymbol{x}(t) + H_{\boldsymbol{u}}^*\delta\boldsymbol{u}(t) - \boldsymbol{\lambda}^T(t)\delta\dot{\boldsymbol{x}}(t)]dt$$
$$= 0 \tag{3.55}$$

where we retain $H^* - \boldsymbol{\lambda}^T\dot{\boldsymbol{x}}^*$ in lieu of L^*.

After integration by parts [Eq. (3.44)], we obtain

$$dJ|_{\varepsilon=0} = d\phi^* + (H^* - \boldsymbol{\lambda}^T\dot{\boldsymbol{x}}^*)\Big|_{t_f^*} dt_f - \boldsymbol{\lambda}^T(t_f^*)\delta\boldsymbol{x}(t_f^*)$$
$$+ \int_{t_o}^{t_f^*} [H_{\boldsymbol{x}}^*\delta\boldsymbol{x}(t) + H_{\boldsymbol{u}}^*\delta\boldsymbol{u}(t) + \dot{\boldsymbol{\lambda}}^T(t)\delta\boldsymbol{x}]dt$$
$$= 0 \tag{3.56}$$

where we have used $\delta\boldsymbol{x}(t_o) = 0$, as before.

After replacing $\delta\boldsymbol{x}(t_f^*)$ with $\delta\boldsymbol{x}(t_f^*) = d\boldsymbol{x}_f - \dot{\boldsymbol{x}}_f^* dt_f$ we have the analog to Eq. (3.46):

$$dJ\big|_{\varepsilon=0} = d\phi^* + H_f^* dt_f - \boldsymbol{\lambda}^T(t_f^*)d\boldsymbol{x}_f$$
$$+ \int_{t_o}^{t_f^*} [(H_{\boldsymbol{x}}^* + \dot{\boldsymbol{\lambda}}^T)\delta\boldsymbol{x} + H_{\boldsymbol{u}}^*\delta\boldsymbol{u}]dt$$
$$= 0 \tag{3.57}$$

which, as illustrated in Fig. 3.7, is subject to the final boundary conditions:

$$d\boldsymbol{\Psi} = \boldsymbol{\Psi}_{t_f}^* dt_f + \boldsymbol{\Psi}_{x_f} d\boldsymbol{x}_f = 0 \tag{3.58}$$

We note that in this formulation most authors assume that $\boldsymbol{\Psi}$ is a p-vector where

$$1 \leq p \leq n+1 \tag{3.59}$$

so that up to $n+1$ constraints can be enforced, allowing all the final states and the final time to be specified. Equations (3.58) and (3.59) are the basis of the *alternate form of the transversality condition*. Rather than being a q-vector, where $0 \leq q \leq n$, $\boldsymbol{\Psi}$ is now a p-vector, where $1 \leq p \leq n+1$. Thus, there is always at least one terminal constraint. *If the value of the final time is specified, it becomes a component of the* $\boldsymbol{\Psi}$ *vector*. By contrast, in Sect. 3.3.1 a specified final time is treated as a separate constraint independent of the $\boldsymbol{\Psi}$ vector. If the final time is unspecified, some component of the $\boldsymbol{\Psi}$ vector acts as a "stopping condition" that determines the value of the final time, e.g. final altitude, final velocity, etc.

The variables p and q are related in a simple manner: $p = q$ unless the final time is specified, in which case $p = q + 1$. Then the first component, Ψ_1, is equal to $t_f - t_f^*$.

As in Eq. (3.47), we select

$$\dot{\boldsymbol{\lambda}}^T(t) = -H_{\boldsymbol{x}}[t, \boldsymbol{x}^*(t), \boldsymbol{u}^*(t), \boldsymbol{\lambda}(t)] \tag{3.60}$$

which eliminates the first parenthetical term in the integrand of Eq. (3.57). In order to eliminate the non-integral terms in Eq. (3.57):

$$H_f^* dt_f - \boldsymbol{\lambda}^T(t_f^*) d\boldsymbol{x}_f + d\phi^* = 0 \tag{3.61}$$

subject to the terminal constraint:

$$d\boldsymbol{\Psi} = 0 \tag{3.62}$$

The terminal cost term in Eq. (3.61) can be expanded as

$$d\phi = \left(\frac{\partial \phi}{\partial \boldsymbol{x}_f}\right) d\boldsymbol{x}_f + \left(\frac{\partial \phi}{\partial t_f}\right) dt_f \tag{3.63a}$$

Substituting Eq. (3.63a) into (3.61) yields

$$\left(\frac{\partial \phi^*}{\partial \boldsymbol{x}_f} - \boldsymbol{\lambda}^T(t_f^*)\right) d\boldsymbol{x}_f + \left(\frac{\partial \phi^*}{\partial t_f} + H_f^*\right) dt_f = 0 \tag{3.63b}$$

Because $d\boldsymbol{x}_f$ and dt_f (if it is non-zero) are independent and arbitrary, choose

$$\boldsymbol{\lambda}^T(t_f^*) = \frac{\partial \phi^*}{\partial \boldsymbol{x}_f} \tag{3.63c}$$

and

$$\frac{\partial \phi^*}{\partial t_f} + H_f^* = 0 \tag{3.63d}$$

Equation (3.63c) provides the boundary condition for Eqs. (3.60) and (3.63d) applies only if $dt_f \neq 0$ (t_f unspecified). In combination with $d\boldsymbol{\Psi} = \boldsymbol{0}$, Eq. (3.63c) is equivalent to Eqs. (3.53b) and (3.63d) is equivalent to Eq. (3.54). (See Exercise 4.)

Thus, the only term remaining in Eq. (3.57) is:

$$dJ = \int_{t_o}^{t_f^*} H_{\boldsymbol{u}}^* \delta \boldsymbol{u} dt$$

$$= 0 \tag{3.64}$$

so we conclude as before that a necessary condition is

$$H_{\boldsymbol{u}}^* = \boldsymbol{0}^T \tag{3.65}$$

which is valid only for unconstrained control.

To summarize the Euler-Lagrange theorem in which the transversality condition is expressed in differential form, the following set of necessary conditions must be met:

$$\dot{\boldsymbol{x}} = \boldsymbol{f}(t, \boldsymbol{x}, \boldsymbol{u}) = H_{\boldsymbol{\lambda}}^T \tag{3.66a}$$

$$\dot{\boldsymbol{\lambda}} = -H_{\boldsymbol{x}}^T \tag{3.66b}$$

$$H_{\boldsymbol{u}} = \boldsymbol{0}^T \tag{3.66c}$$

where Eqs. (3.66a) and (3.66b) each represent n differential equations and Eq. (3.66c) represents m algebraic equations. In addition we have the end conditions:

$$\boldsymbol{x}(t_o) = \boldsymbol{x}_o, \quad t_o \text{ specified} \tag{3.67a}$$

$$\boldsymbol{\Psi}(\boldsymbol{x}_f, t_f) = \boldsymbol{0} \tag{3.67b}$$

where Eq. (3.67a) represents $n + 1$ initial conditions and Eq. (3.67b) represents p terminal conditions where $1 \leq p \leq n+1$. Finally we have the *differential form of the transversality condition*:

$$H_f dt_f - \boldsymbol{\lambda}_f^T d\boldsymbol{x}_f + d\phi = 0 \tag{3.68}$$

which provides $n + 1 - p$ boundary conditions. As previously mentioned, Eq. (3.68) can be separated into Eqs. (3.63c) and (3.63d). [That is, in this counting scheme we need $2n + 2$ B.C.s to solve for $2n$ differential equations, Eqs. (3.66a) and (3.66b).

Since Eqs. (3.67a) and (3.67b) provide $n+1+p$ B.C.s, Eq. (3.68) provides the remaining conditions: $2n + 2 - (n + 1 + p) = n + 1 - p$.] Examples of how the Euler-Lagrange theorem leads to a two-point boundary-value problem using Eq. (3.68) are given in Chap. 4.

3.4. Summary

The problem posed by Johann Bernoulli inspired Lagrange to develop a new mathematical tool, the calculus of variations, to find a function which minimizes (or extremizes) a functional. Lagrange's method led to the Euler-Lagrange equation, which converts the problem of minimizing the integral into a problem of solving a differential equation.

In the optimization of space trajectories a similar problem arises, but is complicated by a choice of controls (such as choosing the steering law for the thrust vector) and by the possibility of having free final boundary conditions. However, the variational approach of Lagrange once again yields a set of differential equations that must apply for an optimal trajectory.

The Problem of Bolza is specified in Eq. (3.32) and consists of a terminal cost, which depends on the final time and final state, and a path cost, which depends on an integral over the path. The Bolza problem includes a set of process equations (usually the equations of motion) along with a specified set of initial and final boundary conditions, Eqs. (3.33).

The Euler-Lagrange theorem states that if $\boldsymbol{u}^*(t)$ is the optimal control (which is assumed to be continuous and unconstrained) that minimizes J of the Bolza problem, then a set of differential equations and algebraic equations, Eqs. (3.35), must be satisfied. Thus, along with the process (or state) equations, Eqs. (3.33a), the differential equations for the Lagrange multipliers (or costates), Eqs. (3.35a), must also be solved. These costate equations arose from the introduction of the Lagrange multipliers into the proof, and have an interesting and useful physical interpretation. A set of algebraic equations, Eqs. (3.35c), provides control laws for $\boldsymbol{u}(t)$. Equation (3.35d), which is called the transversality condition, provides the additional necessary boundary conditions to solve for the differential equations, Eqs. (3.33a) and (3.35a), as we will show later.

The proof of the Euler-Lagrange theorem depends on the definition of the Hamiltonian, Eq. (3.34a), and the use of the one-parameter families for the control and for the state, $\boldsymbol{u}(t, \varepsilon)$ and $\boldsymbol{x}(t, \varepsilon)$, respectively. The proof requires the application of Leibniz' rule and integration by parts. The arbitrary nature of the Lagrange multipliers permits selection of these functions and their final boundary conditions to simplify the proof. The proof consists of Eqs. (3.36) through (3.51).

It is important to note the limitations of the proof. We have assumed that the control is unconstrained (unbounded) and is continuous, and that the state is continuously differentiable. At this stage we have no theorem for the cases where the control is bounded or piecewise continuous. We will need other theorems discussed in later chapters for such problems.

One final remark will be made before moving on to applications of the Euler-Lagrange theorem. Are there other approaches besides the calculus of variations to solve the optimal control problem? With the availability of computational techniques, direct methods of solving Bolza's problem now exist. By direct, we mean the numerical construction of a control law and trajectory path which can be inserted into the integrand so that a numerical value of the cost can be calculated. Then a number of nearby trajectories can be submitted to calculate their costs. Another approach, discussed by Lanczos [1986], is to represent the path by a truncated Fourier series in which the coefficients are solved. This method changes the optimal control problem into a parameter optimization problem, but requires a small error to be accepted due to the truncation. Lanczos credits Hilbert for originating this concept of "function space." Further discussion of direct methods is outside the scope of this text. The goal of the present work is to provide an introduction to the classical indirect method—the use of the calculus of variations—to solve the Problem of Bolza.

3.5. Exercises

1. State and prove the Euler-Lagrange theorem using the same assumptions and techniques presented in Sect. 3.3.1 except for the following differences:

   1a. The initial state $\boldsymbol{x}(t_o)$ is free and the initial time t_o is free,

   1b. The boundary conditions are expressed as

$$\boldsymbol{\Psi}(t_o, \boldsymbol{x}_o, t_f, \boldsymbol{x}_f) = \boldsymbol{0} \tag{3.69}$$

   (i.e., in terms of initial and final conditions), and

   1c. $J = \phi(t_o, \boldsymbol{x_o}, t_f, \boldsymbol{x}_f) + \int_{t_o}^{t_f} L(t, \boldsymbol{x}, \boldsymbol{u})dt$.

   Hint: you should find that the additional necessary condition is $\boldsymbol{\lambda}^T(t_o) = -\partial\Phi/\partial\boldsymbol{x}_o$ and $\partial\Phi/\partial t_o - H(t_o) = 0$.

2. Repeat Exercise 1 using the assumptions of Sect. 3.3.3 in which the terminal costs are *not* adjoined to the cost functional.
   Hint: you should find that the new transversality conditions are $[Hdt - \boldsymbol{\lambda}^T d\boldsymbol{x}]\Big|_{t_o}^{t_f} + d\phi = 0$ subject to $d\boldsymbol{\Psi} = \boldsymbol{0}$.
3. Derive Eq. (3.49) by substituting Eq. (3.48) into Eq. (3.46).
4. Show that, as mentioned after Eq. (3.49), the variable Ω can be calculated as $\frac{\partial\Phi^*}{\partial t_f} + H^*(t_f^*) = 0$.
5. The terminal constraint, Eq. (3.53c), can be generalized to be $\boldsymbol{\Psi}(t_f, \boldsymbol{x}_f, \boldsymbol{c}) = \boldsymbol{0}$, when $\boldsymbol{c}$ is a vector of constants, such as the specified final altitude, etc. Show that the vector $\boldsymbol{\nu}$ introduced in Eq. (3.34b), has the physical interpretation of relating a change in the cost due to a small change in $\boldsymbol{c}$, namely $\frac{\partial J}{\partial \boldsymbol{c}} = \boldsymbol{\nu}^T\boldsymbol{\Psi}_c$. Also show that in component form we have $\frac{\partial J}{\partial c_i} = \sum_{j=1}^{q} \nu_j \frac{\partial \Psi_j}{\partial c_i}$.
   Hint: $\boldsymbol{\nu}^T d\boldsymbol{\Psi} = \boldsymbol{\nu}^T\left[\boldsymbol{\Psi}_{t_f} dt_f + \boldsymbol{\Psi}_{x_f} d\boldsymbol{x}_f + \boldsymbol{\Psi}_c d\boldsymbol{c}\right]$.

6. Zermelo's Problem
We wish to minimize the time for a boat to cross a river (see Fig. 2.10). As mentioned earlier, Zermelo's problem has an interesting aerospace application: that of an aircraft flying in a crosswind.

$$\text{Min } J = \int_{t_o}^{t_f} dt \tag{3.70}$$

subject to

$$\dot{x} = V\cos\theta + u(x, y) \tag{3.71a}$$

$$\dot{y} = V\sin\theta + v(x, y) \tag{3.71b}$$

where t_o, x_o, y_o, x_f, y_f are given and V is constant. Let u and v represent strong currents which may depend on location.

6a. Make use of the fact that since the Hamiltonian is not an explicit function of time, it is a constant throughout the motion (we will prove this in Sect. 4.8). Show that

$$\lambda_x = \frac{-\cos\theta}{V + u\cos\theta + v\sin\theta} \tag{3.72a}$$

$$\lambda_y = \frac{-\sin\theta}{V + u\cos\theta + v\sin\theta} \tag{3.72b}$$

6b. We note that if u and v are constants, then θ is a constant and the minimum paths are straight lines. Show that if

$$u = u(y) \tag{3.73a}$$

$$v = v(y) \tag{3.73b}$$

then

$$\frac{\cos\theta}{V + u(y)\cos\theta + v(y)\sin\theta} = \text{constant} \tag{3.74}$$

Equation (3.74) implies that the heading angle, θ, changes with local current velocities in direct analogy with Snell's law of optics.

6c. Suppose

$$u = -Vy/h \tag{3.75a}$$

$$v = 0 \tag{3.75b}$$

where h is constant. Show that

$$\cos\theta = \frac{\cos\theta_f}{\left[1 + \frac{y}{h}\cos\theta_f\right]} \tag{3.76}$$

for $(x_f, y_f) = (0, 0)$.

References

A.E. Bryson Jr., Y.C. Ho, *Applied Optimal Control* (Hemisphere Publishing, Washington, D.C., 1975)

S.J. Citron, *Elements of Optimal Control* (Holt, Rinehart, and Winston, New York, 1969)

D.T. Greenwood, *Classical Dynamics* (Dover, New York, 1997)

M.R. Hestenes, *Calculus of Variations and Optimal Control Theory* (Wiley, New York, 1966)

P. Kenneth, R. McGill, Two-point boundary value problem techniques, in *Advances in Control Systems: Theory and Applications*, vol. 3, chap. 2, ed. by C.T. Leondes (Academic, New York, 1966)

C. Lanczos, *The Variational Principles of Mechanics*, 4th edn. (Dover, New York, 1986)

D.A. Pierre, *Optimization Theory with Applications* (Wiley, New York, 1969)

Chapter 4

Application of the Euler-Lagrange Theorem

4.1. Introduction

In this chapter we will look at some applications of the Euler-Lagrange theorem. The theorem transforms the Problem of Bolza into a set of differential equations and attendant boundary conditions. In some cases, simple closed-form solutions are available which completely solve the problem. In other cases, numerical methods are required to solve the "two-point boundary-value problem." In some instances we find that the Euler-Lagrange theorem does not supply enough conditions to determine the optimal control law. In such cases we appeal to another theorem (the Weierstrass condition or Minimum Principle, discussed in Chap. 5) to solve the problem.

4.2. Two-Point Boundary-Value Problem (TPBVP)

Before defining a two-point boundary-value problem (TPBVP),we review what is meant by an initial-value problem. The initial-value problem is the typical problem students encounter in their first course on linear, ordinary differential equations. Suppose

$$\dot{\boldsymbol{x}} = \boldsymbol{f}(t, \boldsymbol{x}) \tag{4.1}$$

is an initial-value problem. Because $\boldsymbol{x}$ is an N-vector, we know that Eq. (4.1) represents N scalar equations. Then $N+1$ numbers $(t_o, x_{1o}, \ldots, x_{No})$ are needed to obtain a unique solution. We imagine propagating the solution forward in time (by numerical methods) from the initial conditions (I.C.s). (Whether an analytical solution exists is immaterial to the present discussion.) Because we have the entire initial state $(x_{1o}, \ldots, x_{No})$ and the initial time (t_o), we can determine the state at an infinitesimal time step in the future. In numerical methods we take small, finite steps in time, but in principle we can combine these steps (making them as small as necessary) to create a complete, highly accurate solution at a future time.

Example 4.1 Split boundary conditions (B.C.s).

Consider the scalar differential equation

$$\ddot{x} + x = 0 \tag{4.2}$$

J.M. Longuski et al., *Optimal Control with Aerospace Applications*,
Space Technology Library 32, DOI 10.1007/978-1-4614-8945-0_4,

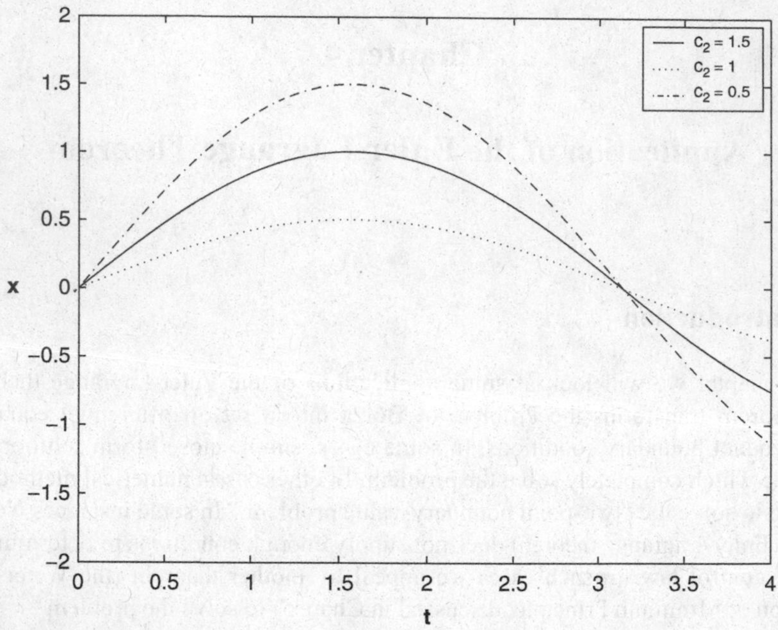

Figure 4.1. Example of split B.C.s where I.C. is given but final B.C. is not, $x = c_2 \sin t$.

In this particular case we have the analytical solution:

$$x = c_1 \cos t + c_2 \sin t \tag{4.3}$$

To be an initial value problem three numbers $(N + 1)$ must be specified, i.e., $[t_o, x(t_o), \dot{x}(t_o)]$. Now suppose that t_o and $x(t_o)$ are given but not $\dot{x}(t_o)$.

In particular, let us assume:

$$t_o = 0$$

$$x(t_o) = 0 \tag{4.4}$$

so that

$$x = c_2 \sin t \tag{4.5}$$

This is an example of split B.C.s (where not all conditions are I.C.s). In this case of split B.C.s, we require $N + 2$ boundary conditions. Figure 4.1 illustrates the situation. Suppose $t_f = \frac{\pi}{2}$. Then we also need $x(\frac{\pi}{2})$. Say, for example, that $x(\frac{\pi}{2}) = 1$. Then we have the case of $c_2 = 1$, as shown in Fig. 4.1. Here we note that:

$$t_o = 0 \tag{4.6a}$$

$$x_o = 0 \tag{4.6b}$$

$$t_f = \frac{\pi}{2} \tag{4.6c}$$

$$x_f = 1 \tag{4.6d}$$

give the required $N+2$ B.C.s for a well-defined TPBVP. This example also illustrates that it is possible to have an ill-defined TPBVP: if $t_f = \pi$ then we learn nothing new about the solution because all curves pass through zero at this point [$x(\pi) = 0$]. Such a case is called a conjugate point. A similar case arises when we seek the shortest path from the Earth's North Pole to the South Pole: all meridians give the same answer, so no proper minimum exists. Conjugate points are discussed in more detail in Sect. 6.7 and in Bryson and Ho [1975], Citron [1969], Hull [2003], and Vagners [1983].

4.3. Two Approaches to Terminal Constraints

There are two approaches for handling terminal B.C.s which appear in the literature and are discussed in Chap. 3. These methods are distinguished by whether the terminal B.C.s are multiplied by a constant multiplier vector, $\boldsymbol{\nu}$, and adjoined to the cost functional, J in Eq. (3.32), or not. We will refer to these two approaches as the *adjoined method* and the *un-adjoined method*. In both methods the number of equations to be solved is the same.

Most books on optimal control concentrate exclusively on one method or the other. Bryson and Ho [1975] and subsequent works overwhelmingly favor the adjoined method which leads to the algebraic form of the transversality condition. Prior to 1975 authors (e.g. Hestenes [1966]) most often employed the un-adjoined approach resulting in the differential form of the transversality condition, Eq. (3.68).

Citron [1969] is one of the few authors who presents both approaches. He demonstrates the equivalence of both methods by example and indicates that it is up to the user to determine which method works best for a particular problem.

In the un-adjoined method the differential form of the transversality condition is combined with the differentials of the terminal constraints to provide natural boundary conditions. This method is a concise technique that is very efficient in obtaining a TPBVP. Because this approach is not addressed in modern texts, we are providing numerous examples in the present work.

The adjoined technique is well documented in Bryson and Ho [1975] and in many modern texts that have followed it. Since there are numerous works emphasizing the adjoined method, we provide fewer examples than for the un-adjoined method—but certainly sufficient examples to make the method clear and accessible to the reader. We also demonstrate in a number of examples that the two methods can be combined.

The reader, on a first reading, may elect to concentrate on only one method and in either case the text will provide a thorough introduction on how to setup a TPBVP for optimal control problems.

The main advantage of the adjoined method is due to the elegant mathematical structure which is readily translated into a numerical algorithm. However, we do have to solve explicitly for the $\boldsymbol{\nu}$ q-vector, but with this step comes a great benefit: the ν_i

components provide a numerical sensitivity of the optimal cost due to small changes in the terminal constraints (as discussed in Exercise 5 of Chap. 3). A similar observation is made in Sect. 3.3.1 about how each $\lambda_i(t_o)$ provides the gradient of the cost J due to a differential change in $x_i(t_o)$.

In the un-adjoined method we have $2n$ differential equations (for $\boldsymbol{x}$ and for $\boldsymbol{\lambda}$ which in this counting scheme require $2n+2$ B.C.s including t_o and possibly t_f). The terminal B.C.s give p equations (where $1 \le p \le n+1$)

$$\boldsymbol{\Psi}(t_f, \boldsymbol{x}_f) = \boldsymbol{0} \tag{4.7}$$

and we have $n+1$ I.C.s $(\boldsymbol{x}_o, t_o)$ so we need

$$2n+2-(n+1)-p = n+1-p \tag{4.8}$$

additional B.C.s. These $n+1-p$ B.C.s are obtained through the differential form of the transversality condition, Eq. (3.68)

$$H_f dt_f - \boldsymbol{\lambda}_f^T d\boldsymbol{x}_f + d\phi = 0 \tag{4.9}$$

which has $n+1$ differential terms $(d\boldsymbol{x}_f, dt_f)$ which, in general, are not independent. There are p constraints, $\boldsymbol{\Psi}(t_f, \boldsymbol{x}_f) = \boldsymbol{0}$, which after elimination from Eq. (4.9) reduces the differential form of the transversality condition to $n+1-p$ independent equations, thus providing the remaining B.C.s required for a well-defined TPBVP.

The main disadvantage of the un-adjoined method is that the differential constraints on the transversality condition must be eliminated by hand—which for simple problems (the kind most often found in textbooks)—is usually easy. However, it is not always possible to perform the elimination, so the method is highly specific. Citron [1969] points out that this method does not provide an explicit set of equations with which to work until the problem of interest has been defined, which is another reason that we provide more examples of the un-adjoined method than of the adjoined. On the other hand, Citron also recognizes that it is useful to have explicit equations for the general case (i.e. the adjoined method).

Here we note that in the adjoined method $\boldsymbol{\Psi}$ is a q-vector (where $0 \le q \le n$) as in Bryson and Ho [1975]. In this counting scheme $\boldsymbol{\Psi}$ does not include t_o and t_f. The initial time, t_o, is ignored and the final time, t_f if specified, is treated as a separate constraint independent of $\boldsymbol{\Psi}$. If the final time is unspecified, some component of $\boldsymbol{\Psi}$ acts as stopping condition as noted in Sect. 3.3.3. In this approach the $2n$ system of equations (for $\dot{\boldsymbol{x}}$ and $\dot{\boldsymbol{\lambda}}$) are required to have $2n$ B.C.s (as opposed to $2n+2$ B.C.s in the un-adjoined method).

In the examples throughout the book we always make clear whether $\boldsymbol{\Psi}$ is a p vector $(1 \le p \le n+1)$ or a q vector $(0 \le q \le n)$, corresponding to the un-adjoined or adjoined method, respectively.

4.4. Transversality Condition

In the proof of the Euler-Lagrange theorem for the un-adjoined method (in Chap. 3) we chose final boundary conditions for the Lagrange multipliers to satisfy:

$$H_f dt_f - \boldsymbol{\lambda}_f^T d\boldsymbol{x}_f + d\phi = 0 \tag{4.10}$$

and we were given p B.C.s:

$$d\boldsymbol{\Psi} = \mathbf{0} \tag{4.11}$$

Eq. (4.10) is called "the transversality condition." Because the λ's were originally arbitrary functions, we were able to choose these functions and their B.C.s to simplify the proof. Thus, we have a total of $2n$ differential equations (i.e. N differential equations using the notation of Sect. 4.2) plus m algebraic equations:

$$\dot{\boldsymbol{x}} = \boldsymbol{f} \tag{4.12a}$$

$$\dot{\boldsymbol{\lambda}} = -H_{\boldsymbol{x}}^T \tag{4.12b}$$

$$H_{\boldsymbol{u}} = \mathbf{0}^T \tag{4.12c}$$

where $H_{\boldsymbol{x}}$ and $H_{\boldsymbol{u}}$ denote the vectors that contain the derivatives of H with respect to the x's and u's.

In general, we have $2n$ differential equations given by Eqs. (4.12a) and (4.12b). Solution of these equations provides $\boldsymbol{x}$ and $\boldsymbol{\lambda}$ (each are n-vectors). The $\boldsymbol{x}$ is the state vector that usually has explicit physical meaning. The $\boldsymbol{\lambda}$ is the costate vector (or the Lagrange multiplier) that also has physical meaning, as the change in the cost due to a small change in the state vector. To solve the $2n$ differential equations we need $2n + 2$ initial and final boundary conditions. We already know t_0, $\boldsymbol{x}_0$, and $\boldsymbol{\Psi}$, which provide a total of $n + 1 + p$ B.C.s. Then, for the general problem we need $(2n + 2) - (n + 1 + p) = n + 1 - p$ additional boundary conditions. We will discuss this problem in more detail after first looking at some special cases.

4.4.1. Case 1 Final Time Specified

Suppose the final time is given:

$$\Psi_1 = t_f - T = 0 \tag{4.13}$$

so that $p = 1$, as illustrated in Fig. 4.2. Then

$$d\Psi_1 = dt_f = 0 \tag{4.14}$$

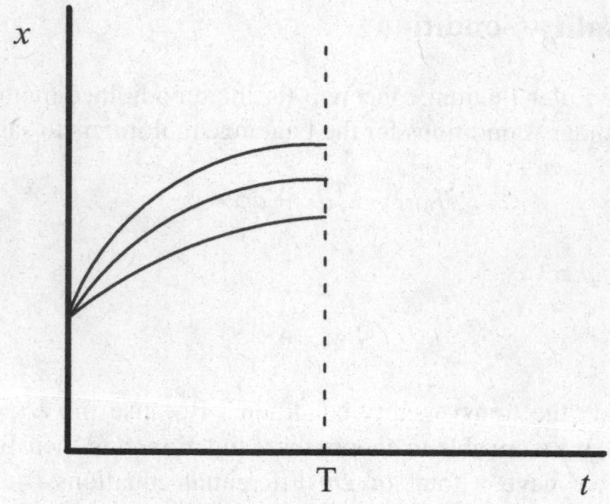

Figure 4.2. Fixed final time problem.

In this case for a well-defined TPBVP we need $n + 1 - p = n$ more B.C.s. assuming the un-adjoined method. From the differential form of the transversality condition we have[1]:

$$H_f dt_f - \boldsymbol{\lambda}_f^T d\mathbf{x}_f + \phi_{t_f} dt_f + \phi_{\mathbf{x}_f} d\mathbf{x}_f = 0 \tag{4.15}$$

subject to

$$dt_f = 0 \tag{4.16}$$

we have,

$$(-\boldsymbol{\lambda}_f^T + \phi_{\mathbf{x}_f}) d\mathbf{x}_f = 0 \tag{4.17}$$

and

$$\lambda_{if} = \frac{\partial \phi}{\partial x_{if}} \qquad (i = 1, \ldots, n) \tag{4.18}$$

Thus, the $2n + 2$ B.C.s are given by t_o, $\mathbf{x}_o$, $t_f = T$, and $\lambda_{1f} = \frac{\partial \phi}{\partial x_{1f}}, \ldots, \lambda_{nf} = \frac{\partial \phi}{\partial x_{nf}}$. Note that as mentioned near the end of Sect. 3.3.1, the dependence of the cost on a small change in the value of T is determined by $\frac{\partial J}{\partial T} = H_f + \phi_{t_f}$ (because $\Phi = \phi$ in this example).

[1] Recall that $\phi_{\mathbf{x}_f}$ is a row vector.

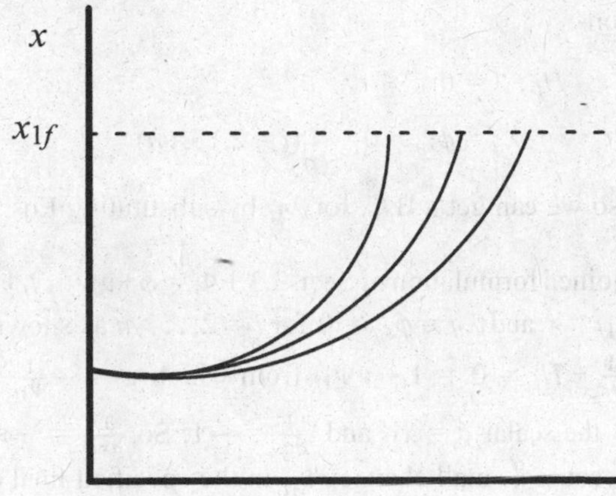

Figure 4.3. One final state specified.

4.4.2. Case 2 Final State Specified

Suppose one component of the final state is specified:

$$\Psi_1 = x_1(t_f) - x_{1f} = 0 \tag{4.19}$$

so that $p = 1$.

Example 4.2 Minimum time-to-climb problem.

Consider the case in which we minimize the time for an aircraft to climb to a specified altitude, as illustrated in Fig. 4.3. Let $\phi = t_f$. Then, from Eq. (4.19) $d\Psi = dx_{1f} = 0$. Following the un-adjoined method, we need n more B.C.s for a well-defined TPBVP (i.e., $n + 1 - p = n + 1 - 1 = n$). From the transversality condition:

$$H_f dt_f - \boldsymbol{\lambda}_f^T d\boldsymbol{x}_f + \phi_{t_f} dt_f + \phi_{x_f}^T d\boldsymbol{x}_f = 0 \tag{4.20}$$

Applying the final boundary condition

$$dx_{1f} = 0 \tag{4.21}$$

the transversality condition becomes:

$$(H_f + \phi_{t_f})dt_f + \sum_{i=2}^{n}(-\lambda_{if} + \phi_{x_{if}})dx_{if} = 0 \tag{4.22}$$

and we have n equations:

$$H_f + 1 = 0 \tag{4.23}$$
$$\lambda_{if} = \phi_{x_{if}} = 0 \qquad (i = 2, \ldots, n)$$

Recall $H = L + \boldsymbol{\lambda}^T \boldsymbol{f}$, so we can get a B.C. for λ_{1f} by substituting Eqs. (4.23) into the Hamiltonian.

Note that in the adjoined formulation of Sect. 3.3.1 $\Phi = \phi + \nu\Psi = t_f + \nu\left[x_1(t_f) - x_{1f}\right]$ and $\boldsymbol{\lambda}_f^T = \frac{\partial\Phi}{\partial x_f}$. Thus $\lambda_{1f} = \nu$ and $\lambda_{if} = \phi_{x_{if}} = 0$ for $i = 2, \ldots, n$ as shown in Eq. (4.23). Also, $\frac{\partial\Phi}{\partial t_f} = 1$ and $\frac{\partial\Phi}{\partial t_f} + H_f = 0 = 1 + \nu\Psi_{1f}$ from which $\nu = -\frac{1}{\Psi_{1f}} = -\frac{1}{\dot{x}_{1f}}$. From Exercise 5 in Chap. 3 the scalar $c = x_{1f}$ and $\frac{\partial\Psi}{\partial c} = -1$. So $\frac{\partial J}{\partial x_{1f}} = -\nu = \frac{1}{\dot{x}_{1f}} = \frac{dt_f}{dx_{1f}}$, which makes sense because a small change dx_{1f} in the specified final altitude results in the small change in the cost dt_f.

4.4.3. Case 3 Final Endpoint Specified

Suppose we have a fixed endpoint:

$$\boldsymbol{\Psi}(t_f, \boldsymbol{x}_f) = \begin{bmatrix} t_f - T \\ x_1(t_f) - x_{1f} \\ \vdots \\ x_n(t_f) - x_{nf} \end{bmatrix} = 0 \tag{4.24}$$

(where $T, x_{1f}, \ldots, x_{nf}$ are given) then,

$$dt_f = 0 \tag{4.25a}$$
$$dx_{1f} = 0 \tag{4.25b}$$
$$\vdots$$
$$dx_{nf} = 0 \tag{4.25c}$$

and we have $p = n + 1$ equations from the final boundary conditions. Since we already have $n + 1$ equations from the I.C.s, we have the required conditions for a well-defined TPBVP. In this case the transversality condition provides no additional boundary conditions, since $n + 1 - p = 0$.

4.5. General Case of Supplying Needed B.C.s

We can now summarize how the two forms of the transversality condition can be used to supply the needed B.C.s in the general case in which the Euler-Lagrange theorem applies. (A summary of the Euler-Lagrange theorem is given in Sect. 3.3.2.)

The m-vector control $\boldsymbol{u}$, is solved for through Eq. (3.52c) and the result is substituted into Eqs. (3.52a) and (3.52b) which provide a system of $2n$ differential equations.

4.5.1. Adjoined Method

In the adjoined method we recall that $\boldsymbol{\Psi}$ is a q-vector (where $0 \leq q \leq n$) that provides the terminal B.C.s, $\boldsymbol{\Psi} = \boldsymbol{0}$, as given in Eq. (3.53c). In this method $\boldsymbol{\Psi}$ is multiplied by a constant vector $\boldsymbol{\nu}$, and adjoined to the cost functional J in Eq. (3.32). In the counting scheme for the adjoined method (the same scheme used by Bryson and Ho [1975]), the number of B.C.s needed for the system of $2n$ differential equations is $2n$. Equation (3.53a) provides n I.C.s and Eq. (3.53b) provides n B.C.s (which are treated as independent) where the terminal function $\Phi(t_f, \boldsymbol{x}_f)$ is given by Eq. (3.34b). The q-vector, $\boldsymbol{\nu}$, which appears in the definition of Φ is solved through the use of Eq. (3.53c).

In the case where the final time is unspecified (i.e. $dt_f \neq 0$) we have the algebraic form of the transversality condition, Eq. (3.54)

$$\Omega = L_f + \frac{d\Phi}{dt_f} = 0 \tag{4.26}$$

4.5.2. Un-adjoined Method

In the un-adjoined method we write the terminal B.C.s, $\boldsymbol{\Psi} = \boldsymbol{0}$ (which is now a p-vector) as

$$d\boldsymbol{\Psi} = \boldsymbol{\Psi}_{t_f} dt_f + \boldsymbol{\Psi}_{x_f} d\boldsymbol{x}_f = \boldsymbol{0} \tag{4.27}$$

to solve for p differentials in terms of the remaining $n + 1 - p$ differentials. That is, substitute the known final B.C.s into Eq. (4.27) to obtain p independent equations.

Next we substitute for the selected p differentials into

$$H_f dt_f - \boldsymbol{\lambda}_f^T d\boldsymbol{x}_f + d\phi_f = 0 \tag{4.28}$$

which originally contains $n + 1$ differential terms. Thus, we have an expression in $n + 1 - p$ independent differentials.

Finally we set the coefficients of the remaining $n + 1 - p$ differentials to zero:

$$N_i(t_f, \boldsymbol{x}_f, \boldsymbol{\lambda}_f) = 0, \quad i = 1, \ldots, n + 1 - p \tag{4.29}$$

To summarize, we have $n + 1$ I.C.s (t_0, x_0), p B.C.s Ψ_i, and $n + 1 - p$ differential expressions, N_i.

4.6. Examples

Example 4.3 Minimum effort, minimum control problem.

Consider the following:

Minimize:

$$J = \frac{1}{2}\int_0^1 (x^2 + Pu^2)dt \tag{4.30}$$

Here we are trying to minimize the square of the error, x^2, and of the control effort, u^2, over the time interval from 0 to 1, subject to:

$$\dot{x} = u \tag{4.31a}$$

$$x(0) = 1 \tag{4.31b}$$

where x and u are scalars. This is the state regulator problem (see Bryson and Ho [1975]) where P is a weighting factor on the control effort. Note that if $P = 0$ there is no solution, as discussed in Chap. 3. On the other hand, as $P \rightarrow \infty$ the curve is flattened. See Fig. 4.4. In a typical application, P is a finite, non-zero value that specifies the relative cost of the control effort compared to the error in maintaining the state near zero.

We note that when solving an optimization problem, it is often a good practice to guess what the solution might be and to analyze the extreme cases.

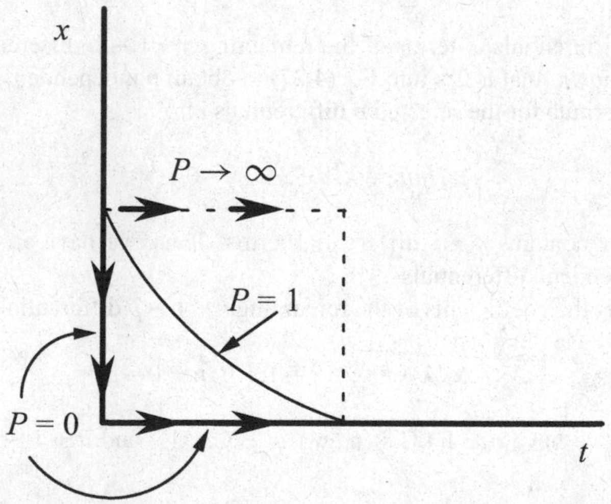

Figure 4.4. Error and control effort minimization depends on the control-effort weighting factor P, in Eq. (4.30).

Letting $P = 1$, we solve our problem in a step-by-step process:

1. Form the Hamiltonian:

$$H = \frac{1}{2}x^2 + \frac{1}{2}u^2 + \lambda u \tag{4.32}$$

2. Write the Euler-Lagrange equations $\dot{\lambda} = -H_x$, $H_u = 0$:

$$\dot{\lambda} = -\frac{\partial H}{\partial x} = -x \tag{4.33}$$

$$\frac{\partial H}{\partial u} = u + \lambda = 0 \tag{4.34}$$

so that

$$u = -\lambda \tag{4.35}$$

3. For the differential form of the transversality condition we have

$$H_f dt_f - \boldsymbol{\lambda}_f^T d\mathbf{x_f} + d\phi_f = 0 \tag{4.36}$$

subject to:

$$d\boldsymbol{\Psi} = 0 \tag{4.37}$$

In this case $\phi = 0$ and $\Psi = t_f - 1 = 0$, so $d\Psi = dt_f = 0$ and

$$\lambda_f = \lambda(t_f) = 0 \tag{4.38}$$

Using the adjoined method Eq. (4.38) results from $\lambda_f = \lambda(t_f) = \frac{\partial \phi}{\partial x(t_f)} = 0$. Our well-defined TPBVP is (for both methods):

$$\dot{x} = -\lambda \tag{4.39a}$$

$$\dot{\lambda} = -x \tag{4.39b}$$

$$t_o = 0,\ x(0) = 1 \tag{4.39c}$$

$$t_f = 1,\ \lambda_f(1) = 0 \tag{4.39d}$$

where we have two first-order differential equations ($2n = 2$) and $2n + 2 = 2 + 2 = 4$ B.C.s.

Let us put our system in state-variable form by defining:

$$z \equiv \begin{bmatrix} z_1 \\ z_2 \end{bmatrix} \equiv \begin{bmatrix} x \\ \lambda \end{bmatrix} \tag{4.40}$$

and, therefore

$$\dot{\boldsymbol{z}} = \begin{bmatrix} \dot{z}_1 \\ \dot{z}_2 \end{bmatrix} = \begin{bmatrix} -z_2 \\ -z_1 \end{bmatrix} \tag{4.41}$$

$$z_1(0) = 1 \tag{4.42a}$$

$$z_2(1) = 0 \tag{4.42b}$$

Eqs. (4.40)–(4.42) are in a form suitable for numerical integration (as a TPBVP). In this particular case, an analytical solution exists. We can write

$$\ddot{z_1} = -\dot{z_2} = z_1 \tag{4.43}$$

so that

$$\ddot{z_1} - z_1 = 0 \tag{4.44}$$

The solution has the form:

$$z_1 = c_1 e^t + c_2 e^{-t} \tag{4.45a}$$

$$z_2 = -\dot{z_1} = -c_1 e^t + c_2 e^{-t} \tag{4.45b}$$

From the B.C.s:

$$z_1(0) = c_1 + c_2 = 1 \tag{4.46a}$$

$$z_2(1) = -c_1 e^1 + c_2 e^{-1} = 0 \tag{4.46b}$$

we can solve for c_1 and c_2:

$$c_1 = \frac{1}{1+e^2} \tag{4.46c}$$

and

$$c_2 = \frac{e^2}{1+e^2} \tag{4.46d}$$

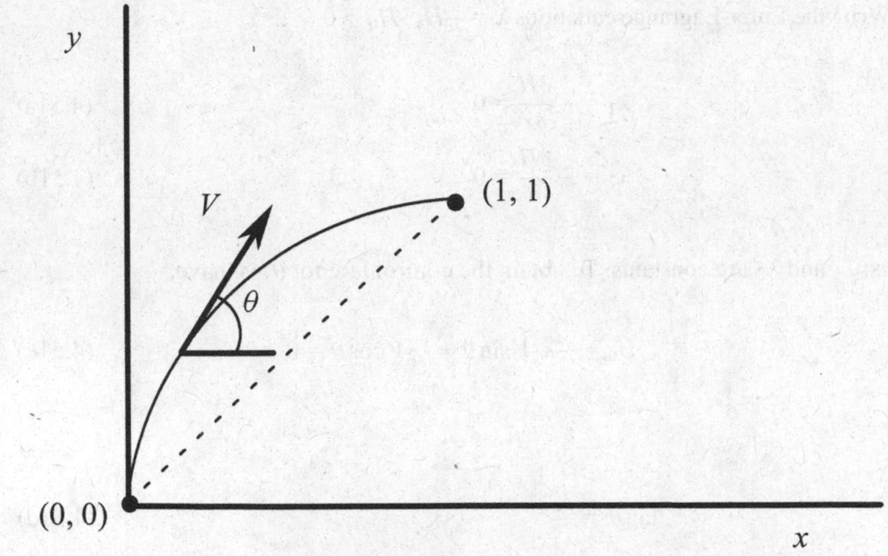

Figure 4.5. Zermelo's problem in which a boat crosses a river in minimum time.

Thus, for our optimal solution we obtain:

$$x^* = \quad z_1 \quad = \frac{e^t + e^2 e^{-t}}{1 + e^2} \tag{4.47a}$$

$$u^* = -\lambda = -z_2 = \frac{e^t - e^2 e^{-t}}{1 + e^2} \tag{4.47b}$$

Example 4.4 Zermelo's problem.

Minimize:

$$J = t_f \tag{4.48}$$

subject to:

$$\dot{x} = V \cos\theta \tag{4.49a}$$

$$\dot{y} = V \sin\theta \tag{4.49b}$$

where V is a given constant and $x(t_f) = y(t_f) = 1$. See Fig. 4.5. We already know that for no current the solution is the line $\theta = 45$ deg, as we found in Chap. 2 and illustrated in Fig. 2.11. Let us follow our systematic procedure on this simple problem:

1. Form the Hamiltonian:

$$H = \lambda_1 V \cos\theta + \lambda_2 V \sin\theta \tag{4.50}$$

2. Write the Euler-Lagrange equations $\dot{\boldsymbol{\lambda}} = -H_{\boldsymbol{x}}, H_{\boldsymbol{u}} = 0$:

$$\dot{\lambda}_1 = -\frac{\partial H}{\partial x} = 0 \tag{4.51a}$$

$$\dot{\lambda}_2 = -\frac{\partial H}{\partial y} = 0 \tag{4.51b}$$

Thus, λ_1 and λ_2 are constants. To obtain the control law for θ we have,

$$H_u = -\lambda_1 V \sin\theta + \lambda_2 V \cos\theta = 0 \tag{4.51c}$$

and

$$\tan\theta = \frac{\lambda_2}{\lambda_1} \tag{4.51d}$$

so that

$$\theta = \text{constant} = \theta_o \tag{4.51e}$$

Solving the state equations, Eqs. (4.49) with both V and θ constant provides

$$x(t) = x_o + (V\cos\theta_o)t \tag{4.52a}$$

$$y(t) = y_o + (V\sin\theta_o)t \tag{4.52b}$$

Evaluating at t_f:

$$x(t_f) = 1 = 0 + (V\cos\theta_o)t_f \tag{4.53a}$$

$$y(t_f) = 1 = 0 + (V\sin\theta_o)t_f \tag{4.53b}$$

so

$$\tan\theta_o = 1 \tag{4.53c}$$

and

$$\theta_o^* = 45^o \tag{4.53d}$$

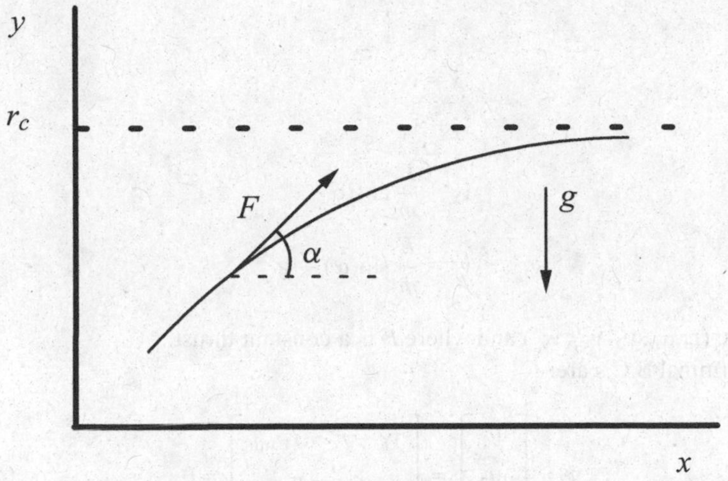

Figure 4.6. Flat-Earth problem: launch of a satellite into circular orbit in minimum time.

as before. For the optimal time we obtain:

$$t_f^* = \frac{1}{V\cos 45°} = \frac{\sqrt{2}}{V} \tag{4.53e}$$

Example 4.5 Launch into circular orbit from flat-Earth.

In Fig. 4.6 we illustrate the problem of launching a satellite into circular orbit where we assume a uniform gravitational field, g = constant. Our assumptions that the x and y axes are rectilinear and that the gravitational acceleration is constant are equivalent to assuming that the Earth is flat. Although these assumptions may seem to be oversimplifying, it is interesting to note that the Saturn V launch vehicle (which launched men to the Moon) used a guidance system based on the control law resulting from the flat-Earth model, with frequent updates of the control parameters.

We base our model on the one introduced in Chap. 2. [See Eqs. (2.1)–(2.8).] Since we are trying to maximize the payload delivered into circular orbit, and since we assume a constant burn rate for the propellant, we note that minimizing the time to reach orbit is an equivalent goal.

So, the problem can be stated as:

Minimize:

$$J = t_f \tag{4.54}$$

subject to:

$$\dot{x} = v_x \tag{4.55a}$$

$$\dot{y} = v_y \tag{4.55b}$$

$$\dot{v}_x = \frac{F}{m}\cos(\alpha) \tag{4.55c}$$

$$\dot{v}_y = \frac{F}{m}\sin(\alpha) - g \tag{4.55d}$$

with I.C.s: $(t_o, x_o, y_o, v_{x_o}, v_{y_o})$ and where F is a constant thrust.

The terminal B.C.s are:

$$\boldsymbol{\Psi} = \begin{bmatrix} \Psi_1 \\ \Psi_2 \\ \Psi_3 \end{bmatrix} = \begin{bmatrix} y_f - r_c + r_{\text{Earth}} \\ v_{xf} - v_c \\ v_{yf} - 0 \end{bmatrix} = 0 \tag{4.56}$$

Next, we set up the necessary conditions from the Euler-Lagrange theorem:

1. Form the Hamiltonian: $H = L + \boldsymbol{\lambda}^T \boldsymbol{f}$

$$H = \lambda_1 v_x + \lambda_2 v_y + \lambda_3 \frac{F}{m}\cos\alpha + \lambda_4 \left(\frac{F}{m}\sin\alpha - g\right) \tag{4.57}$$

2. Write the Euler-Lagrange equations: $\dot{\boldsymbol{\lambda}} = -H_{\boldsymbol{x}}, H_{\boldsymbol{u}} = 0$. We have, for the costate equations:

$$\dot{\lambda_1} = -\frac{\partial H}{\partial x} = 0 \tag{4.58a}$$

$$\dot{\lambda_2} = -\frac{\partial H}{\partial y} = 0 \tag{4.58b}$$

Thus

$$\lambda_1 = c_1 \tag{4.58c}$$

and

$$\lambda_2 = c_2 \tag{4.58d}$$

so that

$$\dot{\lambda_3} = -\frac{\partial H}{\partial v_x} = -\lambda_1 \tag{4.58e}$$

$$\dot{\lambda}_4 = -\frac{\partial H}{\partial v_y} = -\lambda_2 \tag{4.58f}$$

Since λ_1 and λ_2 are constant,

$$\lambda_3 = -c_1 t + c_3 \tag{4.58g}$$

and

$$\lambda_4 = -c_2 t + c_4 \tag{4.58h}$$

The control equation is found from:

$$\frac{\partial H}{\partial \alpha} = -\lambda_3 \frac{F}{m} \sin\alpha + \lambda_4 \frac{F}{m} \cos\alpha = 0 \tag{4.59}$$

so we have

$$\tan\alpha = \frac{\pm\lambda_4}{\pm\lambda_3} \tag{4.60}$$

Thus, we have the *bi-linear tangent steering law*:

$$\tan\alpha = \frac{\lambda_4}{\lambda_3} = \frac{-c_2 t + c_4}{-c_1 t + c_3} \tag{4.61}$$

which is an important result for flight control.

Inspection of Eq. (4.61) reveals that it also obeys:

$$\tan\alpha = \frac{\lambda_4}{\lambda_3} = \frac{-\lambda_4}{-\lambda_3} \tag{4.62}$$

which means that the control law does not distinguish the quadrant in which α resides. The values of $\cos\alpha$ and $\sin\alpha$ are undetermined:

$$\cos\alpha = \frac{\pm\lambda_3}{\sqrt{\lambda_3^2 + \lambda_4^2}} \tag{4.63a}$$

$$\sin\alpha = \frac{\pm\lambda_4}{\sqrt{\lambda_3^2 + \lambda_4^2}} \tag{4.63b}$$

where we must take both plus signs or both minus signs. (We assume the positive sign for the square root.) This problem is solved by an additional necessary condition, which will be stated and used now (and will be proven later).

The additional necessary condition we need is called the Minimum Principle (discussed in Chap. 6) which states that the Hamiltonian must be minimized with respect to the control. (The precise meaning of this statement will be explained later.) A more restricted version of this principle is the Weierstrass condition (discussed in Chap. 5) where the control must be continuous. In our particular problem we assume that the control is continuous and unbounded. The additional necessary condition is

$$\left[\left. \frac{\partial^2 H}{\partial \boldsymbol{u}^2} \right|^* \right] \geq 0 \tag{4.64}$$

That is, the matrix, $H^*_{\boldsymbol{uu}}$ must be positive semi-definite. Since our control is scalar for our flat-Earth problem, we have:

$$\frac{\partial^2 H}{\partial \alpha^2} = -\lambda_3 \frac{F}{m} \cos\alpha - \lambda_4 \frac{F}{m} \sin\alpha \tag{4.65a}$$

$$= \frac{F}{m} \left[-\lambda_3 \left(\frac{\pm\lambda_3}{\sqrt{\lambda_3^2 + \lambda_4^2}} \right) - \lambda_4 \left(\frac{\pm\lambda_4}{\sqrt{\lambda_3^2 + \lambda_4^2}} \right) \right] \geq 0 \tag{4.65b}$$

We pick the minus sign to obtain:

$$\frac{\partial^2 H}{\partial \alpha^2} = \frac{F}{m} \left[\frac{\lambda_3^2 + \lambda_4^2}{\sqrt{\lambda_3^2 + \lambda_4^2}} \right] \geq 0 \tag{4.66}$$

Eq. (4.66) is satisfied because F, m, and the square root are all positive.

Alternatively, there is a more direct way to determine the correct algebraic signs in Eq. (4.63) and in all similar problems. First, let

$$\boldsymbol{\mu} = \begin{bmatrix} \lambda_3 \\ \lambda_4 \end{bmatrix} \tag{4.67}$$

and

$$\boldsymbol{\sigma} = \begin{bmatrix} \cos\alpha \\ \sin\alpha \end{bmatrix} \tag{4.68}$$

where $\boldsymbol{\sigma}$ is a unit vector. Then, Eq. (4.57) can be written as:

$$H = \lambda_1 v_x' + \lambda_2 v_y - \lambda_4 g + \frac{F}{m}\boldsymbol{\mu}^T\boldsymbol{\sigma} \tag{4.69}$$

To minimize H choose $\boldsymbol{\sigma}$ to be antiparallel to $\boldsymbol{\mu}$, so that

$$\boldsymbol{\mu}^T\boldsymbol{\sigma} = -\mu \tag{4.70}$$

for which

$$\boldsymbol{\sigma} = \frac{-\boldsymbol{\mu}}{\mu}; \quad \mu = \sqrt{\lambda_3^2 + \lambda_4^2} \tag{4.71}$$

and note that the vector equation Eq. (4.71) is the same as Eqs. (4.63) with the minus signs.

3. Transversality condition for flat-Earth problem:
Apply the differential form of the transversality condition to obtain a well-defined TPBVP. To solve the differential equations (the state and costate equations), $2n+2$ conditions are needed. In this case $n = 4$, so 10 B.C.s are required. We have five initial conditions $(t_o, x_o, y_o, v_{xo}, v_{yo})$ and three final conditions (y_f, v_{xf}, v_{yf}). We are missing two terms (associated with t_f, x_f).

Since we know:

$$dy_f = 0 \tag{4.72a}$$

$$dv_{xf} = 0 \tag{4.72b}$$

$$dv_{yf} = 0 \tag{4.72c}$$

the transversality condition becomes:

$$H_f dt_f - \lambda_{1f} dx_f + \phi_{t_f} dt_f + \phi_{x_f} dx_f = 0 \tag{4.73}$$

But dt_f and dx_f are independent differentials. We have $\phi = t_f$ so:

$$\frac{\partial \phi}{\partial t_f} = 1 \tag{4.74a}$$

$$\frac{\partial \phi}{\partial x_f} = 0 \tag{4.74b}$$

Therefore

$$H_f dt_f - \lambda_{1f} dx_f + dt_f = 0 \tag{4.75a}$$

so that

$$(H_f + 1)dt_f - \lambda_{1f} dx_f = 0 \tag{4.75b}$$

and

$$H_f = -1, \lambda_{1f} = 0 \tag{4.75c}$$

Using the adjoined method

$$\begin{aligned}\Phi &= \phi + \boldsymbol{\nu}^T \boldsymbol{\Psi} \\ &= t_f + \nu_1(y_f - r_c + r_{\text{Earth}}) + \nu_2(v_{xf} - v_c) + \nu_3(v_{yf})\end{aligned} \tag{4.76}$$

Then, consistent with Eqs. (4.58c) and (4.75c),

$$\lambda_{1f} = \frac{\partial \Phi}{\partial x_f} = 0 \tag{4.77}$$

and

$$\lambda_{2f} = \frac{\partial \Phi}{\partial y_f} = \nu_1 \tag{4.78}$$

$$\lambda_{3f} = \frac{\partial \Phi}{\partial v_{xf}} = \nu_2 \tag{4.79}$$

$$\lambda_{4f} = \frac{\partial \Phi}{\partial v_{yf}} = \nu_3 \tag{4.80}$$

where the final values λ_{1f}, λ_{2f}, λ_{3f}, and λ_{4f} are the sensitivities of the cost to small changes in x_f, y_f, v_{xf}, and v_{yf}. The sensitivity to a small change in x_f is zero. Also, $\frac{\partial \phi}{\partial t_f} + H_f = 0$, which yields $H_f = -1$ as in Eq. (4.75c). In this case then (with the negative signs chosen):

$$\tan \alpha = \frac{c_2 t - c_4}{c_1 t - c_3} \tag{4.81}$$

But since $\lambda_1 = c_1$ and $\lambda_{1f} = 0$, we have $c_1 = 0$. Therefore, we obtain the *linear tangent steering law*

$$\tan \alpha = \frac{c_2 t}{-c_3} + \frac{-c_4}{-c_3} = at + b \tag{4.82}$$

This law was used in the Apollo guidance system with updates for a and b each second.

We can now specify a well-defined TPBVP. For our state equations we have

$$\dot{x} = v_x \tag{4.83a}$$

$$\dot{y} = v_y \tag{4.83b}$$

$$\dot{v}_x = \frac{F}{m_o + \dot{m}t}\left(\frac{-\lambda_3}{\sqrt{\lambda_3^2 + \lambda_4^2}}\right) \tag{4.83c}$$

$$\dot{v}_y = \frac{F}{m_o + \dot{m}t}\left(\frac{-\lambda_4}{\sqrt{\lambda_3^2 + \lambda_4^2}}\right) - g \tag{4.83d}$$

For our costate equations we have

$$\dot{\lambda}_1 = 0 \tag{4.83e}$$

$$\dot{\lambda}_2 = 0 \tag{4.83f}$$

$$\dot{\lambda}_3 = -\lambda_1 \tag{4.83g}$$

$$\dot{\lambda}_4 = -\lambda_2 \tag{4.83h}$$

thus there are eight differential equations, namely,

$$\begin{aligned} \dot{\boldsymbol{x}} &= \boldsymbol{f}(t, \boldsymbol{x}, \boldsymbol{\lambda}) \\ \dot{\boldsymbol{\lambda}} &= \tilde{\boldsymbol{f}}(t, \boldsymbol{x}, \boldsymbol{\lambda}) \end{aligned} \tag{4.84}$$

Now how do we deal with the split B.C.s? We have five initial conditions and five final conditions. One numerical approach that has been used is called the *shooting method*. While the shooting method is not necessarily the best numerical method, it has the additional advantage of serving as a good "thought experiment" and so contributes to understanding the TPBVP.

In our example, the shooting method could be done in the following steps.

1. Guess $\lambda_{1o}, \lambda_{2o}, \lambda_{3o}, \lambda_{4o}, t_f$.
2. Integrate $\dot{\boldsymbol{x}} = \boldsymbol{f}, \dot{\boldsymbol{\lambda}} = \tilde{\boldsymbol{f}}$ forward to $t = t_f$.
3. Compute the final conditions:

$$\Psi_1 = y_f - r_c + r_{\text{Earth}} \tag{4.85}$$

$$\Psi_2 = v_{xf} - v_c \tag{4.86}$$

$$\Psi_3 = v_{yf} \tag{4.87}$$

$$\Psi_4 = \lambda_{1f} \tag{4.88}$$

$$\Psi_5 = H_f + 1 \tag{4.89}$$

which change as $(\boldsymbol{\lambda}_o, t_f)$ are changed iteratively.

4. Converge on $\Psi_i(\boldsymbol{\lambda}_o, t_f) = 0$ for $i = 1, \ldots, 5$.

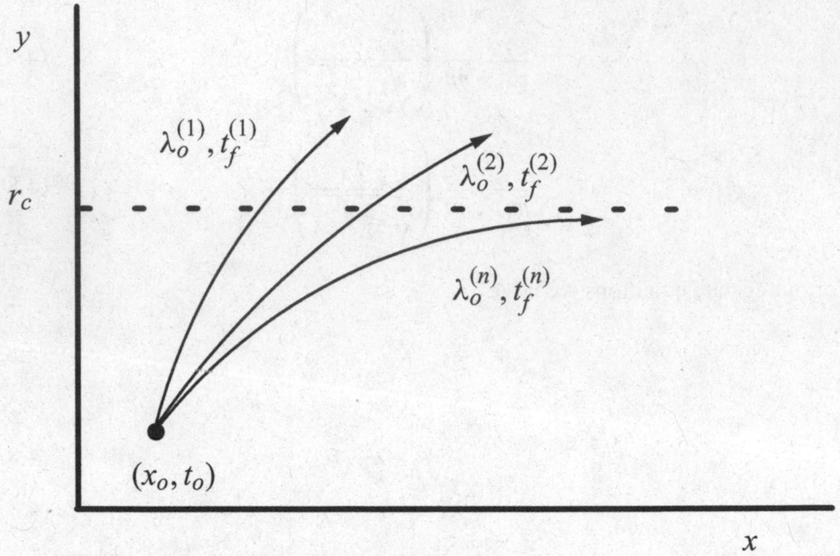

Figure 4.7. Shooting method for flat-Earth problem.

The technique is illustrated in Fig. 4.7. A discussion on the shooting method can be found in Citron [1969].

4.7. A "Cookbook" for Optimization Problems

In the examples, we have developed a step-by-step procedure to reach a well-defined TPBVP based on the Euler-Lagrange theorem and the Minimum Principle. We can now outline this systematic method in "cookbook" form. While it would not be wise to blindly apply such an approach to all problems, this cookbook should prove useful in many cases. As we shall see, however, there are a number of examples in the literature of trajectory optimization which require innovative and creative thinking beyond the rote application of these four steps.

Step 1
Form the Hamiltonian:

$$H = L + \boldsymbol{\lambda}^T \boldsymbol{f} \tag{4.90}$$

Step 2
Write the Euler-Lagrange equations:

$$\dot{\lambda}_i = -\frac{\partial H}{\partial x_i}, \quad i = 1, \ldots, n \tag{4.91}$$

$$H_{\boldsymbol{u}} = 0 \tag{4.92}$$

When possible, use Eq. (4.92) to find the control $\boldsymbol{u}$ (an m-vector). When a definitive control is not obtained from Eq. (4.92), use step 4 (the Minimum Principle). In either case substitute the resulting control into the state equation:

$$\dot{\boldsymbol{x}} = \boldsymbol{f}(t, \boldsymbol{x}, \boldsymbol{u}) \tag{4.93}$$

and into Eq. (4.91)

Step 3: Adjoined Method
If the final time is unspecified (ie. $dt_f \neq 0$), use the algebraic form of the transversality condition Eq. (3.54)

$$\Omega = L_f + \frac{d\Phi}{dt} = 0 \tag{4.94}$$

where $\Phi(t_f, \boldsymbol{x}_f)$ is obtained from Eq. (3.34b):

$$\Phi(t_f, \boldsymbol{x}_f) \equiv \phi(t_f, \boldsymbol{x}_f) + \boldsymbol{\nu}^T \boldsymbol{\Psi}(t_f, \boldsymbol{x}_f) \tag{4.95}$$

and where $\boldsymbol{\Psi}$ is a q-vector ($0 \leq q \leq n$) that specifies the terminal constraints

$$\boldsymbol{\Psi}(t_f, \boldsymbol{x}_f) = \boldsymbol{0} \tag{4.96}$$

If the final time is specified then Eq. (4.94) is invalid. Note that in the counting scheme for the adjoined method the $2n$ differential equations, Eqs. (4.91) and (4.93), require $2n$ B.C.s which are provided by Eqs. (3.53a) and (3.53b). Treat the components of $\boldsymbol{\lambda}(t_f)$ as independent. Solve for the q-vector, $\boldsymbol{\nu}$ which appears in Φ using Eq. (3.53c).

Step 3: Un-Adjoined Method
Write the terminal B.C.s

$$\boldsymbol{\Psi}(t_f, \boldsymbol{x}_f) = \boldsymbol{0} \tag{4.97}$$

as a p-vector where $1 \leq p \leq n + 1$. In the counting scheme for the un-adjoined method, the $2n$ differential equations, Eqs. (4.91) and (4.93) require $2n + 2$ B.C.s. The I.C.s $(t_o, \boldsymbol{x})$ provide $n + 1$ conditions and Eq. (4.97) provide p conditions. The number of additional conditions needed are

$$2n + 2 - (n + 1) - p = n + 1 - p \tag{4.98}$$

Use the differential form of the transversality condition:

1. Use $d\boldsymbol{\Psi} = 0$ to solve for p differentials in terms of the remaining $n + 1 - p$ differentials.
2. Substitute these p differentials into:

$$H_f dt_f - \boldsymbol{\lambda}_f^T d\boldsymbol{x}_f + d\phi = 0 \tag{4.99}$$

3. Obtain an expression involving $n + 1 - p$ "independent" differentials and equate their coefficients to zero:

$$N_i(t_f, \boldsymbol{x}_f, \boldsymbol{\lambda}_f) = 0, \qquad i = 1, \ldots, n + 1 - p \tag{4.100}$$

Step 4
Apply the Minimum Principle ("minimize H with respect to $\boldsymbol{u}$") to minimize J. If $\boldsymbol{u}^*$ is an *interior control* (not on control boundary) then,

$$\left[\left. \frac{\partial^2 H}{\partial \boldsymbol{u}^2} \right|^* \right] \geq 0 \tag{4.101}$$

which is to say that the matrix is positive semi-definite.

We note that $H_{\boldsymbol{uu}}^* > 0$ is called the convexity condition (or the strengthened Legendre-Clebsch condition), a sufficient condition for a minimum of H and part of the sufficient conditions for a minimum of the cost J. These concepts are developed in Chaps. 5 and 6. The main purpose of the Minimum Principle is to provide additional conditions on the control law so that a specific control law may be found and then used in Step 2.

4.7.1. Examples of Step 4

Example 4.6 Unbounded control.

Suppose u is a scalar and the Hamiltonian is given by

$$H = \lambda_1 \cos u + \lambda_2 \sin u \tag{4.102}$$

Following the procedure described after Eq. (4.66) write the Hamiltonian in Eq. (4.102) as

$$H = \boldsymbol{\mu}^T \boldsymbol{\sigma} \tag{4.103}$$

where

$$\boldsymbol{\mu} = \begin{bmatrix} \lambda_1 \\ \lambda_2 \end{bmatrix} \tag{4.104}$$

and

$$\boldsymbol{\sigma} = \begin{bmatrix} \cos u \\ \sin u \end{bmatrix} \tag{4.105}$$

where $\boldsymbol{\sigma}$ is a unit vector. Then H is minimized by choosing

$$\boldsymbol{\sigma} = \frac{-\boldsymbol{\mu}}{\mu} \tag{4.106}$$

for which

$$\cos u = \frac{-\lambda_1}{\sqrt{\lambda_1^2 + \lambda_2^2}}, \tag{4.107}$$

$$\sin u = \frac{-\lambda_2}{\sqrt{\lambda_1^2 + \lambda_2^2}} \tag{4.108}$$

Example 4.7 Bounded control.

Suppose u is a scalar, u is bounded such that $|u| \leq 1$, and the Hamiltonian is:

$$H = H_o(t, \boldsymbol{x}, \boldsymbol{\lambda}) + H_1(t, \boldsymbol{x}, \boldsymbol{\lambda})u \tag{4.109}$$

Since H is linear in u

$$\frac{\partial H}{\partial u} = H_1 \tag{4.110}$$

In this case we apply the Minimum Principle differently from Example 4.6 because now the control is bounded. We are seeking a control law such that the Hamiltonian given in Eq. (4.109) is minimized by the value we select for u. The Minimum Principle says that we should minimize H with respect to u, so we must choose the u that makes the value of the Hamiltonian the smallest it can be. Our choice of u cannot influence $H_0(t, \boldsymbol{x}, \boldsymbol{\lambda})$, so we can ignore this term. If $H_1(t, \boldsymbol{x}, \boldsymbol{\lambda})$ is a positive number, then to minimize H we should pick the lowest value of u, i.e., $u = -1$ (which is on the lower bound of the control). Similarly, if $H_1 < 0$ then we pick $u = 1$. If $H_1 = 0$ the Minimum Principle does not provide us with a specific control law. In summary, the Minimum Principle gives us:

$$u = \begin{cases} -1 & \text{if} \quad H_1 > 0 \\ \text{undetermined} & \text{if} \quad H_1 = 0 \\ 1 & \text{if} \quad H_1 < 0 \end{cases} \tag{4.111}$$

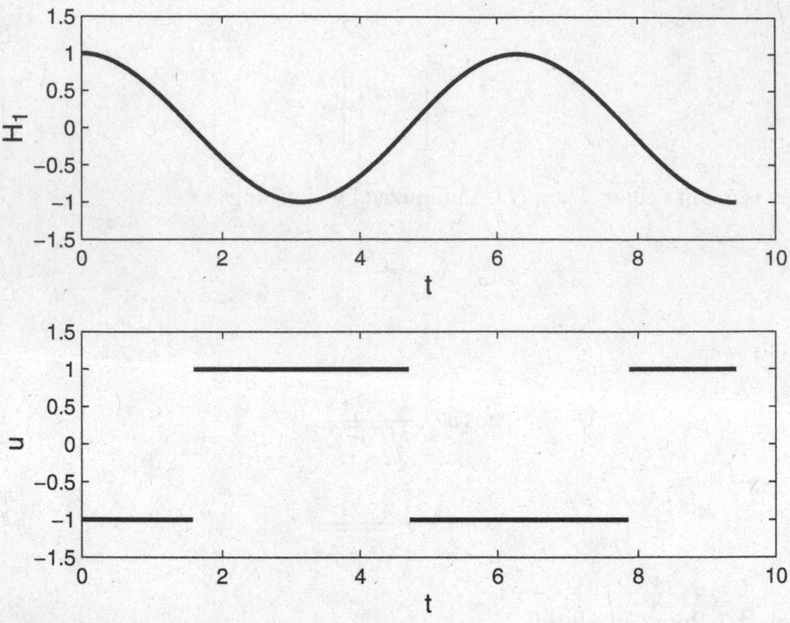

Figure 4.8. The switching function and the bang-bang control given by Eq. (4.111).

The coefficient $H_1(t, \boldsymbol{x}, \boldsymbol{\lambda})$ is called the *switching function*. Figure 4.8 shows an example where H_1 oscillates through positive and negative values (moving instantaneously through zero). In such a case, the control law switches back and forth between its minimum and maximum values and is called a "bang-bang" control.

In the case where $H_1(t, \boldsymbol{x}, \boldsymbol{u}) \equiv 0$ on some non-zero interval, say $[t_1, t_2] \leq [t_o, t_f]$, then the subarc $[t_1, t_2]$ is called a singular subarc. When singular subarcs appear we need yet another theorem called the Generalized Legendre-Clebsch condition to determine the control law. (See Chap. 9.)

Example 4.8 Bounded control problem.

Consider the following:

Minimize:

$$J = \frac{1}{2}\int_0^3 x^2 dt \tag{4.112}$$

subject to:

$$\dot{x} = u \tag{4.113}$$

with B.C.s:

$$x(0) = 1 \tag{4.114}$$

$$x(3) = 1 \tag{4.115}$$

and with the control constraint:

$$|u| \leq 1 \tag{4.116}$$

The solution will involve analysis of the switching function. We solve it in the following steps.

1. Form the Hamiltonian:

$$H = \frac{1}{2}x^2 + \lambda u \tag{4.117}$$

2. Write the Euler-Lagrange equations:

$$\dot{\lambda} = -H_x = -x \tag{4.118}$$

$$H_u = \lambda = H_1 \tag{4.119}$$

Thus, we have a switching function so the control law is:

$$u = \begin{cases} -1 & \text{if} \quad H_1 > 0 \\ \text{undetermined} & \text{if} \quad H_1 = 0 \\ 1 & \text{if} \quad H_1 < 0 \end{cases} \tag{4.120}$$

3. The differential form of the transversality condition is:

$$H_f dt_f - \boldsymbol{\lambda}_f^T d\mathbf{x}_f + d\phi = 0 \tag{4.121}$$

In this case x_f and t_f have been specified and $\phi = 0$, thus $dt_f = dx_f = d\phi = 0$ and the transversality condition provides no additional information. This can be verified by noting that there are $2n+2 = 2(1)+2 = 4$ conditions already specified. They are t_o, t_f, x_o, and x_f.

However, using the adjoined method one can generalize the terminal constraint from $\Psi = x(3) - 1 = 0$ to be $\Psi = x(3) - x_f = 0$. Then use Exercise 5 of Chap. 3 to obtain $\frac{\partial J}{\partial x_f} = -\nu = -\lambda(3)$, which provides the change in the cost due to a small change in x_f without explicitly recalculating the cost using Eq. (4.112). The value of λ is determined using the second equation in the system of equations below:

$$\dot{x} = u \tag{4.122}$$

$$\dot{\lambda} = -x \tag{4.123}$$

with B.C.s:

$$x(0) = 1 \tag{4.124}$$

$$x(3) = 1 \tag{4.125}$$

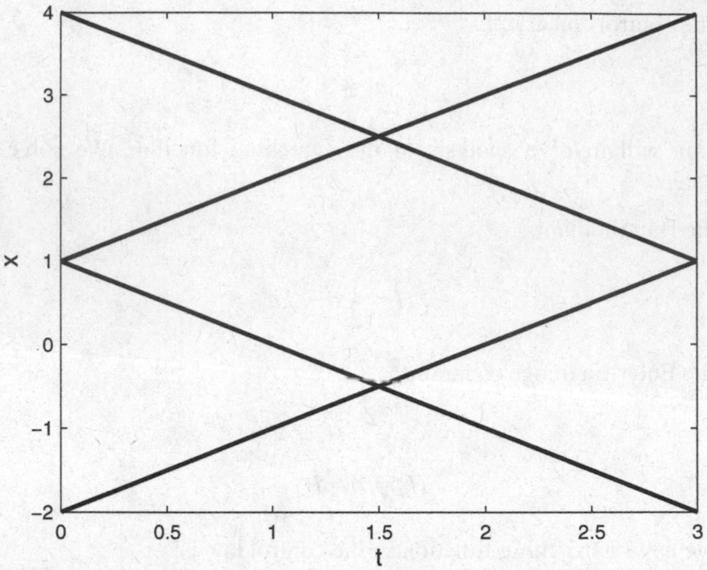

Figure 4.9. Possible paths for constant *u* from Eqs. (4.126)–(4.129).

With these two boundary conditions, four possible solutions for x appear, depending on whether $u = \pm 1$ at the initial and final times.

For $x(0) = 1$ we have:

$$x = +t + 1 \ (u = +1) \tag{4.126}$$

$$x = -t + 1 \ (u = -1) \tag{4.127}$$

For $x(3) = 1$ we have:

$$x = +t - 2 \ (u = +1) \tag{4.128}$$

$$x = -t + 4 \ (u = -1) \tag{4.129}$$

Plotting these four solutions and looking at the possible paths from the initial condition to the final condition, we obtain Fig. 4.9.

Considering the functional we are trying to minimize, it is clear that the area under the curve must be as small as possible. As a result, we obtain the solution shown in Fig. 4.10, where we see that $u = -1$ from $t = 0$ to $t = 1$ at which point x becomes zero and the control is turned off. Then at $t = 2$ the control is set to maximum, $u = +1$, to drive x up to the final condition $x(3) = 1$. As mentioned earlier, this behavior is called a bang-bang control.

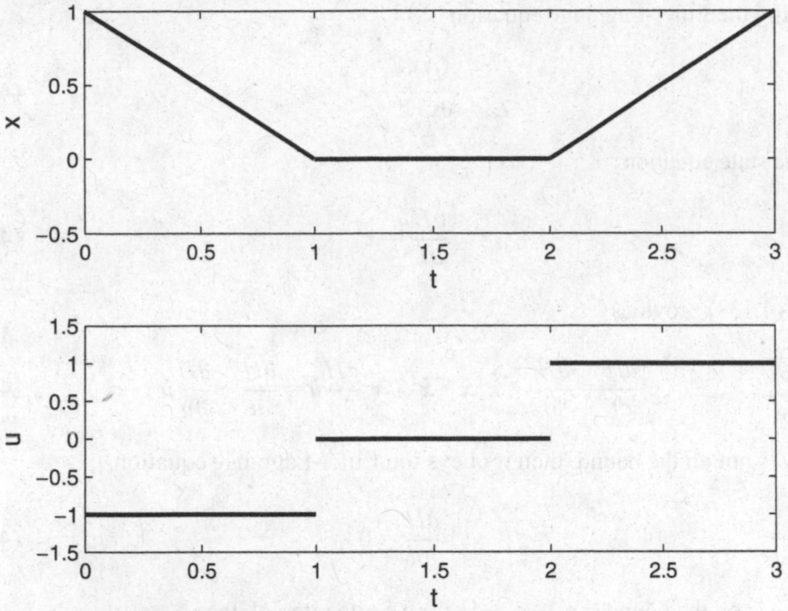

Figure 4.10. Solution for bounded control problem, example 4.8. The cost functional, process equation, and B.C.s are given by Eqs.(4.112)–(4.115). The control is bounded by Eq. (4.116).

4.8. Constant Hamiltonian

Consider an optimal control problem with Hamiltonian $H(t, \boldsymbol{x}, \boldsymbol{\lambda}, u)$ and bounded, scalar control $|u| \leq 1$. Then

$$\frac{dH}{dt} = \frac{\partial H}{\partial t} \tag{4.130}$$

so if H does not contain time explicitly, $\frac{\partial H}{\partial t} = 0$ and

$$H = \text{constant} \tag{4.131}$$

We show this by considering the Hamiltonian:

$$H = L + \boldsymbol{\lambda}^T \boldsymbol{f} \tag{4.132}$$

Taking the total time derivative we have:

$$\frac{dH}{dt} = \frac{\partial H}{\partial t} + \frac{\partial H}{\partial \boldsymbol{x}}\dot{\boldsymbol{x}} + \frac{\partial H}{\partial \boldsymbol{\lambda}}\dot{\boldsymbol{\lambda}} + \frac{\partial H}{\partial u}\dot{u} \tag{4.133}$$

Using the Euler-Lagrange equation

$$\frac{\partial H}{\partial \boldsymbol{x}} = -\dot{\boldsymbol{\lambda}}^T \tag{4.134}$$

and the state equation

$$\frac{\partial H}{\partial \boldsymbol{\lambda}} = \dot{\boldsymbol{x}}^T \tag{4.135}$$

in Eq. (4.133) provides

$$\frac{dH}{dt} = \frac{\partial H}{\partial t} - \dot{\boldsymbol{\lambda}}^T\dot{\boldsymbol{x}} + \dot{\boldsymbol{x}}^T\dot{\boldsymbol{\lambda}} + \frac{\partial H}{\partial u}\dot{u} = \frac{\partial H}{\partial t} + \frac{\partial H}{\partial u}\dot{u} \tag{4.136}$$

If u is not on the bound, then it obeys the Euler-Lagrange equation

$$\frac{\partial H}{\partial u} = 0 \tag{4.137}$$

and if u is on the bound (i.e., $u = \pm 1$) for a finite interval, then

$$\dot{u} = 0 \tag{4.138}$$

Thus Eq. (4.136) becomes

$$\frac{dH}{dt} = \frac{\partial H}{\partial t} \tag{4.139}$$

and if the Hamiltonian is not an explicit function of time then it is a constant.

The observation that H is constant in certain optimization problems will prove to be important in their solution.

4.9. Summary

Many technical challenges arise in the application of the Euler-Lagrange theorem. While the theorem converts the Problem of Bolza into a set of differential equations (for the states and the costates), the boundary conditions are split. That is, we have some of the conditions set as initial conditions and some as final conditions. Mathematically the two-point boundary-value problem (TPBVP) can be well defined, but the engineer may have to resort to numerical methods to solve the problem. In some cases analytical solutions exist, which can provide insight into particular cases.

A "cookbook" is presented to make the application of the Euler-Lagrange theorem as systematic as possible. However, we soon find that there are several obstacles. In obtaining necessary conditions for the minimum time to launch a satellite into orbit, we find that the Euler-Lagrange theorem fails to uniquely specify the optimal control law (to steer the thrust). We must invoke another theorem called the Minimum Principle. A more restricted version of the Minimum Principle, the Weierstrass condition, is discussed and proven in Chap. 5. (Chapter 6 states the more general Minimum Principle without proof.)

We also find an example where the Minimum Principle fails to provide the control law in the case of a singular subarc. Chapter 9 presents a theorem to surmount this difficulty, called the Generalized Legendre-Clebsch condition.

Clearly the optimization of trajectories cannot be completely solved in "cookbook" form. There exist deep mathematical and numerical challenges. Each new application may require new techniques or special handling.

4.10. Exercises

1. Solve for the optimal control, $u^*(t)$, and the optimal trajectory, $x^*(t)$:

$$\text{Min}\, J = \frac{1}{2}\int_0^1 (x^2 + u^2)dt$$

subject to

$$\dot{x} = -x + u$$

$$x(0) = 1,\ x(1) = 0$$

2. Pose a well-defined TPBVP for the following problem. (Do not attempt to solve the differential equations.)

Given a constant-thrust rocket engine, T = Thrust, operating for a given length of time, t_f, we wish to find the thrust-direction history, $\phi(t)$, to transfer a rocket vehicle from a given initial circular orbit to the largest possible circular orbit, where:

r = radial distance of spacecraft from attracting center,

u = radial component of velocity,

v = tangential component of velocity,

m = mass of spacecraft,

$\dot{m}$ = fuel consumption rate (constant),

ϕ = thrust direction angle, and

μ = gravitational constant of attracting center.

The problem may be stated as follows (Fig. 4.11).

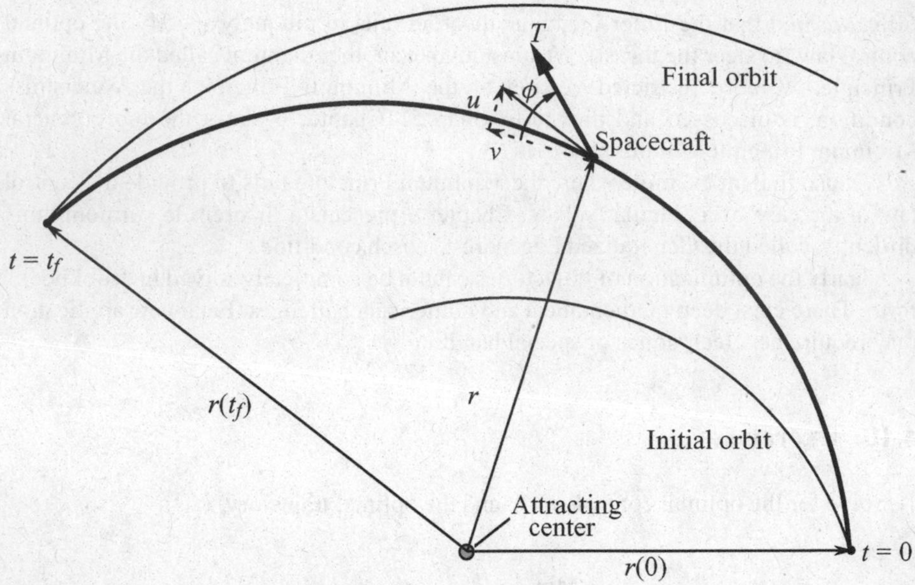

Figure 4.11. Maximum radius orbit transfer in a given time (Adapted from Bryson and Ho [1975]).

Find $\phi(t)$ to maximize $r(t_f)$ subject to

$$\dot{r} = u,$$
$$\dot{u} = \frac{v^2}{r} - \frac{\mu}{r^2} + \frac{T \sin\phi}{m_0 - |\dot{m}|t},$$
$$\dot{v} = -\frac{uv}{r} + \frac{T\cos\phi}{m_0 - |\dot{m}|t},$$

and

$$r(0) = r_0,\ u(0) = 0,\ v(0) = \sqrt{\frac{\mu}{r_0}},$$
$$\psi_1 = u(t_f) = 0,\ \psi_2 = v(t_f) - \sqrt{\frac{\mu}{r(t_f)}} = 0.$$

Note: Explicitly state all the equations and all of the boundary conditions (i.e., initial and final) for a well-defined TPBVP.

3. Zermelo's Problem
A particle must travel through a region in which its instantaneous velocity magnitude is given as a function of position, $V(x, y)$.

$$\text{Min } J = \int_{t_o}^{t_f} dt$$

subject to

$$\dot{x} = V(x, y) \cos\theta$$

$$\dot{y} = V(x, y) \sin\theta$$

where t_o, x_o, y_o are given and

$$x(t_f) = 0, \; y(t_f) = 0$$

The control variable is $\theta(t)$.

3a. Show that, along a minimum-time path, $\theta(t)$ must satisfy the differential equation

$$\dot{\theta} = \frac{\partial V}{\partial x} \sin\theta - \frac{\partial V}{\partial y} \cos\theta \tag{4.140}$$

Hints: The Hamiltonian is constant. Substitute solutions for λ_x and λ_y into $\dot{\lambda}_x = -H_x$ and $\dot{\lambda}_y = -H_y$ to find an equation in $\dot{\theta}$. Note that

$$\frac{d}{dt} V(x, y) = \frac{\partial V}{\partial x}\dot{x} + \frac{\partial V}{\partial y}\dot{y}.$$

Equation (4.140) indicates that if V is constant then the minimum-time paths are straight lines.

3b. Consider the special case where

$$V = V(y).$$

Show that

$$\frac{\cos\theta}{V(y)} = \text{constant} \tag{4.141}$$

Eq. (4.141) is known as Snell's law.

4. Solve Example 4.3, retaining P as an arbitrary constant (instead of setting P to 1).
5. Briefly explain why $H_{\boldsymbol{u}}^T = 0$ is not a necessary condition for a minimum of J if H is a *linear* function of $\boldsymbol{u}$.

References

A.E. Bryson, Jr., Y.C. Ho, *Applied Optimal Control* (Hemisphere, Washington, D.C., 1975)
S.J. Citron, *Elements of Optimal Control* (Holt, Rinehart, and Winston, New York, 1969)
D.G. Hull, *Optimal Control Theory for Applications* (Springer, New York, 2003)
J. Vagners, Optimization techniques. in *Handbook of Applied Mathematics*, 2nd edn., ed. by C.E. Pearson (Van Nostrand Reinhold, New York, 1983) pp. 1140–1216

Chapter 5

The Weierstrass Condition

5.1. Introduction

In Chap. 4 we noted that there are optimization problems that cannot be resolved by the Euler-Lagrange theorem alone. Pontryagin's Minimum Principle often provides an additional condition that leads to a specific control law and to the solution of the problem. The most general form of the Minimum Principle is stated in Chap. 6 without proof.

In this chapter we state and prove the Weierstrass necessary condition. At first glance the Weierstrass condition appears to be identical to Pontryagin's Minimum Principle. The main difference is that the former requires the control to be a continuous, unbounded function of time while the latter is far more general in allowing the control to be a "measurable function" (which includes piecewise continuous, bounded functions of time). We are interested in applications in which the control may be a piecewise continuous function (e.g., bang-bang control in which a thruster is turned on and off). So we will make use of the most general form of the Minimum Principle, but for the level of this text we do not have sufficient mathematical tools to prove Pontryagin's Minimum Principle.

5.2. Statement of the Weierstrass Necessary Condition

The Weierstrass condition is as follows.

If $\boldsymbol{x}^*(t) \in C^1[t_o, t_f]$ and $\boldsymbol{u}^*(t) \in C^0[t_o, t_f]$—i.e., if the optimal trajectory is continuously differentiable and the optimal control is continuous over the closed interval of time from t_0 to t_f—such that they minimize the Problem of Bolza:

$$J = \phi(t_f, \boldsymbol{x}_f) + \int_{t_o}^{t_f} L(t, \boldsymbol{x}, \boldsymbol{u})dt \tag{5.1}$$

subject to:

$$\dot{\boldsymbol{x}} = \boldsymbol{f}(t, \boldsymbol{x}, \boldsymbol{u}) \tag{5.2}$$

$$\boldsymbol{x}(t_o) = \boldsymbol{x}_o \tag{5.3}$$

$$\boldsymbol{\Psi}(t_f, \boldsymbol{x}_f) = \boldsymbol{0} \tag{5.4}$$

J.M. Longuski et al., *Optimal Control with Aerospace Applications*,
Space Technology Library 32, DOI 10.1007/978-1-4614-8945-0_5,

then for all $t \in [t_o, t_f]$ it is necessary that:

$$H[t, \boldsymbol{x}^*(t), \boldsymbol{u}^*(t), \boldsymbol{\lambda}(t)] \leq H[t, \boldsymbol{x}^*(t), \boldsymbol{u}(t), \boldsymbol{\lambda}(t)] \tag{5.5}$$

for all admissible (unbounded control) $\boldsymbol{u}(t)$. [If $\boldsymbol{u}(t)$ is bounded, then the Weierstrass condition no longer holds.]

Equation (5.5) is the Weierstrass necessary condition. On the left-hand side of the equation we have the Hamiltonian computed on the optimal trajectory, $\boldsymbol{x}^*(t)$, with the optimal control, $\boldsymbol{u}^*(t)$. On the right-hand side we have the Hamiltonian computed on the optimal trajectory at the same time but for a non-optimal control $\boldsymbol{u}(t)$. For any non-optimal control, the Hamiltonian must be greater than or equal to the Hamiltonian computed for the optimal control (again assuming the optimal trajectory is always used).

In 1939, E.J. McShane found a way to express Eq. (5.5) in simple English as quoted in Bryson and Ho [1975], "The Hamiltonian must be minimized over the set of all admissible controls" (which applies to the more general Minimum Principle, as well). The meaning of this principle will become clear when we present applications in Chaps. 6 and 7. We have already seen an example in Chap. 4 where, for unbounded control, $H_{\boldsymbol{u}} = 0$ and $H_{\boldsymbol{uu}} \geq 0$ which is a special case of minimizing H with respect to $\boldsymbol{u}$.

5.3. Proof Outline of the Weierstrass Necessary Condition

We will use a special variation in the proof outline of this theorem. The control blip $\boldsymbol{U}(t)$ disturbs the state from the optimal by $\boldsymbol{X}(t)$. The opposite blip $\boldsymbol{u}(t, e)$ brings the state back to $\boldsymbol{x}^*(t)$ via $\boldsymbol{x}(t, e)$, as illustrated in Fig. 5.1. Let $t \in (t_o, t_f)$. Let $e > 0$ be a small parameter. Implicit in this method of proof is the assumption that it is possible to return to the optimal path.

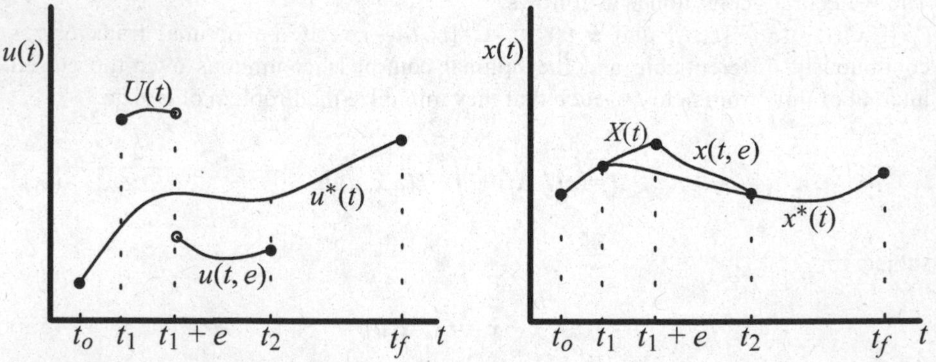

Figure 5.1. Control and state history. The nonoptimal control blip, $U(t)$, disturbs the state from its optimal path. The opposite blip $u(t, e)$, restores the optimal path at time t_2.

Note that while holding t_2 fixed,

$$\lim_{e\to 0} \boldsymbol{x}(t, e) = \boldsymbol{x}^*(t) \tag{5.6}$$

$$\lim_{e\to 0} \dot{\boldsymbol{x}}(t, e) = \dot{\boldsymbol{x}}^*(t) \tag{5.7}$$

$$\lim_{e\to 0} \boldsymbol{u}(t, e) = \boldsymbol{u}^*(t) \tag{5.8}$$

Now we define our special variation, $\tilde{\boldsymbol{x}}$, of the state variable:

$$\tilde{\boldsymbol{x}}(t) \equiv \begin{cases} \boldsymbol{x}^*(t) \;\; \text{on } [t_o, t_1] \text{ and on } [t_2, t_f] & \textit{No variations} \\ \boldsymbol{X}(t) \;\; \text{on } [t_1, t_1 + e] & \textit{Strong variation} \\ \boldsymbol{x}(t, e) \; \text{on } [t_1 + e, t_2] & \textit{Weak variation} \end{cases} \tag{5.9}$$

A similar variation is used by Citron [1969], Ewing [1985], and Vagners [1983]. The terms *weak* and *strong* will be discussed in greater detail later on (in Sect. 6.4). We note that in the strong case the slope of the variation is different from that of the optimal path. The discontinuous jump in control, $\boldsymbol{U}(t)$, can be arbitrarily large, as long as $\boldsymbol{u}(t, e)$ can bring the trajectory back to $\boldsymbol{x}^*(t)$. (Of course e must be infinitesimally small.) Hull [2003] points out that "Even though the minimum control is assumed to be continuous, controls in general can be discontinuous so that the admissible comparison control can be continuous or discontinuous."

Let $J(e)$ be the cost of the trajectory defined by $\tilde{\boldsymbol{x}}(t)$. The basic concept of the proof of Eq. (5.5) is to presume:

$$J(e) - J^* \geq 0 \tag{5.10}$$

as illustrated in Fig. 5.2. Expanding J about $e = 0$ we have:

$$\left[J(0) + \left.\frac{dJ}{de}\right|_{e=0} e + \mathcal{O}(e^2) \right] - J^* \geq 0 \tag{5.11}$$

Ignoring the higher order terms we obtain:

$$\left.\frac{dJ}{de}\right|_{e=0} \geq 0 \tag{5.12}$$

(where we have used $J(0) = J^*$ and $e = de$.) It is then clear that, in the proof we are trying to show, J^* is indeed the optimal solution. The rest involves Leibniz' rule:

$$\frac{d}{d\varepsilon}\left(\int_{a(\varepsilon)}^{b(\varepsilon)} f(x, \varepsilon) dx \right) = f[b(\varepsilon), \varepsilon]\frac{db(\varepsilon)}{d\varepsilon} - f[a(\varepsilon), \varepsilon]\frac{da(\varepsilon)}{d\varepsilon} + \int_{a(\varepsilon)}^{b(\varepsilon)} \frac{\partial f(x, \varepsilon)}{\partial \varepsilon} dx \tag{5.13}$$

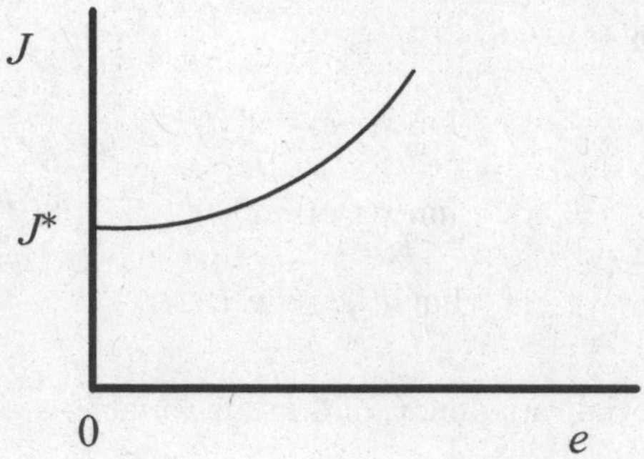

Figure 5.2. Basic concept of the proof of the Weierstrass condition: $J(e) \geq J^*$.

Let us write the expression for the functional:

$$
\begin{aligned}
J(e) = \phi(t_f^*, \boldsymbol{x}_f^*) &+ \int_{t_o}^{t_1} \{H^* - \boldsymbol{\lambda}^T \dot{\boldsymbol{x}}^*(t)\} dt \\
&+ \int_{t_1}^{t_1+e} \{H[t, \boldsymbol{X}(t), \boldsymbol{U}(t), \boldsymbol{\lambda}(t)] - \boldsymbol{\lambda}^T \dot{\boldsymbol{X}}(t)\} dt \\
&+ \int_{t_1+e}^{t_2} \{H[t, \boldsymbol{x}(t,e), \boldsymbol{u}(t,e), \boldsymbol{\lambda}(t)] - \boldsymbol{\lambda}^T \dot{\boldsymbol{x}}(t,e)\} dt \\
&+ \int_{t_2}^{t_f} \{H^* - \boldsymbol{\lambda}^T \dot{\boldsymbol{x}}^*(t)\} dt
\end{aligned}
\tag{5.14}
$$

where $H^* = H[t, \boldsymbol{x}^*(t), \boldsymbol{u}^*(t), \boldsymbol{\lambda}(t)]$.

To guide our analysis, we note in Eq. (5.14) that:

1. The first term $\phi(t_f^*, \boldsymbol{x}_f^*)$ does not contain e,
2. The second and fifth terms have no terms in e,
3. The third term depends on e outside the integrand, and
4. The fourth term depends on e both outside and inside the integrand.

Therefore, only the third and fourth terms will contribute to $\frac{dJ}{de}$. Applying Leibniz' rule we obtain:

$$
\begin{aligned}
\frac{dJ}{de} = &\{H[t, \boldsymbol{X}(t), \boldsymbol{U}(t), \boldsymbol{\lambda}(t)] - \boldsymbol{\lambda}^T \dot{\boldsymbol{X}}(t)\}|_{t_1+e} \frac{d(t_1+e)}{de} \\
&- \{H[t, \boldsymbol{x}(t,e), \boldsymbol{u}(t,e), \boldsymbol{\lambda}(t)] - \boldsymbol{\lambda}^T \dot{\boldsymbol{x}}(t,e)\}|_{t_1+e} \frac{d(t_1+e)}{de} \\
&+ \int_{t_1+e}^{t_2} \left[H_{\boldsymbol{x}}(t,e) \frac{\partial \boldsymbol{x}(t,e)}{\partial e} + H_{\boldsymbol{u}}(t,e) \frac{\partial \boldsymbol{u}(t,e)}{\partial e} - \boldsymbol{\lambda}^T \frac{\partial \dot{\boldsymbol{x}}(t,e)}{\partial e} \right] dt
\end{aligned}
\tag{5.15}
$$

Since we require $\left.\frac{dJ}{de}\right|_{e=0} \geq 0$ and noting that $\frac{d(t_1+e)}{de} = 1$, we have:

$$\begin{aligned}\left.\frac{dJ}{de}\right|_{e=0} &= H[t_1, \boldsymbol{X}(t_1), \boldsymbol{U}(t_1), \boldsymbol{\lambda}(t_1)] - \boldsymbol{\lambda}^T(t_1)\dot{\boldsymbol{X}}(t_1) \\ &- H[t_1, \boldsymbol{x}^*(t_1), \boldsymbol{u}^*(t_1), \boldsymbol{\lambda}(t_1)] + \boldsymbol{\lambda}^T(t_1)\dot{\boldsymbol{x}}^*(t_1) \\ &+ \int_{t_1}^{t_2} \left[H_{\boldsymbol{x}}^* \left.\frac{\partial \boldsymbol{x}(t,e)}{\partial e}\right|_{e=0} + H_{\boldsymbol{u}}^* \left.\frac{\partial \boldsymbol{u}(t,e)}{\partial e}\right|_{e=0} - \boldsymbol{\lambda}^T \left.\frac{\partial \dot{\boldsymbol{x}}(t,e)}{\partial e}\right|_{e=0} \right] dt \geq 0\end{aligned} \tag{5.16}$$

For the last term (inside the integral) we apply integration by parts:

$$\int_{t_1}^{t_2} -\boldsymbol{\lambda}^T \frac{\partial \dot{\boldsymbol{x}}(t,e)}{\partial e} dt = \left[-\boldsymbol{\lambda}^T \frac{\partial \boldsymbol{x}(t,e)}{\partial e} \right]_{t_1}^{t_2} + \int_{t_1}^{t_2} \dot{\boldsymbol{\lambda}}^T \frac{\partial \boldsymbol{x}(t,e)}{\partial e} dt \tag{5.17}$$

Thus, we have:

$$\begin{aligned}\left.\frac{dJ}{de}\right|_{e=0} &= H[t_1, \boldsymbol{X}(t_1), \boldsymbol{U}(t_1), \boldsymbol{\lambda}(t_1)] - H^*(t_1) \\ &- \boldsymbol{\lambda}^T(t_1)\dot{\boldsymbol{X}}(t_1) + \boldsymbol{\lambda}^T(t_1)\dot{\boldsymbol{x}}^*(t_1) \\ &- \boldsymbol{\lambda}^T(t_2) \left.\frac{\partial \boldsymbol{x}(t,e)}{\partial e}\right|_{\substack{e=0 \\ t=t_2}} + \boldsymbol{\lambda}^T(t_1) \left.\frac{\partial \boldsymbol{x}(t,e)}{\partial e}\right|_{\substack{e=0 \\ t=t_1}} \\ &+ \int_{t_1}^{t_2} \left[(H_{\boldsymbol{x}}^* + \dot{\boldsymbol{\lambda}}^T) \left.\frac{\partial \boldsymbol{x}(t,e)}{\partial e}\right|_{e=0} + (H_{\boldsymbol{u}}^*) \left.\frac{\partial \boldsymbol{u}(t,e)}{\partial e}\right|_{e=0} \right] dt \geq 0\end{aligned} \tag{5.18}$$

Now, the following term vanishes:

$$-\boldsymbol{\lambda}^T(t_2) \left.\frac{\partial \boldsymbol{x}(t,e)}{\partial e}\right|_{\substack{e=0 \\ t=t_2}} = 0 \tag{5.19}$$

since we are back to the optimal at t_2 [i.e., $x(t_2, e) = x^*(t_2)$] and

$$H_{\boldsymbol{x}}^* + \dot{\boldsymbol{\lambda}}^T = 0 \tag{5.20}$$

$$H_{\boldsymbol{u}}^* = 0 \tag{5.21}$$

from the Euler-Lagrange theorem. Figure 5.3 provides a closer look at the varied path, for clarity. Thus,

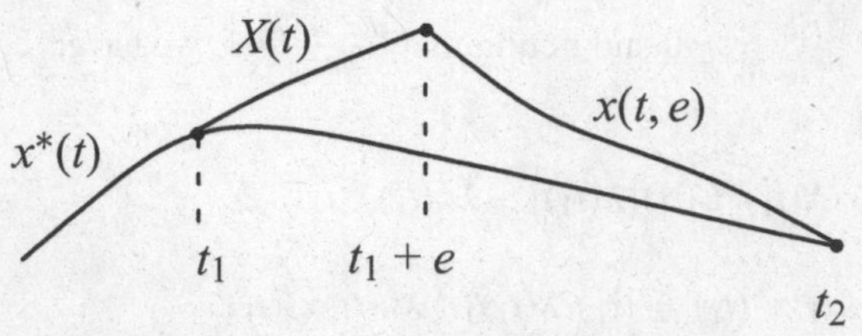

Figure 5.3. A closer look at the special variation given by Eq. (5.9).

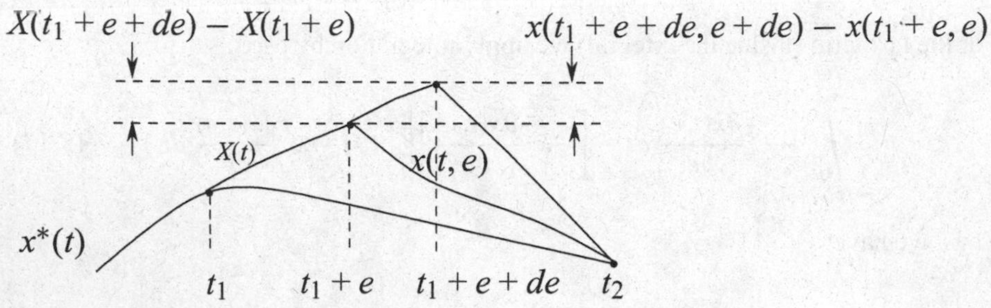

Figure 5.4. Characterization of $\left.\frac{\partial x(t,e)}{\partial e}\right|_{\substack{e=0\\ t=t_1}}$ which appears in Eq. (5.22).

$$\left.\frac{dJ}{de}\right|_{e=0} = H[t_1, \boldsymbol{X}(t_1), \boldsymbol{U}(t_1), \boldsymbol{\lambda}(t_1)] - H^*(t_1)$$
$$- \boldsymbol{\lambda}^T(t_1)\dot{\boldsymbol{X}}(t_1) + \boldsymbol{\lambda}^T(t_1)\dot{\boldsymbol{x}}^*(t_1) + \boldsymbol{\lambda}^T(t_1) \left.\frac{\partial \boldsymbol{x}(t,e)}{\partial e}\right|_{\substack{e=0\\ t=t_1}}$$
$$\geq 0 \tag{5.22}$$

We will use Fig. 5.4 in order to characterize $\left.\frac{\partial x(t,e)}{\partial e}\right|_{\substack{e=0\\ t=t_1}}$, which is not zero.

Let us compute the total derivative $\left.\frac{dx}{de}\right|_{\substack{e=0\\ t=t_1}}$:

$$\frac{d[\boldsymbol{x}(t_1+e,e)]}{de} = \frac{\partial \boldsymbol{x}}{\partial t}\frac{d(t_1+e)}{de} + \frac{\partial \boldsymbol{x}(t_1+e,e)}{\partial e} \tag{5.23}$$

so that

$$\left.\frac{d\boldsymbol{x}}{de}\right|_{\substack{e=0\\ t=t_1}} = \dot{\boldsymbol{x}}^*(t_1) + \left.\frac{\partial \boldsymbol{x}}{\partial e}\right|_{\substack{e=0\\ t=t_1}} \tag{5.24}$$

From the definition of a derivative we have:

$$\left.\frac{d\boldsymbol{x}}{de}\right|_{\substack{e=0\\t=t_1}} = \lim_{de\to 0}\left.\left[\frac{\boldsymbol{x}(t_1+e+de, e+de)-\boldsymbol{x}(t_1+e, e)}{de}\right]\right|_{e=0}$$
$$= \lim_{de\to 0}\left.\left[\frac{\boldsymbol{X}(t_1+e+de)-\boldsymbol{X}(t_1+e)}{de}\right]\right|_{e=0} \tag{5.25}$$

therefore:

$$\left.\frac{d\boldsymbol{x}}{de}\right|_{\substack{e=0\\t=t_1}} \equiv \dot{\boldsymbol{X}}(t_1) \tag{5.26}$$

and we obtain

$$\dot{\boldsymbol{X}}(t_1) = \dot{\boldsymbol{x}}^*(t_1) + \left.\frac{\partial \boldsymbol{x}}{\partial e}\right|_{\substack{e=0\\t=t_1}} \tag{5.27}$$

Equivalent expressions are given by Ewing [1985] and Vagners [1983]. From Eq. (5.27) we have:

$$\left.\frac{\partial \boldsymbol{x}}{\partial e}\right|_{\substack{e=0\\t=t_1}} = \dot{\boldsymbol{X}}(t_1) - \dot{\boldsymbol{x}}^*(t_1) \tag{5.28}$$

which, after substituting into Eq. (5.22), implies:

$$H - H^* - \boldsymbol{\lambda}^T(t_1)\dot{\boldsymbol{X}}(t_1) + \boldsymbol{\lambda}^T(t_1)\dot{\boldsymbol{x}}^*(t_1) + \boldsymbol{\lambda}^T(t_1)[\dot{\boldsymbol{X}}(t_1) - \dot{\boldsymbol{x}}^*(t_1)] \geq 0 \tag{5.29}$$

Thus, from Eq. (5.29) we obtain

$$H - H^* \geq 0 \tag{5.30}$$

or,

$$H[t, \boldsymbol{x}^*(t), \boldsymbol{u}^*(t), \boldsymbol{\lambda}(t)] \leq H[t, \boldsymbol{X}(t), \boldsymbol{U}(t), \boldsymbol{\lambda}(t)] \tag{5.31}$$

for $t = t_1 \in (t_o, t_f)$, where

$$\boldsymbol{X}(t_1) = \boldsymbol{x}^*(t_1) \tag{5.32}$$

Thus, Eq. (5.31) provides the Weierstrass necessary condition. In the proof we assumed $t = t_1 \in (t_o, t_f)$. Later it will be shown that H is continuous, which implies that the condition must also hold at the endpoints. (By continuity $H - H^* \geq 0$ for $t_1 \in [t_o, t_f]$; otherwise, if $H < H^*$ at t_o, there is a contradiction.)

5.4. Summary

The Weierstrass condition, Eq. (5.5), has been proved using a special variation, Eq. (5.9). A more general condition, the Minimum Principle, is presented in Chap. 6 without proof. The Weierstrass condition applies when the control is unbounded and continuous, whereas the Minimum Principle applies when the control may be piecewise continuous and bounded (or more generally, measurable). The Minimum Principle is employed in Step 4 of our "Cookbook" for optimization problems presented in Chap. 4. The meaning of this principle will become clearer in Chaps. 6 and 7 where we give its precise statement and several examples of its application.

5.5. True or False Quiz for Chaps. 1–5

Answer the following questions in the context of the material presented in Chaps. 1–5. Circle the correct answer.

1. The problems of Bolza, Lagrange, and Mayer are interchangeable.

   True False

2. The Dirac delta function, $\delta(t)$, is piecewise continuous.

   True False

3. All optimization problems lead to admissible controls.

   True False

4. In the proof of the Euler-Lagrange theorem we make use of the expression

$$d\boldsymbol{x}_f = \delta\boldsymbol{x}(t_f^*) + \dot{\boldsymbol{x}}^*(t_f^*)dt_f$$

   True False

5. In the Euler-Lagrange theorem, if $\dot{\boldsymbol{x}} = \boldsymbol{f}(t, \boldsymbol{x}, \boldsymbol{u})$ is an n-vector, then the number of boundary conditions required for a well-defined TPBVP is $n + 2$.

   True False

6. The Minimum Principle can be stated as: "H^* must be minimized over the set of all admissible u."

   True False

7. In the un-adjoined approach, the transversality condition can be ignored if all the initial and terminal boundary conditions are specified.

   True False

8. The step function, $h(t)$, is piecewise continuous.

   True False

9. When $(r_f/r_0) > 11.9$, a three-impulse orbital transfer can be more economical than a two-impulse transfer.

True False

10. The Minimum Principle is more general than the Weierstrass condition.

True False

Solution: 1T, 2F, 3F, 4T, 5F, 6T, 7T, 8T, 9T, 10T.

References

A.E. Bryson Jr., Y.C. Ho, *Applied Optimal Control* (Hemisphere Publishing, Washington, D.C., 1975)
S.J. Citron, *Elements of Optimal Control* (Holt, Rinehart, and Winston, New York, 1969)
G.M. Ewing, *Calculus of Variations with Applications* (Dover, New York, 1985)
D.G. Hull, *Optimal Control Theory for Applications* (Springer, New York, 2003)
J. Vagners, Optimization techniques, in *Handbook of Applied Mathematics*, 2nd edn., ed. by C.E. Pearson (Van Nostrand Reinhold, New York, 1983), pp. 1140–1216

Chapter 6

The Minimum Principle

6.1. Statement of the Minimum Principle

The Weierstrass condition, which requires the Hamiltonian to be minimized over the set of all admissible controls, is a powerful tool for solving a class of optimization problems that do not immediately yield to our familiar algorithm with the Euler-Lagrange equations and the transversality condition. However, the Weierstrass condition's "set of all admissible controls" is limited to continuously differentiable, unbounded functions, which are by no means the only feasible controls in practice or in principle. For example, "bang-bang" or "on-off" control schemes are frequently employed in everyday engineering applications, but these controls do not fall within the Weierstrass condition's set.

There exists a rigorous proof of a more general theorem by Pontryagin et al. [1962], usually stated in the literature as the Minimum Principle, which expands the Weierstrass condition's set of controls to include all "measurable" functions. (What constitutes a measurable function is beyond the scope of this text; however, piecewise continuous functions like the bang-bang control scheme are included.) Pontryagin's Minimum Principle is a much stronger statement than the Weierstrass condition because it provides the most general continuity restrictions on the control and on the functions of the Bolza problem. McShane (as reported in Bryson and Ho [1975]) expressed the principle in the wonderfully concise and clear verbal statement: "H^* must be minimized over the set of all admissible $\boldsymbol{u}$."

6.1.1. Problem Statement

For the optimal control problem:

Minimize:

$$J = \lambda_o \phi(t_o, \boldsymbol{x}_o, t_f, \boldsymbol{x}_f) + \lambda_o \int_{t_o}^{t_f} L(t, \boldsymbol{x}, \boldsymbol{u})dt \tag{6.1}$$

subject to:

$$\dot{\boldsymbol{x}} = \boldsymbol{f}(t, \boldsymbol{x}, \boldsymbol{u}) \tag{6.2}$$

$$\boldsymbol{x}(t_o) = \boldsymbol{x}_o \tag{6.3}$$

J.M. Longuski et al., *Optimal Control with Aerospace Applications*, Space Technology Library 32, DOI 10.1007/978-1-4614-8945-0_6,

where $\boldsymbol{x}$ is an n-vector and $\boldsymbol{u}$ is an m-vector, and where the terminal constraints are

$$\boldsymbol{\Psi}(t_f, \boldsymbol{x}_f) = \boldsymbol{0} \tag{6.4}$$

and where

$$\boldsymbol{u} \in U \tag{6.5}$$

is the set of admissible controls.

The set U is an arbitrary set in m-dimensional Euclidean space and $\boldsymbol{u}(t)$ may be a "measurable" function on $[t_o, t_f]$ whose values lie in U. Some authors (e.g. Vagners [1983]) say Pontryagin's Minimum Principle is applicable to any "topological control space" U. Specifically, U includes the case where the control $\boldsymbol{u}(t)$ may assume a number of discrete levels $u_1, \ldots, u_k$. The definitions of measurable functions and of topological control space are outside the scope of this text (see Ewing [1985]). In applications it is usually sufficient to assume that $\boldsymbol{u}(t)$ is a bounded P.C. function on $[t_o, t_f]$ (see Berkovitz [1974]).

6.1.2. Pontryagin's Minimum Principle

For the problem and assumptions stated above, suppose $\boldsymbol{u}^*(t) \in U$ causes J to be minimized. Let us write the Hamiltonian as

$$H(t, \boldsymbol{x}, \boldsymbol{u}, \boldsymbol{\lambda}) \equiv \lambda_o L(t, \boldsymbol{x}, \boldsymbol{u}) + \boldsymbol{\lambda}^T \boldsymbol{f}(t, \boldsymbol{x}, \boldsymbol{u}) \tag{6.6}$$

According to the multiplier rule (see Hestenes [1966] and Ewing [1985]) there exist continuous costate functions $\lambda_i(t)$ on $[t_o, t_f]$ for $i = 0, 1, \ldots, n$ such that:

$$\dot{\lambda}_i = -H^*_{x_i} \qquad (i = 1, \ldots, n) \tag{6.7}$$

at each $t \in [t_o, t_f]$ on arcs where $\boldsymbol{u}(t)$ is continuous, and λ_o is a constant such that:

$$[\lambda_o, \lambda_1(t), \ldots, \lambda_n(t)] \neq \boldsymbol{0}. \tag{6.8}$$

at each $t \in [t_o, t_f]$. (Nevertheless, some components may be zero. When $\lambda_o = 1$ the problem is "normal." Examples of normal and abnormal problems are provided in Sect. 6.1.3.) For the conditions stated above [Eqs. (6.1)–(6.8)] we have Pontryagin's Minimum Principle:

$$H[t, \boldsymbol{x}^*(t), \boldsymbol{u}^*(t), \boldsymbol{\lambda}(t)] \leq H[t, \boldsymbol{x}^*(t), \boldsymbol{u}(t), \boldsymbol{\lambda}(t)] \tag{6.9}$$

where $\boldsymbol{u} \in U$.

Pierre [1969] emphasized that Eq. (6.9) *does not imply* that:

$$H[t, \boldsymbol{x}(t), \boldsymbol{u}^*(t), \boldsymbol{\lambda}(t)] \leq H[t, \boldsymbol{x}(t), \boldsymbol{u}(t), \boldsymbol{\lambda}(t)] \tag{6.10}$$

for arbitrary (i.e., non-optimal) $\boldsymbol{x}(t)$. Proof of Pontryagin's Minimum Principle is notoriously difficult. (See Pontryagin et al. [1962], Hestenes [1966], and Berkovitz [1974].) In his discussion of the Minimum Principle, Pierre notes that Athans and Falb [1966] provide a heuristic proof in forty pages of text! In his introductory text Ross [2009] points out that while proving Pontryagin's Minimum Principle is not easy, *applying* the principle is *very easy*. He then proceeds to give several important applications and avoids dealing with the proof entirely. In Chap. 2 on Pontryagin's Principle, Ross jokingly adds, "You want proof? You can't handle the proof!" (For more Curious Quotations see appendix D of the present text.) As mentioned earlier (in Chap. 5), we also make use of Pontryagin's Principle without proof.

A problem is said to be *abnormal* if $\lambda_o = 0$ and the conditions of the Minimum Principle are satisfied. Otherwise, $\lambda_o = +1$ and the problem is normal (the usual case). Of course, any positive constant for λ_0 could be used in the normal case since its particular value does not change the optimal trajectory and optimal control—the choice of unity is merely a convenience.

6.1.3. Examples

Example 6.1 A non-optimization problem.

Consider the following:

Minimize:

$$J = \lambda_o f(x, y) \tag{6.11}$$

subject to:

$$S(x, y) = x^2 + y^2 = 0 \tag{6.12}$$

We see immediately that Eq. (6.12) implies that

$$x^* = y^* = 0 \tag{6.13}$$

so, this is not an optimization problem to begin with!

Example 6.2 Abnormal and normal problems.

Consider the following problem.

Minimize:

$$J = \lambda_o \int_0^1 x^2 dt \tag{6.14}$$

subject to:

$$\dot{x} = u \tag{6.15}$$

$$x(0) = 1 \tag{6.16}$$

$$|u| \leq 1 \tag{6.17}$$

where x and u are scalars. The Hamiltonian is:

$$H = \lambda_o x^2 + \lambda_1 u \tag{6.18}$$

Let us consider *two cases*:

1. $x(1) = 0$,
2. $x(1)$ is free.

For both cases, let us set $\lambda_o = 0$ to find out if they are abnormal cases (i.e., that we can ignore the cost functional, J).

The costate equation becomes

$$\dot{\lambda}_1 = -H_x = 0 \tag{6.19}$$

so that

$$\lambda_1 = \text{constant.} \tag{6.20}$$

The control equation is found from:

$$H_u = \lambda_1 \tag{6.21}$$

where λ_1 is the switching function:

$$\begin{aligned} \lambda_1 > 0 &\Rightarrow u = -1 \\ \lambda_1 = 0 &\Rightarrow u = \text{undetermined} \\ \lambda_1 < 0 &\Rightarrow u = +1 \end{aligned} \tag{6.22}$$

Equations (6.22) are, of course, the direct result of minimizing H with respect to u. The differential form of the transversality condition (see Sect. 3.5, Exercise 2) is:

$$[H^* dt - \lambda_1 dx]\Big|_{t_o}^{t_f} + \lambda_o d\phi = 0 \tag{6.23}$$

and since $d\phi = 0$ we have

$$H_f^* dt_f - H_o^* dt_o - \lambda_{1f} dx_f + \lambda_{1o} dx_o = 0, \tag{6.24}$$

where all the differentials vanish except

$$\lambda_{1f} dx_f = 0 \tag{6.25}$$

For case 1 the final state is specified:

$$x(1) = 0 \tag{6.26}$$

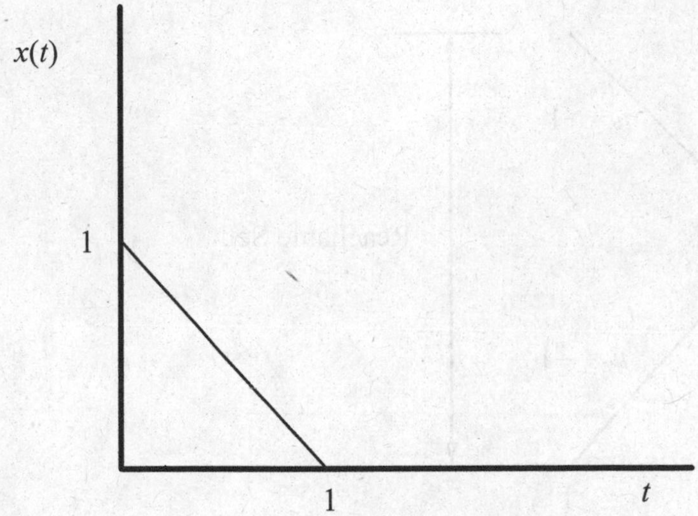

Figure 6.1. Case 1 where the final state is specified.

so

$$dx_f = 0 \tag{6.27}$$

The transversality condition drops out. Thus, we have a well-defined TPBVP and conclude that if $\lambda_o = 0$, then $u = -1$ and (consequently) $\lambda_1 > 0$ (i.e., the switching function must be positive) which allows us to satisfy the necessary conditions (as illustrated in Fig. 6.1). It is seen that $(\lambda_o, \lambda_1) = (0, \lambda_1) \neq \mathbf{0}$ and we have satisfied the conditions of the Minimum Principle. Therefore, this is an abnormal problem. In an abnormal problem we obtain a solution in which the cost functional, in this case Eq. (6.14), is ignored—so we are not really optimizing anything.

For case 2 the final state is free (as shown in Fig. 6.2) so

$$dx_f \neq 0 \tag{6.28}$$

The differential form of the transversality condition is

$$\lambda_{1f} dx_f = 0 \tag{6.29}$$

so that

$$\lambda_1(t_f) = 0 \tag{6.30}$$

Thus, we also have

$$\lambda_1(t) = 0 \tag{6.31}$$

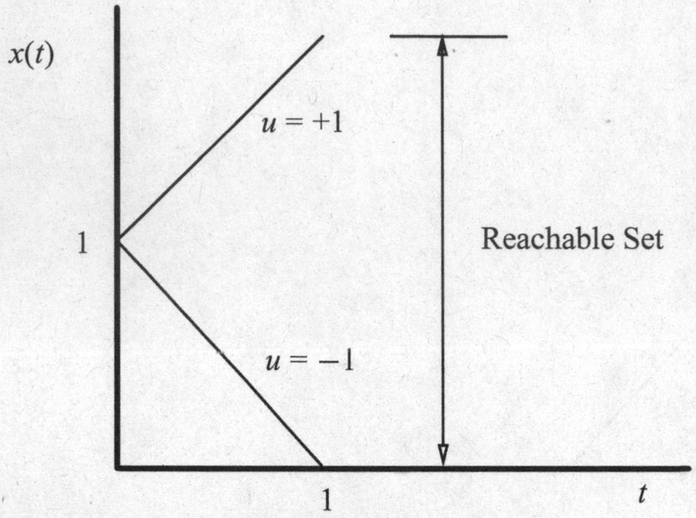

Figure 6.2. Case 2 where final state is free.

since $\dot{\lambda}_1 = 0$. Therefore $(\lambda_o, \lambda_1) = \mathbf{0}$, which violates the first statement [Eq. (6.8)] of the Minimum Principle. Thus, the necessary conditions are not satisfied.

Letting $\lambda_0 = 1$, we obtain:

$$H = x^2 + \lambda_1 u \tag{6.32}$$

$$\dot{x} = u \tag{6.33}$$

with I.C.s $x(0) = 1$ and

$$\dot{\lambda}_1 = -H_x = -2x \tag{6.34}$$

so that

$$\lambda_1(t) \neq 0 \tag{6.35}$$

The transversality condition is:

$$\lambda_{1f} dx_f = 0 \tag{6.36}$$

so

$$\lambda_1(t_f) = 0 \tag{6.37}$$

Thus, we have a well-defined TPBVP:

$$\dot{x} = u, \quad t_o = 0, \quad x(t_o) = 1 \tag{6.38}$$

$$\dot{\lambda}_1 = -2x, \quad t_f = 1, \quad \lambda_1(t_f) = 0 \tag{6.39}$$

For $u = 1$ we obtain:

$$\dot{x} = 1 \tag{6.40}$$

since $x(t_0) = 1$, the solution for x is

$$x(t) = t + 1 \tag{6.41}$$

and thus

$$\dot{\lambda}_1 = -2t - 2 \tag{6.42}$$

The solution for λ_1, using $\lambda(t_f) = 0$ is

$$\lambda_1(t) = -t^2 - 2t + 3 \tag{6.43}$$

For $u = -1$

$$\dot{x} = -1 \tag{6.44}$$

so with $x(t_0) = 1$ we have $x(t) = -t + 1$. Also,

$$\dot{\lambda}_1 = 2t - 2 \tag{6.45}$$

and when we apply $\lambda_1(t_f) = 0$ we obtain $\lambda_1(t) = t^2 - 2t + 1$.

To choose between these two solutions ($u = 1$ or $u = -1$), we could evaluate J; nevertheless, by inspecting the cost functional $J = \int_0^1 x^2 dt$, it is clear that the $u = -1$ solution provides the lowest cost:

$$x(t) = 1 - t \tag{6.46}$$

$$\lambda_1(t) = (1 - t)^2 \tag{6.47}$$

$$u(t) = -1 \tag{6.48}$$

That is, we see from Fig. 6.2 that when $u(t) = -1$ the value of x decreases at the quickest possible rate to make the cost functional, Eq. (6.14), the lowest possible value.

6.2. Legendre-Clebsch Necessary Condition

We have made use of the Legendre-Clebsch necessary condition in Step 4 of our Cookbook (in Chap. 4). We now formally state and prove this condition through use of the Minimum Principle.

If H is second-order differentiable in $\boldsymbol{u}$, the optimal control at t is interior (not on the control boundary), and $\boldsymbol{u}(t)$ is "infinitesimally close" to $\boldsymbol{u}^*(t)$, i.e., a weak variation, then:

$$H_{\boldsymbol{u}}^*(t) = 0,\ H_{\boldsymbol{uu}}^*(t) \geq 0 \tag{6.49}$$

which means the matrix $H_{\boldsymbol{uu}}$ must be positive semi-definite.

The proof follows from the Minimum Principle:

$$H^*[t, \boldsymbol{x}^*(t), \boldsymbol{u}^*(t), \boldsymbol{\lambda}(t)] \leq H[t, \boldsymbol{x}^*(t), \boldsymbol{u}(t), \boldsymbol{\lambda}(t)] \tag{6.50}$$

Using a Taylor series to expand about $\boldsymbol{u}^*$ to second-order in $\boldsymbol{u}$ we obtain:

$$H^* \leq H^* + H_{\boldsymbol{u}}^*(\boldsymbol{u} - \boldsymbol{u}^*) + \frac{1}{2}(\boldsymbol{u} - \boldsymbol{u}^*)^T H_{\boldsymbol{uu}}^*(\boldsymbol{u} - \boldsymbol{u}^*) + \mathcal{O}(\Delta \boldsymbol{u}^3) \tag{6.51}$$

but $H_{\boldsymbol{u}}^* = 0$, from the Euler-Lagrange theorem since we assume the optimal control is not on the bound. Rearranging Eq. (6.51) and ignoring $\mathcal{O}(\Delta \boldsymbol{u}^3)$ we deduce using Eq. (6.50)

$$H_{\boldsymbol{uu}}^* \geq 0. \tag{6.52}$$

which is the Legendre-Clebsch necessary condition.

6.3. Notes on Necessary and Sufficient Conditions

Example 6.3 Necessary and sufficient condition problem.

Consider Fig. 6.3. Suppose we have a set A with the property:

$$\{(x, y) \in A\}. \tag{6.53}$$

A sufficient condition guarantees that (x, y) is in the set A but is usually too restrictive to uniquely define all (x, y) in set A. A necessary condition applies to every element in the set A but is usually too loose to define all (x, y) in A uniquely.

Now let us consider particular statements that pertain to the set A in Fig. 6.3. For example:

$$x = \frac{1}{2}a,\ y = \frac{1}{2}b \tag{6.54}$$

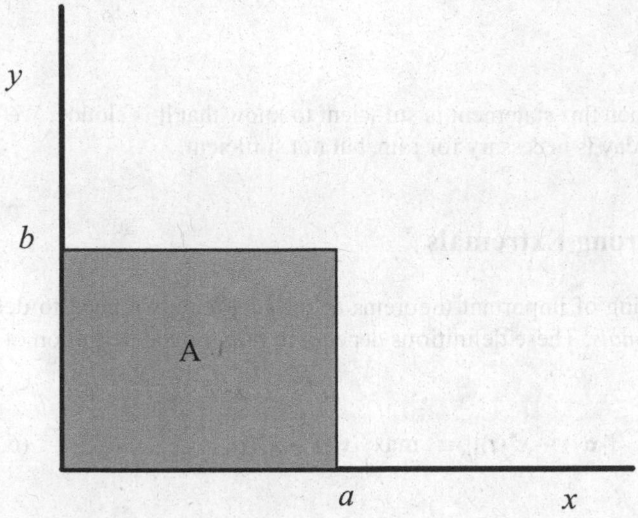

Figure 6.3. $(x, y) \in A$.

Equations (6.54) provide a sufficient condition since they guarantee that $(x, y) \in A$. On the other hand:

$$0 \leq x \leq a \tag{6.55}$$

is a necessary but not sufficient condition that $(x, y) \in A$. Similarly

$$0 \leq y \leq b \tag{6.56}$$

is a necessary condition but it is not sufficient. Finally, we see that

$$0 \leq x \leq a, \; 0 \leq y \leq b \tag{6.57}$$

is a necessary and sufficient condition.

We note that the combination of two necessary conditions [Eqs. (6.55) and (6.56)] give a necessary and sufficient condition [Eq. (6.57)]. We also note that the combination of a necessary condition [say Eq. (6.55)] and a sufficient condition [say Eq. (6.54)] might not even come close to providing a necessary and sufficient condition.

Following Hull [2003], we can generalize the concepts of necessary conditions and sufficient conditions. Let A and B be mathematical statements and let the symbol "$\Rightarrow$" mean "implies" and the symbol "$\Leftarrow$" mean "is implied by". We can then write

$A \Rightarrow B$ means A is sufficient for B
$A \Leftarrow B$ means that A is necessary for B
$A \Leftrightarrow B$ means that A is necessary and sufficient for B

For example

rain $\Rightarrow$ clouds

clouds $\Leftarrow$ rain

So, if it is raining then this statement is sufficient to know that it is cloudy. We can also say that a cloudy day is necessary for rain, but not sufficient.

6.4. Weak and Strong Extremals

To describe the meaning of important theorems in the literature, we need to define *weak* and *strong extremals*. These definitions depend, in turn, on the definition of the *norm*:

$$||\boldsymbol{x}(t) - \boldsymbol{x}^*(t)|| \equiv \max_{t\in[t_o,t_f]} |\boldsymbol{x}(t) - \boldsymbol{x}^*(t)| \tag{6.58}$$

Next, we define N_1 and N_2 *neighborhoods*:

$$N_1(\boldsymbol{x}^*, \boldsymbol{u}^*) \equiv \{\ (\boldsymbol{x}, \boldsymbol{u}) \mid ||\boldsymbol{x} - \boldsymbol{x}^*|| < \varepsilon\ \}, \tag{6.59}$$

$$N_2(\boldsymbol{x}^*, \boldsymbol{u}^*) \equiv \{\ (\boldsymbol{x}, \boldsymbol{u}) \mid ||\boldsymbol{x} - \boldsymbol{x}^*|| < \varepsilon,\ \ ||\boldsymbol{u} - \boldsymbol{u}^*|| < \delta\ \} \tag{6.60}$$

where $\varepsilon > 0$ and $\delta > 0$, and both are sufficiently small.

We can now define weak and strong extremals as follows:

1. $\boldsymbol{x}^*(t)$ is called a *weak extremal* if $J(\boldsymbol{x}^*) \le J(\boldsymbol{x})$ for all $(\boldsymbol{x}, \boldsymbol{u}) \in N_2(\boldsymbol{x}^*, \boldsymbol{u}^*)$ which satisfy the constraints.
2. $\boldsymbol{x}^*(t)$ is called a *strong extremal* if $J(\boldsymbol{x}^*) \le J(\boldsymbol{x})$ for all $(\boldsymbol{x}, \boldsymbol{u}) \in N_1(\boldsymbol{x}^*, \boldsymbol{u}^*)$ which satisfy the constraints.

We see from the definitions that the salient characteristic of a strong extremal is that no change in control—no matter how large—can provide a lower cost.

At this point, let us consider a simple analogy to guide the subsequent discussion. For the moment, let us assume that we are solving a parameter optimization problem (instead of an optimal control problem) and that the cost function is $f(x)$, where x is a scalar. Then we can apply the method of Chap. 1 to minimize $f(x)$. For a local minimum (assuming no constraints) we require $f_x = 0$ and $f_{xx} > 0$. If we find a local minimum, then sufficiently small perturbations in x from x^* will satisfy the conditions of a local minimum. This local minimum is analogous to the weak extremal of optimal control. Next we consider that the scalar value of x has the range: $0 \le x \le x_{\max}$. Suppose we search over the full range of x and find the global minimum. This global minimum is analogous to the strong extremal of optimal control because, in the former, no value of $f(x)$ is less than or equal to $f(x^*)$ and, in the latter, no value of $\boldsymbol{u}$ can produce a lower value than $J(x^*)$.

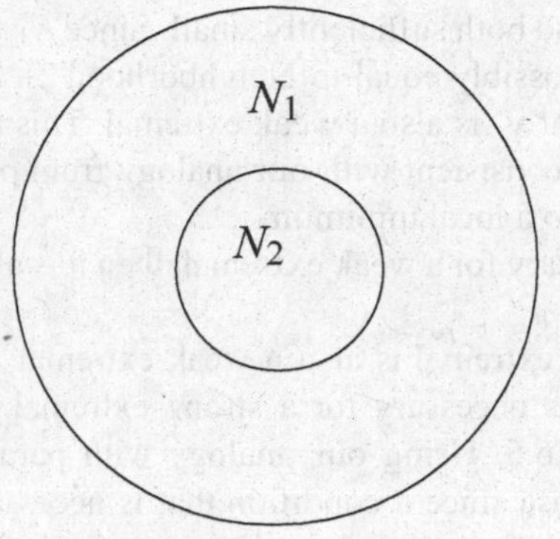

Figure 6.4. Weak and strong neighborhoods.

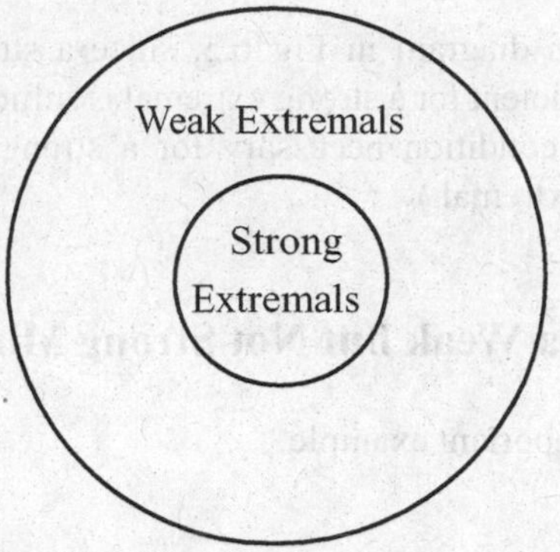

Figure 6.5. Sets of weak and strong extremals.

Several observations can be made about weak and strong extremals as follows.

1. If $\boldsymbol{x}^*(t)$ is a strong extremal, then it is also a weak extremal.
   Suppose $\boldsymbol{x}^*(t)$ is a strong extremal: this implies that

$$J(\boldsymbol{x}^*) \leq J(\boldsymbol{x}) \tag{6.61}$$

   for all $(\boldsymbol{x}, \boldsymbol{u}) \in N_1$.

   We recall that the neighborhood of the strong extremal, N_1, implies

$$||\boldsymbol{x} - \boldsymbol{x}^*|| < \varepsilon \tag{6.62}$$

   while the neighborhood, N_2, implies

$$||\boldsymbol{x} - \boldsymbol{x}^*|| < \varepsilon, \quad ||\boldsymbol{u} - \boldsymbol{u}^*|| < \delta \tag{6.63}$$

for $\varepsilon > 0$ and $\delta > 0$ and both sufficiently small. Since $N_1 \supseteq N_2$ (i.e. Neighborhood 1 is a superset of or possibly equal to Neighborhood 2), then $J(\boldsymbol{x}^*) \leq J(\boldsymbol{x})$ for all $(\boldsymbol{x}, \boldsymbol{u}) \in N_2$ implies that $\boldsymbol{x}^*$ is also a weak extremal. This argument is illustrated in Fig. 6.4. The result is consistent with our analogy from parameter optimization: a global minimum is also a local minimum.

2. If a condition is necessary for a weak extremal, then it is also necessary for a strong extremal.

   Since every strong extremal is also a weak extremal, then anything necessary for a weak extremal is necessary for a strong extremal, as demonstrated by the Venn diagram in Fig. 6.5. Using our analogy with parameter optimization, this observation makes sense since a condition that is necessary for a local minimum would also be necessary for a global minimum.
3. If a condition is sufficient for a strong extremal, then it is sufficient for a weak extremal.

   Again, by the Venn diagram in Fig 6.5, since a strong extremal is a weak extremal, anything sufficient for a strong extremal is sufficient for a weak extremal. (We also note that a condition necessary for a strong extremal is not always necessary for a weak extremal.)

6.5. An Example of a Weak but Not Strong Minimum

Consider the following important example.

Minimize:

$$J = \int_0^a u^3 dt \tag{6.64}$$

subject to:

$$\dot{x} = u \tag{6.65}$$

with B.C.s:

$$x(0) = 0 \tag{6.66}$$

$$x(a) = b \tag{6.67}$$

where $a > 0$ and $b > 0$.

We will show in this problem that:

1. The Legendre-Clebsch condition, $H^*_{\boldsymbol{uu}} \geq 0$, is satisfied.
2. But the Weierstrass condition, $H^* \leq H(\boldsymbol{u})$, is not satisfied.

From the Euler-Lagrange theorem we have:

$$H = u^3 + \lambda u \tag{6.68}$$

$$\dot{\lambda} = -H_x = 0 \tag{6.69}$$

so that

$$\lambda = \text{constant} \tag{6.70}$$

$$H_u = 3u^2 + \lambda = 0 \tag{6.71}$$

so

$$u = \pm\sqrt{\frac{-\lambda}{3}} \tag{6.72}$$

$$H_{uu} = 6u \geq 0 \tag{6.73}$$

and thus u must be positive.

Therefore, we have:

$$u = +\sqrt{\frac{-\lambda}{3}}. \tag{6.74}$$

All of this implies:

$$u = c_1 > 0 \tag{6.75}$$

$$\dot{x} = c_1 \tag{6.76}$$

and

$$x(t) = c_1 t + c_2 \tag{6.77}$$

where c_1 and c_2 are constants.

Then using the B.C's:

$$x(0) = 0, \ x(a) = b \tag{6.78}$$

we find that

$$c_1 = \frac{b}{a}, \ c_2 = 0 \tag{6.79}$$

Thus, the solution is:

$$x^* = \frac{b}{a}t \tag{6.80}$$

$$u^* = \frac{b}{a} \tag{6.81}$$

Again, checking the Legendre-Clebsch condition, we find

$$H^*_{uu} = 6u = 6\frac{b}{a} \geq 0 \tag{6.82}$$

which is satisfied since a and b are positive constants.

Now let us check the Weierstrass condition. Is

$$H^* \leq H(x^*, u) \tag{6.83}$$

for all u?

We note that $\lambda = -3\left(\frac{b}{a}\right)^2$ from $u = +\sqrt{\frac{-\lambda}{3}}$ so that:

$$H^* = u^3 + \lambda u = \left(\frac{b}{a}\right)^3 + (-3)\left(\frac{b}{a}\right)^2\left(\frac{b}{a}\right) = -2\left(\frac{b}{a}\right)^3 \tag{6.84}$$

Is

$$-2\left(\frac{b}{a}\right)^3 \leq u^3 - 3\left(\frac{b}{a}\right)^2 u \tag{6.85}$$

for all u? Clearly, it is possible to select u small enough to reverse the inequality. Consider:

$$u = -10\frac{b}{a} \tag{6.86}$$

Then:

$$H^* = -2\left(\frac{b}{a}\right)^3 > -1{,}000\left(\frac{b}{a}\right)^3 + 30\left(\frac{b}{a}\right)^3 = -970\left(\frac{b}{a}\right)^3 \tag{6.87}$$

Thus, the Weierstrass condition does not hold for all u, since we have:

$$H^* > H \tag{6.88}$$

when $u = -10\left(\frac{b}{a}\right)$.

Next we construct an example of a strong variation for this problem. We will follow the concept illustrated in Fig. 5.1, in which a control blip is applied to the candidate optimal control which causes a variation in the state variable given in Eq. (5.9). We assume that the control blip begins at time t_1 and that the state is brought back to the candidate optimal state by time t_2. In Fig. 6.6 we show the candidate optimal solution expressed by Eqs. (6.80) and (6.81).

In Fig. 6.7 we show the effect that a strong variation may have on our example problem. The main idea is that we use $U(t) = -b/\varepsilon a$ as our control blip, which begins at t_1 and ends at $t_1 + \varepsilon^2 a$. Then, from time $t_1 + \varepsilon^2 a$ to t_2 we assume a constant control, $u(t, \varepsilon) = c$, to restore the state to the candidate optimal solution. Thus, for the strong variation of the control we assume the following:

$$u(t) = \begin{cases} \frac{b}{a} & t \in [0, t_1] \\ -\frac{b}{\varepsilon a} & t \in [t_1, t_1 + \varepsilon^2 a] \\ c & t \in [t_1 + \varepsilon^2 a, t_2] \\ \frac{b}{a} & t \in [t_2, a] \end{cases} \quad (6.89)$$

For the corresponding strong variation of the state we have:

$$x(t) = \begin{cases} \frac{b}{a}t & t \in [0, t_1] \\ \frac{b}{a}t_1 - \frac{b}{\varepsilon a}(t - t_1) & t \in [t_1, t_1 + \varepsilon^2 a] \\ ct + d & t \in [t_1 + \varepsilon^2 a, t_2] \\ \frac{b}{a}t & t \in [t_2, a] \end{cases} \quad (6.90)$$

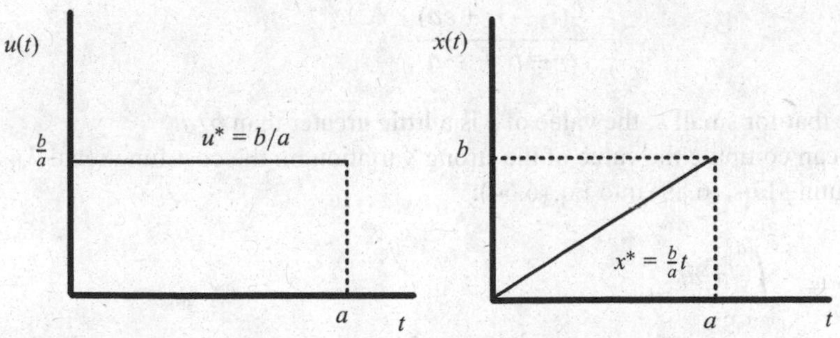

Figure 6.6. Candidate optimal solution from Legendre-Clebsch condition.

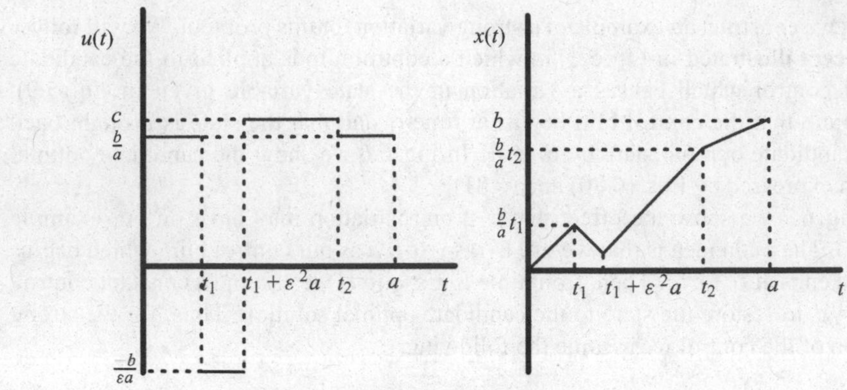

Figure 6.7. Effect of a strong variation on an example.

where d is given by

$$d = \frac{x(t_1 + \varepsilon^2 a)t_2 - x(t_2)(t_1 + \varepsilon^2 a)}{t_2 - (t_1 + \varepsilon^2 a)} \tag{6.91}$$

and where the value of c in Eqs. (6.89) and (6.90) is:

$$\begin{aligned} c &= \frac{x(t_2) - x(t_1 + \varepsilon^2 a)}{t_2 - (t_1 + \varepsilon^2 a)} \\ &= \frac{\frac{b}{a}t_2 - \left[\frac{b}{a}t_1 - \frac{b}{\varepsilon a}(t_1 + \varepsilon^2 a - t_1)\right]}{t_2 - t_1 - \varepsilon^2 a} \\ &= \frac{\frac{b}{a}(t_2 - t_1 + \varepsilon a)}{t_2 - t_1 - \varepsilon^2 a} \end{aligned} \tag{6.92}$$

We see that for small ε, the value of c is a little greater than b/a.

We can compute the value of the strong variation on the cost functional J_{strong}, by substituting Eqs. (6.89) into Eq. (6.64):

$$\begin{aligned} J_{\text{strong}} &= \int_0^a u^3 dt \\ &= \int_0^{t_1} \left(\frac{b}{a}\right)^3 dt + \int_{t_1}^{t_1+\varepsilon^2 a} \left(\frac{-b}{\varepsilon a}\right)^3 dt + \int_{t_1+\varepsilon^2 a}^{t_2} c^3 dt + \int_{t_2}^{a} \left(\frac{b}{a}\right)^3 dt \\ &= \left(\frac{b}{a}\right)^3 t_1 - \frac{1}{\varepsilon^3}\left(\frac{b}{a}\right)^3 (\varepsilon^2 a) + c^3(t_2 - t_1 - \varepsilon^2 a) + \left(\frac{b}{a}\right)^3 (a - t_2) \end{aligned} \tag{6.93}$$

Since $1/\varepsilon$ appears in Eq. (6.93), it represents a large number when ε is small and thus we retain this term but can drop terms of order ε. When ε terms are dropped, Eq. (6.92) gives $c = b/a$, so that Eq. (6.93) becomes:

$$J_{\text{strong}} = \frac{b^3}{a^2}\left(1 - \frac{1}{\varepsilon}\right) \tag{6.94}$$

We can compare the cost of the strong variation, Eq. (6.94), to the cost of the weak variation, J_{weak}, which is [from Eqs. (6.64) and (6.81)]:

$$J_{\text{weak}} = \int_0^a u^3 dt = \int_0^a \left(\frac{b}{a}\right)^3 dt = \frac{b^3}{a^2} \tag{6.95}$$

Of course $\varepsilon < 1$ because we tacitly assumed the time order: $t_1 + \varepsilon^2 a < t_2 < a$ as illustrated in Fig. 6.7. For any $\varepsilon < 1$, J_{strong} is negative, while J_{weak} is always positive, therefore

$$J_{\text{strong}} < J_{\text{weak}} \tag{6.96}$$

Thus, we have shown that the Weierstrass condition (or more generally, the Minimum Principle) is not satisfied by Eqs. (6.80) and (6.81). Furthermore, we have constructed a control which provides a lower cost than that of the weak variation. Here we appeal to the more general Minimum Principle because our control is piecewise continuous. There is no true minimum in this case because the cost approaches an infinitely negative value as ε approaches zero.

6.6. Second-Order Necessary and Sufficient Conditions

Up to now we have determined first-order conditions that are necessary for $\boldsymbol{u}(t)$ to be optimal. In this section, we briefly state a second-order necessary condition and some known conditions that are sufficient.

The second-order necessary condition is the Jacobi (conjugate point) condition [i.e., no points are conjugate on the open interval (t_o, t_f)]. This condition is usually not applied because of the difficulty in doing so. However, a recent simplified procedure has been developed for the case of a continuous control by Prussing and Sandrik [2005]. The Jacobi condition is necessary because if an extremal solution contains a conjugate point, there exists a neighboring solution of lower cost.

One can argue that a necessary condition is more informative than a sufficient condition. If a sufficient condition is not satisfied, the result is inconclusive (see Exercise 2 of Chap. 1). However, if a necessary condition is not satisfied, the result is fatal and the solution is not optimal.

If only one solution satisfies the first necessary conditions (Euler-Lagrange plus transversality), then:

1. It is the optimal solution, or
2. The optimal solution lies outside the class of functions being considered.

If many solutions satisfy the necessary conditions, then further criteria must be developed.

Leitmann [1981] provides sufficient conditions for a proper, weak, relative minimum, as follows.

Let $\boldsymbol{x}^*(t) \in C^2[t_o, t_f]$, $\boldsymbol{u}^*(t) \in C^1[t_o, t_f]$.

If:

1. The Euler-Lagrange equations and the transversality condition are satisfied,
2. The Strengthened Legendre-Clebsch Condition is satisfied (i.e., $H^*_{\boldsymbol{uu}}$ is positive definite: $H^*_{\boldsymbol{uu}} > 0$), and
3. The Strengthened Jacobi (or conjugate point) condition is satisfied (i.e., no points are conjugate on the closed interval $[t_o, t_f]$),

then: $\boldsymbol{x}^*(t)$ is a proper, weak, relative minimum. By *proper* we mean unique in neighborhood N_2.

6.7. Examples Illustrating the Concept of a Conjugate Point

In Chap. 4 (in our discussion of Example 4.1) we mentioned the problem of a conjugate point. If, for example, we consider the shortest path on the surface of the Earth from the North Pole to the South Pole, it is clear that there are infinitely many paths (called meridians) that are equally short. Since these solutions are not unique, none of them are considered to be a proper minimum. From a practical point of view, we wish to identify such problems to avoid attempting their numerical solution (a process unlikely to reach any resolution).

We illustrate the problem of conjugate points with the following classic examples.

Example 6.4 Shortest path to a great circle on a sphere.

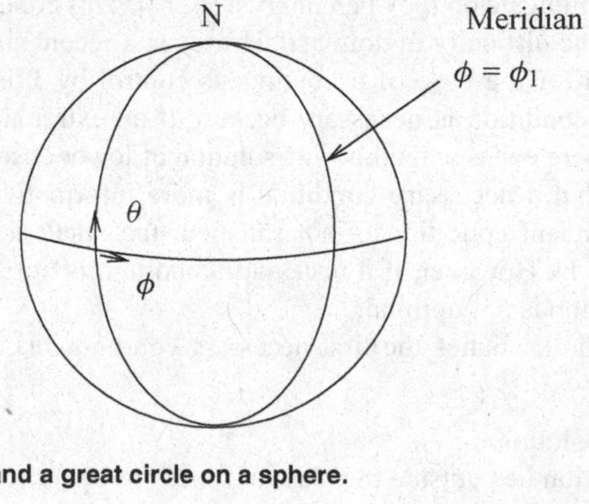

Figure 6.8. A point and a great circle on a sphere.

Consider the shortest path between a point and a great circle on a sphere, as illustrated in Fig 6.8. The element of distance on a sphere is given by:

$$ds = \sqrt{r^2(d\theta)^2 + r^2(\cos\theta)^2(d\phi)^2} \tag{6.97}$$

Without loss of generality, we assume that $r = 1$, that we begin at the origin of Fig. 6.8, and that we wish to find the shortest path to the meridian, ϕ_1, without regard for the final value of latitude. We state our problem as follows (see Bryson and Ho [1975]).

Find $u(\phi)$ to minimize J:

$$J = \int_0^{\phi_1} \sqrt{u^2 + (\cos\theta)^2 d\phi} \tag{6.98}$$

where

$$\frac{d\theta}{d\phi} = u, \quad \theta(0) = 0 \tag{6.99}$$

Let us first approach this problem by using the four steps of the cookbook in Chap. 4.

1. Form the Hamiltonian:

$$H = \sqrt{u^2 + (\cos\theta)^2} + \lambda u \tag{6.100}$$

2. Write Euler-Lagrange equations:

$$\frac{d\lambda}{d\phi} = -\frac{\partial H}{\partial\theta} = \frac{\cos\theta\sin\theta}{\sqrt{u^2 + (\cos\theta)^2}} \tag{6.101}$$

$$\frac{\partial H}{\partial u} = \frac{u}{\sqrt{u^2 + (\cos\theta)^2}} + \lambda = 0 \tag{6.102}$$

so that

$$\lambda = -\frac{u}{\sqrt{u^2 + (\cos\theta)^2}} \tag{6.103}$$

3. Use the differential form of the transversality condition:

$$H_f^* dt_f - \lambda_f dx_f = H_f^* d\phi_f - \lambda_f d\theta_f = 0 \tag{6.104}$$

$$d\phi_f = 0 \tag{6.105}$$

so that

$$\lambda_f = 0 \tag{6.106}$$

We note that $\theta = u = \lambda = 0$ satisfies the above conditions, which corresponds to a path along the equator in Fig. 6.8.

4. Checking the Legendre-Clebsch condition we obtain:

$$\begin{aligned} H_{uu} &= (u^2 + \cos^2\theta)^{-\frac{1}{2}} - \frac{u}{2}(u^2 + \cos^2\theta)^{-\frac{3}{2}} 2u \\ &= (u^2 + \cos^2\theta)^{-\frac{1}{2}}[1 - u^2(u^2 + \cos^2\theta)^{-1}] \\ &= 1 > 0 \end{aligned} \tag{6.107}$$

In the general case, to determine whether a conjugate point exists requires the calculation described in Prussing and Sandrik [2005]. But for this simple example, we can examine the second variation $\delta^2 J$ analytically.

Expanding the cost function to include the second variation yields

$$J = J^* + \delta J + \delta^2 J$$

Because $\delta J = 0$ on the solution that satisfies the first-order necessary conditions, $\delta^2 J > 0$ on neighboring solutions implies that $J > J^*$, i.e., the cost J^* is a local minimum. The classical approach to this analysis is called the *accessory minimum problem*, in which the second variations are minimized to see if that minimum value is positive.

Consider neighboring paths (to $u = 0$, $\theta = 0$). By expanding the index of performance to form the second variation $\delta^2 J$:

$$J = \int_0^{\phi_1} \left[u^2 + \left(1 - \frac{\theta^2}{2!} + \cdots \right)^2 \right]^{\frac{1}{2}} d\phi \tag{6.108}$$

and by ignoring the higher order terms and using a binomial expansion we obtain:

$$\begin{aligned} J &= \int_0^{\phi_1} \left[1 + u^2 - \theta^2 \right]^{\frac{1}{2}} d\phi \\ &\approx \int_0^{\phi_1} \left[1 + \frac{1}{2}(u^2 - \theta^2) \right] d\phi \\ &= \phi_1 + \frac{1}{2} \int_0^{\phi_1} (u^2 - \theta^2) d\phi \end{aligned} \tag{6.109}$$

Thus we have the accessory minimum problem:

Minimize:

$$\delta^2 J = J - \phi_1 = \frac{1}{2} \int_0^{\phi_1} (u^2 - \theta^2) d\phi \tag{6.110}$$

The Hamiltonian is:

$$H = \frac{1}{2}(u^2 - \theta^2) + \lambda u \tag{6.111}$$

and the Euler-Lagrange equations provide:

$$\frac{d\lambda}{d\phi} = -H_\theta = \theta \tag{6.112}$$

$$H_u = u + \lambda = 0 \tag{6.113}$$

$$u = -\lambda \tag{6.114}$$

From the transversality condition we obtain

$$H_f^* d\phi_f - \lambda_f d\theta_f = 0 \tag{6.115}$$

and since

$$d\phi_f = 0 \tag{6.116}$$

we have

$$\lambda_f = \lambda(\phi_1) = 0 \tag{6.117}$$

Now from the process equation we have:

$$\frac{d\theta}{d\phi} = u = -\lambda \tag{6.118}$$

Differentiating and using the costate equation we obtain

$$\frac{d^2\theta}{d\phi^2} = \frac{-d\lambda}{d\phi} = -\theta \tag{6.119}$$

Thus, we have the simple harmonic oscillator equation:

$$\frac{d^2\theta}{d\phi^2} + \theta = 0 \tag{6.120}$$

with B.C.s:

$$\theta(0) = 0 \tag{6.121}$$

$$-\lambda(\phi_1) = \left.\frac{d\theta}{d\phi}\right|_{\phi_1} = 0 \tag{6.122}$$

The neighboring solution for $\theta(0) = 0$ is:

$$\theta = \epsilon \sin\phi, \quad |\epsilon| \ll 1 \tag{6.123}$$

The final B.C. is then:

$$\frac{d\theta}{d\phi}(\phi_1) = \epsilon \cos \phi_1 = 0 \tag{6.124}$$

From the accessory problem we can make the following observations. Analytical solutions are available for the state and costate equations. The final boundary condition is satisfied only by $\epsilon = 0$ if $\phi_1 < \frac{\pi}{2}$, but it is satisfied by any ϵ if $\phi_1 = \frac{\pi}{2}$. Point O (the origin of Fig. 6.8) is said to be a conjugate point to $\phi_1 = \frac{\pi}{2}$. At this point, the second variation vanishes and an infinite number of solutions provide the same minimum distance. No proper minimum exists.

For $\frac{\pi}{2} < \phi_1 < \pi$ the value of $\delta^2 J$ is negative (see Exercise 3) and the extremal solution $\theta = u = 0$ is not a local minimum because the neighboring solution $\theta = \epsilon \sin \phi$ has a lower cost than the extremal solution. Furthermore, this lower cost neighboring solution does not even satisfy the necessary condition, Eq. (6.124).

Example 6.5 Shortest path between two points on a sphere.

Consider the shortest path between two points on a sphere, as illustrated in Fig. 6.9. This is a classical problem of the literature. Here we provide well-known results (as reported by Bryson and Ho [1975], Fox [1987], and Hull [2003]). This problem is similar to Example 6.4 except that $\theta_f = \theta(\phi_1) = 0$ replaces $\lambda_f = 0$ of Eq. (6.106). A conjugate point exists for $\phi_1 = \pi$. The second variation is the same as Eq. (6.110) and a neighboring solution to $\theta = u = 0$ that satisfies the B.C.s is $\theta = \epsilon \sin \pi \frac{\phi}{\phi_1}$ with $|\epsilon| \ll 1$. For the neighboring solution the second variation is equal to $-\frac{\epsilon^2}{4\phi_1}(\phi_1^2 - \pi^2)$, which is

(i) Greater than zero for $\phi_1 < \pi$,
(ii) Equal to zero for $\phi_1 = \pi$, and
(iii) Less than zero for $\phi_1 > \pi$.

As in Example 6.3 if the extremal path $\theta = u = 0$ (which satisfies the first-order necessary conditions) contains a conjugate point (if $\phi_1 > \pi$), it is not a minimizing solution because a neighboring path (that bypasses the conjugate point) has a lower cost.

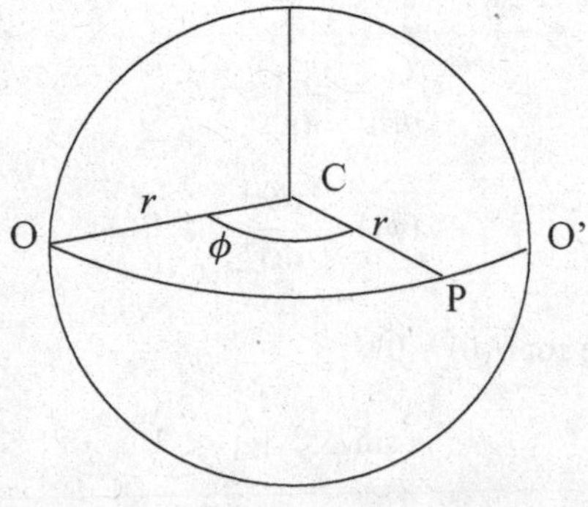

Figure 6.9. Shortest path between two points on a sphere.

6.8. Summary

There are a variety of statements of the Minimum Principle. In words, McShane has given the most succinct, "the Hamiltonian must be minimized over the set of all admissible controls." The Legendre-Clebsch condition states that H_{uu} must be positive semi-definite; it applies only to interior controls and only to weak variations. The Weierstrass condition applies to strong variations (meaning the control can have large changes from the optimal while the state can only have small changes) and where the optimal control must be continuous. Pontryagin provides the most general statement of the Minimum Principle where the control can be piecewise continuous (or more precisely "measurable").

In any case we must bear in mind that the calculus of variations can only provide local optimal solutions because variations of the state must remain infinitesimally small. However, variations of the control may be large. When only small variations of the control are considered, we refer to the variations as weak variations. As we have seen in an example, strong variations can sometimes lead to a lower cost.

In two examples we have illustrated the existence of a conjugate point and demonstrated, by calculating the second variation, that an extremal that contains a conjugate point is not a minimum.

6.9. Exercises

1. An astronaut in space must transfer in minimum time from point A to point B, a distance L. Assume that the astronaut has a thruster which produces a variable thrust F, with a maximum of $F_{\max}$. Also, for simplicity, assume that the total mass of the astronaut, propellant, and thruster remains constant during the maneuver. The initial velocity and final velocity of the astronaut is zero with respect to an inertial frame. There are no other forces acting on the astronaut.

   1a. Set up the minimization problem statement as a Lagrange problem, with $u = F$. (Note: F can be P. C.)

   1b. Show that the switching function is a linear function of time, i.e., $\lambda_2 = c_1 t + c_2$. (Don't evaluate constants c_1 and c_2.)

   1c. Make sketches of the switching function, the thrust profile, and the velocity as functions of time. (Hint: Use engineering judgement to deduce the shapes of these plots. Give correct shape and sign.)

2. A bead can move along a line where its position is given by x. Assume that the bead is initially located at $x = 0$ and that its final position is $x = L$. Assume that the control of the bead is the velocity, V (Fig. 6.10).

   Set up the optimal control problem as a Lagrange problem and work out the following steps.

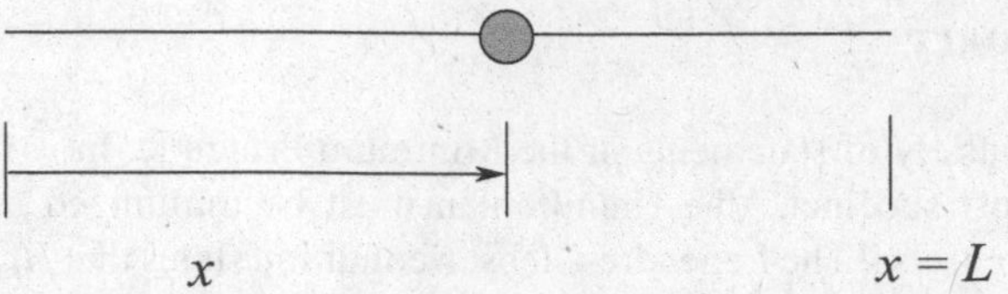

Figure 6.10. Lagrange problem: Bead on a wire.

2a. Assume that we wish to minimize the following:

$$\text{Min } J = \int_{t_0}^{t_f} (V^2 + 1)dt$$

where V is the control. Write the state equation and the initial and final conditions for the state. Assume that $t_0 = 0$ and t_f is free.

2b. Write the Hamiltonian.

2c. Write the costate equation, and give the form of its solution.

2d. Assuming unbounded control, state the control law (for V) in terms of the costate.

2e. Show that your control satisfies the Legendre-Clebsch necessary condition.

Note: you are not required to solve for the constants that may appear in Exercises 2c and 2d.

3. Show that in Example 6.4 the second variation in Eq. (6.110) for the neighboring solution $\theta = \epsilon \sin \phi$ is

(i) > 0 for $\phi_1 < \frac{\pi}{2}$,
(ii) $= 0$ for $\phi_1 = \frac{\pi}{2}$, and
(iii) < 0 for $\frac{\pi}{2} < \phi_1 < \pi$.

4. In Example 6.5 verify that our second variation of Eq. (6.110) for the neighboring path $\theta = \epsilon \sin \pi \frac{\phi}{\phi_1}$ is equal to $-\frac{\epsilon^2}{4\phi_1}(\phi_1^2 - \pi^2)$.

References

M. Athans, P.L. Falb, in *Optimal Control: An Introduction to the Theory and Its Applications* (Dover, New York, 1966)

L.D. Berkovitz, *Optimal Control Theory* (Springer, New York, 1974)

A.E. Bryson Jr., Y.C. Ho, *Applied Optimal Control* (Hemisphere Publishing, Washington, D.C., 1975)

G.M. Ewing, *Calculus of Variations with Applications* (Dover, New York, 1985)

C. Fox, *An Introduction to the Calculus of Variations* (Dover, New York, 1987)

M.R. Hestenes, *Calculus of Variations and Optimal Control Theory* (Wiley, New York, 1966)

D.G. Hull, *Optimal Control Theory for Applications* (Springer, New York, 2003)

G. Leitmann, *The Calculus of Variations and Optimal Control* (Plenum, New York 1981)

D.A. Pierre, *Optimization Theory with Applications* (Wiley, New York, 1969)

L.S. Pontryagin, V.G. Boltyanskii, R.V. Gamkrelidze, E.F. Mishchenko, *The Mathematical Theory of Optimal Processes* (Wiley, New York, 1962)

J.E. Prussing, S.L. Sandrik, Second-order necessary conditions and sufficient conditions applied to continuous-thrust trajectories. J. Guid Control Dyn. **28**(4), 812–816 (2005). Engineering Note

I.M. Ross, *A Primer on Pontryagin's Principle in Optimal Control* (Collegiate Publishers, Carmel, 2009)

J. Vagners, Optimization techniques, in *Handbook of Applied Mathematics*, 2nd edn., ed. by C.E. Pearson (Van Nostrand Reinhold, New York, 1983), pp. 1140–1216

Chapter 7

Some Applications

7.1. Aircraft Performance Optimization

Why discuss aircraft performance in a text chiefly about optimization of spacecraft trajectories? Because aerospace engineers sometimes wish to capture the best of both worlds—as in the examples of Pegasus and SpaceShipOne which use aircraft as first stages and rockets as second stages to send vehicles into space.

Typical flight envelopes for subsonic and supersonic aircraft are shown in Figs. 7.1 and 7.2. When test pilots talk about "pushing the envelope" (as quoted famously in the movie *The Right Stuff*) they are referring to the empirical determination of these flight envelopes. In Figs. 7.1 and 7.2 we have altitude, h, plotted on the vertical axis and speed in Mach number, V, plotted on the horizontal axis. At low speed, aircraft are unable to fly, so the flight envelope starts at $V > 0$ for $h = 0$. As speed increases, the altitude (that the aircraft can reach) increases. Subsonic flight envelopes usually have a simple inverted parabolic shape where the greatest speed occurs at zero altitude, which can be achieved by diving from high altitude (as shown in Fig. 7.1). Supersonic flight envelopes are more complicated (as illustrated in Fig. 7.2) because, in addition to constraints of insufficient lift and insufficient thrust, there may be insufficient structure. That is, at low altitude and high Mach number, the supersonic aircraft may have to throttle back to avoid tearing the wings off in the denser atmosphere.

A classical optimization question is: what is the minimum-time trajectory from P_1 to P_2 (or P_2 to P_3) in Fig. 7.2? This typical aircraft performance question will be addressed in the following analysis.

The free-body diagram for this problem is given in Fig. 7.3 where $\boldsymbol{L}$ and $\boldsymbol{D}$ are the lift and drag vectors and $\boldsymbol{T}$ and $\boldsymbol{W}$ are the thrust and weight vectors, respectively. The inertial velocity vector, $\boldsymbol{V}$, is oriented at flight-path angle, γ, with respect to the local horizontal plane. The angle of attack, α, is the angle between $\boldsymbol{V}$ and the zero-lift axis. By definition, when $\alpha = 0$ the aircraft produces no lift; the zero-lift orientation typically occurs when the nose of the aircraft is pointed slightly downward (relative to $\boldsymbol{V}$). The angle between the zero-lift axis and $\boldsymbol{T}$ is denoted by ε.

Now we will derive the aircraft equations of motion for translation in the vertical plane (the h-x plane of Fig. 7.3). First, we note that lift is perpendicular to velocity. Let us define a frame of reference (fixed at the center of mass of the aircraft) so:

$$\hat{V} = \cos\gamma\,\hat{x} + \sin\gamma\,\hat{h} \tag{7.1a}$$

$$\hat{L} = -\sin\gamma\,\hat{x} + \cos\gamma\,\hat{h} \tag{7.1b}$$

J.M. Longuski et al., *Optimal Control with Aerospace Applications*, Space Technology Library 32, DOI 10.1007/978-1-4614-8945-0_7,

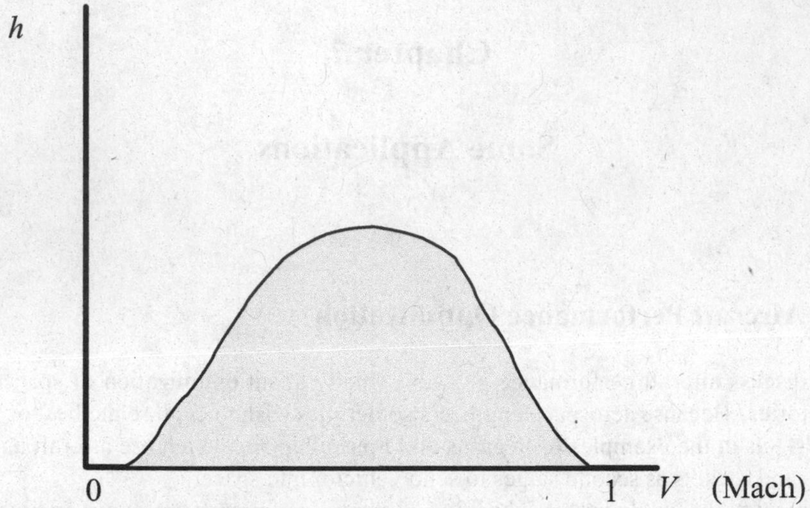

Figure 7.1. Typical flight envelope for subsonic aircraft ($V < 1$).

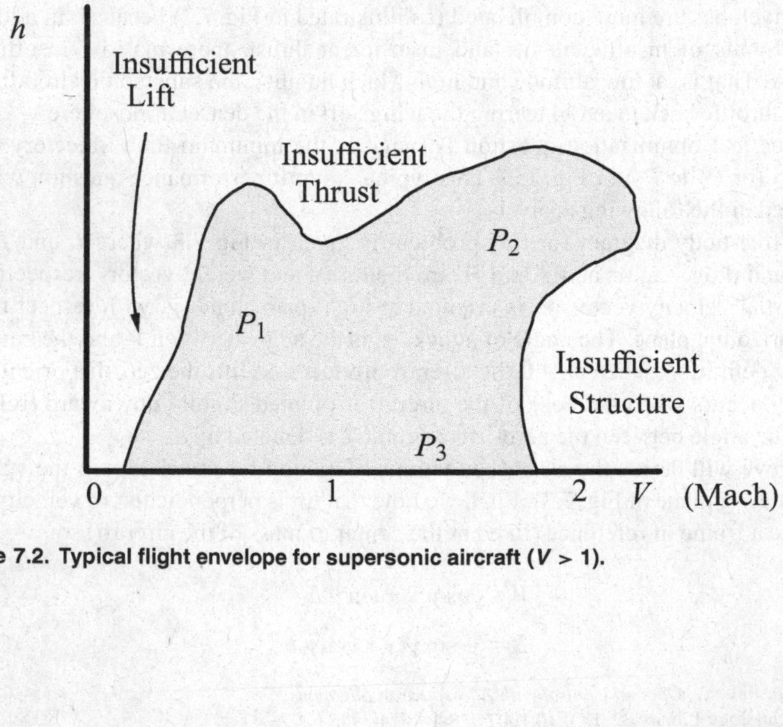

Figure 7.2. Typical flight envelope for supersonic aircraft ($V > 1$).

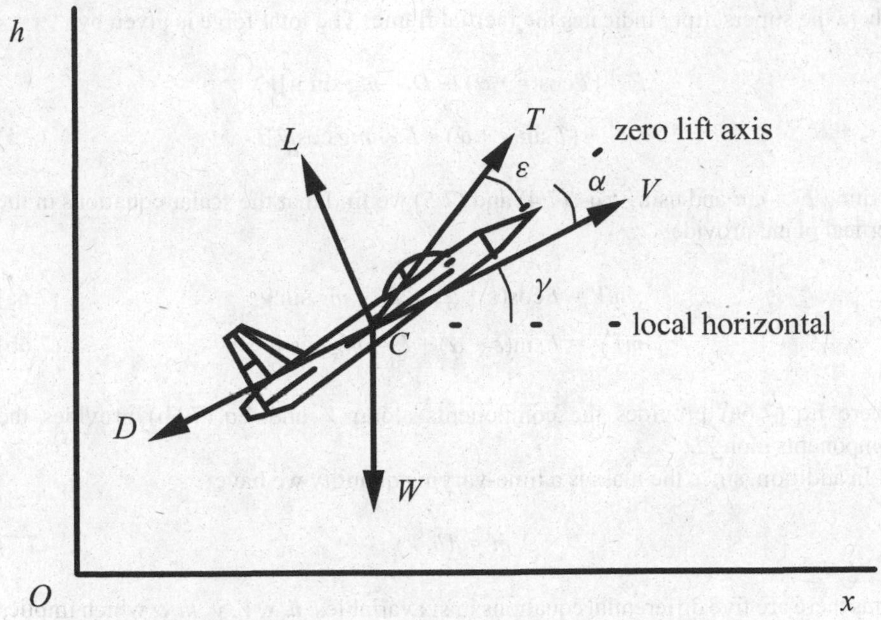

Figure 7.3. Aircraft free-body diagram (Adapted from Vinh [1993]).

where $\hat{V}$ and $\hat{L}$ are unit vectors along the velocity and lift vectors and where $\hat{x}$ and $\hat{h}$ are unit vectors along the x and h axes, respectively. We define $\hat{z}$ as out of the page:

$$\hat{x} \times \hat{h} = \hat{V} \times \hat{L} = \hat{z} \tag{7.2}$$

The scalar velocity components are:

$$\dot{x} = V \cos \gamma \tag{7.3a}$$

$$\dot{h} = V \sin \gamma \tag{7.3b}$$

We derive the equations of motion in the $\hat{V}$-$\hat{L}$ frame by applying Newton's law $\boldsymbol{F} = m\boldsymbol{a}$. The inertial acceleration of the center of mass of the aircraft, C, with respect to the origin, O, of the h-x frame is:

$$\begin{aligned} {}^{i}\boldsymbol{a}^{OC} &= \dot{V}\hat{V} + \dot{\gamma}\hat{z} \times V\hat{V} \\ &= \dot{V}\hat{V} + V\dot{\gamma}\hat{L} \end{aligned} \tag{7.4}$$

where the superscript i indicates the inertial frame. The total force is given by

$$\boldsymbol{F} = [T\cos(\varepsilon+\alpha) - D - mg\sin\gamma]\hat{V} + [T\sin(\varepsilon+\alpha) + L - mg\cos\gamma]\hat{L} \tag{7.5}$$

Writing $\boldsymbol{F} = m\boldsymbol{a}$ and using Eqs. (7.4) and (7.5) we find that the scalar equations in the vertical plane provide:

$$m\dot{V} = T\cos(\varepsilon+\alpha) - D - mg\sin\gamma \tag{7.6a}$$

$$mV\dot{\gamma} = T\sin(\varepsilon+\alpha) + L - mg\cos\gamma \tag{7.6b}$$

where Eq. (7.6a) provides the components along $\hat{V}$ and Eq. (7.6b) provides the components along $\hat{L}$.

In addition, since the mass is a time-varying quantity we have

$$\dot{m} = f(h, V) \tag{7.7}$$

Thus there are five differential equations in six variables: $h, x, V, \gamma, m, \alpha$ which implies that one variable can be chosen as a control variable, leaving five state variables. We assume in this model that $D = D(h, V, \alpha)$ and $L = L(h, V, \alpha)$; drag and lift are known functions of altitude, speed, and angle of attack. We also assume that $T = T(h, V)$ and $\dot{m} = f(h, V)$; thrust and mass rates of change are known functions of altitude and velocity for a given throttle setting.

Next we consider some simplifying assumptions. If we assume the mass is constant and if we drop range considerations (see Miele [1962] for maximum range problem), then we have:

$$\dot{h} = V\sin\gamma \tag{7.8a}$$

$$m\dot{V} = T\cos(\varepsilon+\alpha) - D - mg\sin\gamma \tag{7.8b}$$

$$mV\dot{\gamma} = T\sin(\varepsilon+\alpha) + L - mg\cos\gamma \tag{7.8c}$$

In addition, since $mV\dot{\gamma}$ is usually small (i.e., the acceleration perpendicular to the velocity is negligible):

$$T(h, V)\sin(\varepsilon+\alpha) + L(h, V, \alpha) - W\cos\gamma = 0 \tag{7.9}$$

Equation (7.9) gives the angle of attack, that is,

$$\alpha = \alpha(h, V, \gamma) \tag{7.10}$$

and thus one equation is eliminated.

Before analyzing the problem of unsteady climb, we briefly remark that for steady climb we assume that $\dot{V} = 0$ and usually assume that $\varepsilon + \alpha = 0$. For steady cruise we also set $\gamma = 0$ which implies $L = W$ and $T = D$, resulting in a trivial equilibrium problem.

For unsteady climb analysis, we have two equations, Eqs. (7.8a) and (7.8b), in three variables (h, V, γ). Multiplying Eq. (7.8b) by $V/(mg)$ and adding it to Eq. (7.8a) yields:

$$\dot{h} + \frac{V\dot{V}}{g} = \frac{TV\cos(\varepsilon + \alpha) - DV}{mg} \tag{7.11}$$

Thus, we have eliminated one equation and one variable, γ. Although not necessary, we will gain more insight by assuming $\cos(\varepsilon + \alpha) = 1$, since ε and α are typically small angles, so we have:

$$\dot{h} + \frac{V\dot{V}}{g} = \frac{(T - D)V}{W} \tag{7.12}$$

where the left hand side is the derivative of the total mechanical energy divided by the weight $\frac{d}{dt}[(mgh + \frac{1}{2}mV^2)/W]$.

We now introduce *excess power*, *P*, and *energy height*, h_c, which are defined as follows

$$P(h, V) \equiv \frac{[T(h, V) - D(h, V)]V}{W} \tag{7.13}$$

$$h_c \equiv h + \frac{1}{2}\frac{V^2}{g} \tag{7.14}$$

These definitions are widely used in the *energy method* of evaluating *accelerated climbs* (i.e., unsteady climbs).

From these definitions and Eq. (7.12) we have:

$$\begin{aligned} \dot{h}_c &= P(h, V) \\ &= \frac{[T(h, V) - D(h, V)]V}{W} \\ &= \frac{\{T[h(h_c, V), V] - D[h(h_c, V), V]\}\, V}{W} \end{aligned} \tag{7.15}$$

The rate of change of energy height equals the excess power. We see from Eq. (7.15) that excess power allows us to change the energy height of the aircraft. We view h_c as the state variable, V as the control variable and Eq. (7.15) as the process equation.

In Fig. 7.4 we illustrate a typical altitude-Mach number plot of constant energy height. These curves are universal (i.e., they do not depend on the specific aircraft). An aircraft at A can dive to point B or zoom climb to C in a short time span. For example, an aircraft at 20,000 ft and zero airspeed can dive to zero altitude and pick up a velocity of Mach 1. Similarly, an aircraft at zero altitude traveling at Mach 1 can quickly ascend (zoom climb) to an altitude of 20,000 ft where it loses all its airspeed. (In this simplified analysis we ignore the effect of drag). Thus, when we use energy height as the state variable for unsteady climb analysis, we are viewing the aircraft as a falling body problem which obeys conservation of total energy. A particle dropped

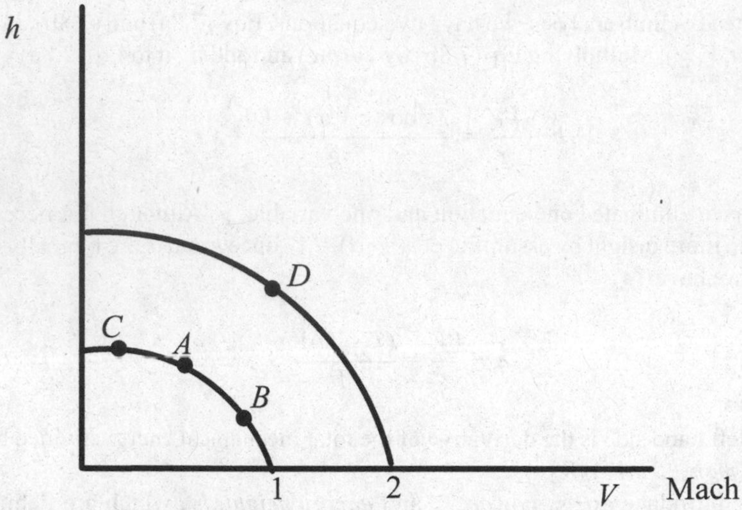

Figure 7.4. **Typical altitude-Mach number plot of constant energy height where energy height, h_c, is defined by Eq. (7.14). Points *A*, *B*, and *C* have the same h_c; point *D* has a higher value.**

from an altitude of 20,000 ft will (again ignoring drag) strike the ground in 125 s. A particle on the ground with an initial vertical speed of Mach 1 will rise to an altitude of 20,000 ft in 125 s. On the other hand, getting from *A* to *D* takes power and time. Airline passengers are used to the slow climb to cruise altitude (typically about 30,000 ft) which takes many minutes—the airliner starts at zero altitude and zero speed and must build up the energy height through the excess power of the engines. As long as the engines can provide more thrust than the drag, the aircraft can continue its accelerated climb according to Eq. (7.15).

Example 7.1 Minimum time to climb.

Minimize the time to climb from (h_o, V_o) to (h_f, V_f) in Fig. 7.5. Note that there is a ground constraint $h \geq 0$, that we should not attempt to violate!

Minimize:

$$J = t_f \tag{7.16}$$

subject to:

$$\dot{h}_c = P(h_c, V) \tag{7.17}$$

with boundary conditions $h_c(0)$ and $h_c(t_f)$.

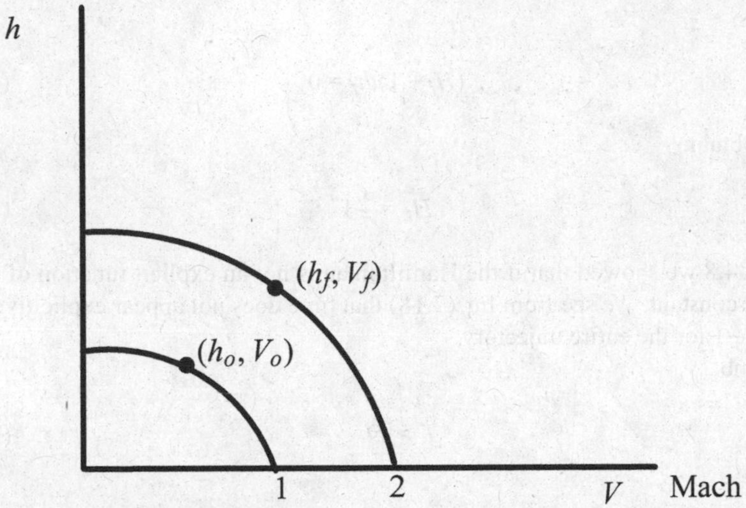

Figure 7.5. **Minimum time-to-climb problem. Aircraft can dive or zoom climb at constant h_c in a matter of seconds but require several minutes to change h_c significantly.**

Once the aircraft achieves the desired energy height (h_c), it can zoom climb (or dive) in virtually no time to get to the desired altitude. We set up the necessary conditions as follows.

1. Form the Hamiltonian:

$$H = \lambda P(h_c, V) \tag{7.18}$$

2. From the Euler-Lagrange theorem and the Legendre-Clebsch condition we find:

$$H_V = 0 \tag{7.19}$$

so

$$\lambda P_V = 0 \tag{7.20}$$

and

$$H_{VV} \geq 0 \tag{7.21}$$

so that

$$\lambda P_{VV} \geq 0 \tag{7.22}$$

3. The differential form of the transversality condition provides:

$$H_f dt_f - \lambda_f dh_c(t_f) + dt_f = 0 \tag{7.23}$$

Thus

$$(H_f + 1)dt_f = 0 \tag{7.24}$$

so we obtain

$$H_f = -1 \tag{7.25}$$

In Sect. 4.8 we showed that if the Hamiltonian is not an explicit function of time, then it is a constant. We see from Eq. (7.18) that time does not appear explicitly; thus $H = \lambda P = -1$ for the entire trajectory.

For climb

$$T \geq D \tag{7.26}$$

so

$$P \geq 0 \tag{7.27}$$

Therefore

$$\lambda \leq 0 \tag{7.28}$$

Clearly $\lambda \neq 0$, since $\lambda P = -1$. Thus, from $\lambda P_V = 0$, $\lambda P_{VV} \geq 0$ and $\lambda < 0$ we have:

$$P_V = 0$$
$$P_{VV} \leq 0 \tag{7.29}$$

which means that *to minimize the time to climb at a given energy height, h_c, pick the velocity, V, to maximize excess power.*

Before we examine this optimal control strategy graphically, let us briefly consider an alternate derivation.

From the process equation:

$$\dot{h}_c = P(V, h_c) = \frac{dh_c}{dt} \tag{7.30}$$

or

$$dt = \frac{dh_c}{P(V, h_c)} \tag{7.31}$$

Integrating, we obtain

$$\int_{t_o}^{t_f} dt = t_f - t_o = \int_{h_{co}}^{h_{cf}} \frac{1}{P(V, h_c)} dh_c \tag{7.32}$$

We see from Eq. (7.32) that to ensure the time is a minimum, the excess power which appears in the denominator of the integrand must be a maximum. Again: to minimize the time to climb at a given height h_c, pick V to maximize excess power.

But how does the pilot pick V? We return to our earlier discussion of dives and zoom climbs. At any given energy height, the pilot can choose from of a range of velocities that he or she can obtain very quickly by diving (to increase V) or by zoom climbing (to decrease V). We assume that the time required for this change of velocity is small (e.g., measured in seconds) compared to the time required to significantly change the energy height by the use of engine excess power (e.g., measured in minutes). In practice the results obtained by ignoring the time for diving and zoom climbing provide very good first-order optimal guesses. Refined optimal solutions for minimum time-to-climb problems can be found using the higher fidelity equations of motion developed earlier in this chapter.

Let us consider Fig. 7.6. For each h_c curve there is a point where P is maximum. In fact, at these points the P curve is tangent to the h_c curve. We observe that there are two local extremals: the arcs A–D and D'–I. To get from the starting point of low energy height and low velocity (h_o, V_o) to the final point of high energy height and high velocity (h_f, V_f) in minimum time, we use the following strategy:

1. Start with low altitude flight to maximize P, 0–0′. During take off, an aircraft accelerates along the runway and maintains low altitude until maximum excess power is reached. The ground constraint $(h \geq 0)$ precludes any option of diving at this time.
2. Fly along the extremal A–D (tangent points between P and h_c contours) where excess power is maximum.
3. Dive through Mach 1 to get to the second extremal D'–I. (Diving takes virtually no time.) Follow the second extremal to E.
4. To reach the final goal of (h_f, V_f), zoom climb on the h_{cf} contour which requires virtually no time.

We note in this supersonic example that two extremal arcs exist, due to two local maxima for P: one at low speed and one at high speed. The excess power contours are, of course, specific to the aircraft. These power contours are typically displayed in the cockpits of high performance aircraft. The energy state approximation not only serves as an initial guess for computation of the true optimum, but also provides valuable insight into the nature of the optimal solution.

The counterintuitive answer that a supersonic aircraft must dive first in order to climb to a given altitude in minimum time was first discovered by Walter Denham and Art Bryson, Jr. See Ho and Speyer [1990] for historical details.

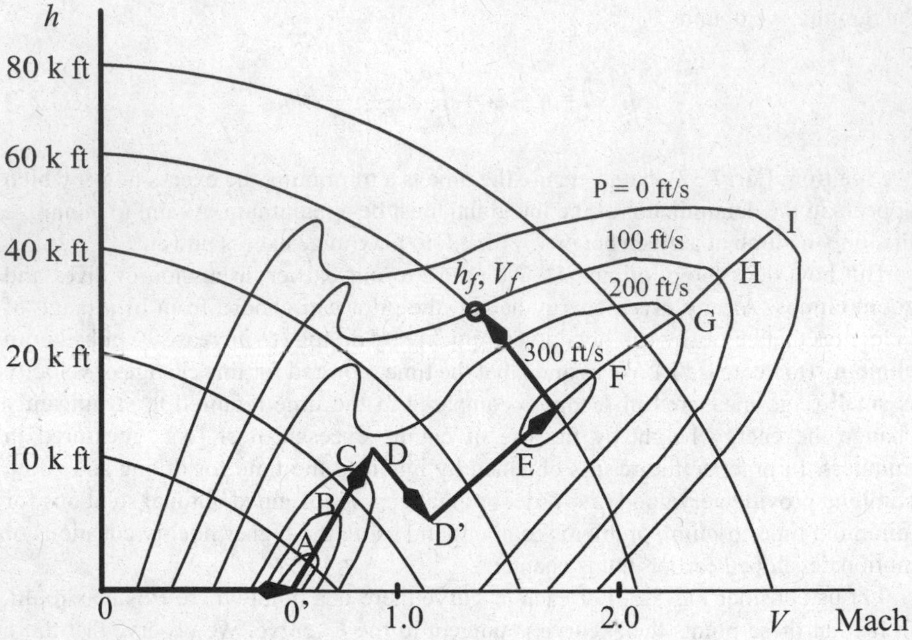

Figure 7.6. **Maximizing excess power (Adapted from Anderson [1999] and Vinh [1993]). In this example the aircraft starts at zero altitude and zero velocity and reaches a desired final altitude and final velocity in minimum time. The power contours range from 500 ft/s (near A) to 0 ft/s (near I).**

7.2. Maximization of the Range of a Rocket

Let us consider the problem of maximizing the range of a rocket, as illustrated in Fig. 7.7. Here we assume a uniform gravity field (with acceleration g) and ignore the effect of drag. We also assume that the thrust acceleration, $f(t)$, is a given positive function of time and that burnout occurs at $t = T$. Thus, our problem is to find the control function $\theta(t)$ to maximize the range, R. That is,
Maximize:

$$J = R \tag{7.33}$$

subject to:

$$\dot{x} = V_x \tag{7.34a}$$

$$\dot{y} = V_y \tag{7.34b}$$

$$\dot{V}_x = f\cos\theta \tag{7.34c}$$

$$\dot{V}_y = f\sin\theta - g \tag{7.34d}$$

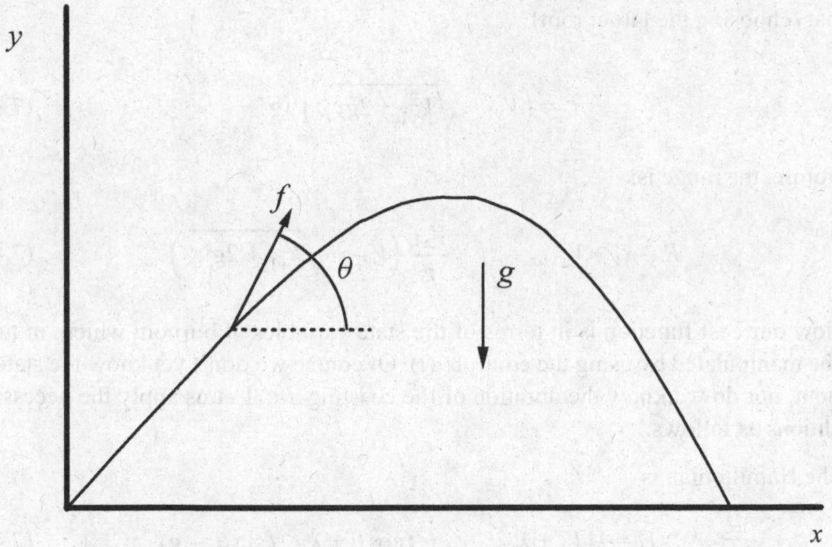

Figure 7.7. Maximization of the range of a rocket.

where

$$f(t) = -\dot{m}I_{spm}/m \tag{7.35}$$

is a predetermined function of time and I_{spm} is specific impulse based on mass flow rate, i.e. effective exhaust velocity.

For the I.C.s we assume:

$$x(0) = y(0) = 0,\ V_x(0) = V_y(0) = 0 \tag{7.36}$$

In the analysis of this problem we follow the approach of Lawden [1963]. If $(x_1, y_1, V_{x1}, V_{y1})$ is the state at burnout (t = T), then the range is found as follows.

To determine the duration of the coasting arc, t_c, we write the familiar equation for the falling body:

$$d = -\frac{1}{2}gt_c^2 + V_{y1}t_c + y_1 \tag{7.37}$$

where d is the altitude of the body above the ground. Setting $d = 0$ for the time of impact we have

$$-\frac{1}{2}gt_c^2 + V_{y1}t_c + y_1 = 0 \tag{7.38}$$

so that (choosing the larger root)

$$t_c = \left(V_{y1} + \sqrt{V_{y1}^2 + 2gy_1}\right)/g \tag{7.39}$$

Therefore, the range is:

$$R = x_1 + V_{x1}t_c = x_1 + \frac{V_{x1}}{g}\left(V_{y1} + \sqrt{V_{y1}^2 + 2gy_1}\right) \tag{7.40}$$

Now our cost function is in terms of the state variables at burnout which, in turn, can be manipulated by using the control $\theta(t)$. Of course we don't yet know the state at burnout, nor do we know the duration of the coasting arc. Let us apply the necessary conditions as follows.

1. The Hamiltonian is

$$H = \lambda_x V_x + \lambda_y V_y + \lambda_{V_x} f\cos\theta + \lambda_{V_y}(f\sin\theta - g) \tag{7.41}$$

2. The Euler-Lagrange equations are

$$\dot{\lambda}_x = -H_x = 0,\ \dot{\lambda}_y = -H_y = 0 \tag{7.42a}$$

so that

$$\lambda_x = c_1,\ \lambda_y = c_2 \tag{7.42b}$$

and

$$\dot{\lambda}_{V_x} = -H_{V_x} = -\lambda_x,\ \dot{\lambda}_{V_y} = -H_{V_y} = -\lambda_y \tag{7.42c}$$

so

$$\lambda_{V_x} = -c_1 t + c_3,\ \lambda_{V_y} = -c_2 t + c_4 \tag{7.42d}$$

where c_1, c_2, c_3, and c_4 are constants.

Following the procedure described after Eq. (4.66) we write the Hamiltonian in Eq. (7.41) as

$$H = \lambda_x V_x + \lambda_y V_y - \lambda_{V_y} g + f\boldsymbol{\mu}^T\boldsymbol{\sigma} \tag{7.43}$$

where

$$\boldsymbol{\mu} = \begin{bmatrix} \lambda_{V_x} \\ \lambda_{V_y} \end{bmatrix} \tag{7.44}$$

and

$$\boldsymbol{\sigma} = \begin{bmatrix} \cos\theta \\ \sin\theta \end{bmatrix} \tag{7.45}$$

where $\boldsymbol{\sigma}$ is a unit vector. Because we want to maximize J (the range) we choose the control to maximize H; so $\boldsymbol{\mu}^T\boldsymbol{\sigma} = \mu$ and

$$\boldsymbol{\sigma} = \frac{\boldsymbol{\mu}}{\mu} \tag{7.46}$$

for which

$$\cos\theta = \frac{\lambda_{V_x}}{\sqrt{\lambda_{V_x}^2 + \lambda_{V_y}^2}} \tag{7.47}$$

$$\sin\theta = \frac{\lambda_{V_y}}{\sqrt{\lambda_{V_x}^2 + \lambda_{V_y}^2}} \tag{7.48}$$

3. From the differential form of the transversality condition we have

$$H_f dt_f - \boldsymbol{\lambda}_f^T d\boldsymbol{x}_f + d\phi = 0 \tag{7.49}$$

Following Lawden [1963], we evaluate the transversality condition at $t_1 = T$:

$$H_1 dt_1 - \boldsymbol{\lambda}_1^T d\boldsymbol{x}_1 + d\phi(\boldsymbol{x}_1) = 0 \tag{7.50}$$

We note that since $t_1 = T$ we have $dt_1 = 0$; we also note that x_1 is free. Recalling that:

$$\phi(\boldsymbol{x}_1) = x_1 + V_{x1}t_c = x_1 + \frac{V_{x1}}{g}\left(V_{y1} + \sqrt{V_{y1}^2 + 2gy_1}\right) \tag{7.51}$$

we have:

$$d\phi(\boldsymbol{x}_1) = \frac{\partial\phi}{\partial x_1}dx_1 + \frac{\partial\phi}{\partial y_1}dy_1 + \frac{\partial\phi}{\partial V_{x1}}dV_{x1} + \frac{\partial\phi}{\partial V_{y1}}dV_{y1} \tag{7.52}$$

Thus from transversality we obtain:

$$\begin{aligned}&\left(-\lambda_{x1}dx_1 - \lambda_{y1}dy_1 - \lambda_{V_{x1}}dV_{x1} - \lambda_{V_{y1}}dV_{y1}\right)\\&+\left\{dx_1 + V_{x1}\left(V_{y1}^2+2gy_1\right)^{-1/2}dy_1 + \frac{1}{g}\left(V_{y1}+\sqrt{V_{y1}^2+2gy_1}\right)dV_{x1}\right.\\&\left.+\frac{V_{x1}}{g}\left[1+V_{y1}\left(V_{y1}^2+2gy_1\right)^{-1/2}\right]dV_{y1}\right\}=0\end{aligned} \tag{7.53}$$

Setting the coefficients to zero:

$$\lambda_{x1} = 1 \tag{7.54a}$$

$$\lambda_{y1} = \frac{V_{x1}}{r} \tag{7.54b}$$

$$\lambda_{V_{x1}} = \frac{V_{y1}+r}{g} \tag{7.54c}$$

$$\lambda_{V_{y1}} = \frac{V_{x1}}{gr}\left(r+V_{y1}\right) \tag{7.54d}$$

where we use

$$r \equiv \sqrt{V_{y1}^2+2gy_1} \tag{7.54e}$$

to simplify the expressions.

Thus, for the control law we have:

$$\tan\theta = \frac{+\lambda_{V_y}}{+\lambda_{V_x}} = \frac{-c_2t+c_4}{-c_1t+c_3} \tag{7.55}$$

where

$$c_1 = \lambda_{x1} = 1 \tag{7.56a}$$

$$c_2 = \lambda_{y1} = \frac{V_{x1}}{r} \tag{7.56b}$$

$$c_3 = \lambda_{V_{x1}} + c_1T = \frac{V_{y1}+r}{g} + T \tag{7.56c}$$

$$c_4 = \lambda_{V_{y1}} + c_2T = \frac{V_{x1}}{gr}(r+V_{y1}) + \frac{V_{x1}}{r}T \tag{7.56d}$$

so that

$$\tan\theta = \frac{-c_2 t + c_4}{-c_1 t + c_3} = \frac{-\frac{V_{x1}}{r} t + \frac{V_{x1}}{r}\left(\frac{r+V_{y1}}{g} + T\right)}{-t + \frac{V_{y1}+r}{g} + T} \tag{7.57}$$

which reduces to the simple form:

$$\tan\theta = \frac{V_{x1}}{r} \tag{7.58}$$

Recalling that r is a constant that depends on V_{y1} and y_1, we see that the maximum range is achieved by θ = constant which means that the thrust must be kept at a constant angle to the horizontal, even though $f = f(t)$.

7.2.1. Integration of Equations of Motion When f Is Constant

Let us consider the special case of constant acceleration. When f is constant, integration of the equations of motion, Eqs. (7.34a)–(7.34d), provides:

$$V_{x1} = f T \cos\theta \tag{7.59a}$$

$$V_{y1} = (f \sin\theta - g)T \tag{7.59b}$$

$$x_1 = \frac{1}{2} f T^2 \cos\theta \tag{7.59c}$$

$$y_1 = \frac{1}{2}(f \sin\theta - g)T^2 \tag{7.59d}$$

for the state at $t = T$. Substituting the solution into the control law, Eq. (7.58), gives:

$$\tan\theta = \frac{f T \cos\theta}{\sqrt{(f\sin\theta - g)^2 T^2 + 2g\frac{1}{2}(f\sin\theta - g)T^2}}$$

$$= \frac{f T \cos\theta}{\sqrt{T^2 f^2(\sin^2\theta - \frac{g}{f}\sin\theta)}}$$

$$= \frac{\cos\theta}{\sqrt{\sin^2\theta - \frac{g}{f}\sin\theta}} \tag{7.60}$$

Equation (7.60) implies that:

$$\sin\theta\sqrt{\sin^2\theta - \frac{g}{f}\sin\theta} = \cos^2\theta$$

$$= 1 - \sin^2\theta \tag{7.61}$$

Squaring both sides of Eq. (7.61) provides:

$$\sin^2\theta(\sin^2\theta - \frac{g}{f}\sin\theta) = 1 - 2\sin^2\theta + \sin^4\theta \tag{7.62}$$

or

$$\frac{g}{f}\sin^3\theta - 2\sin^2\theta + 1 = 0 \tag{7.63}$$

Equation (7.63) is a transcendental equation which we can solve numerically to determine θ. Lawden [1963] summarizes the solutions for θ as follows.

1. If $f > g$, then there are four roots for $0 < \theta < 2\pi$ (one in each quadrant), but only the positive acute solution $\theta < \pi/2$ is meaningful.
2. If $f < g$, then there is insufficient thrust. The solutions for θ are in the third and fourth quadrants, corresponding to non-physical cases of falling below ground level.

7.2.2. The Optimal Trajectory

Once θ has been calculated, we solve for the state at $T(x_1, y_1, V_{x1}, V_{y1})$ from the integrated equations of motion, Eqs. (7.59a)–(7.59d). Thus, the optimal trajectory is given (parametrically) as:

$$\begin{aligned} x &= \frac{1}{2} f t^2 \cos\theta \\ y &= \frac{1}{2}(f\sin\theta - g)t^2 \end{aligned} \tag{7.64}$$

for $0 \leq t \leq T$. Equations (7.64) indicate that the rocket travels along a straight line during thrusting, given by the flight path angle γ (see Fig. 7.8).

$$\tan\gamma = \frac{y}{x} = \tan\theta - \frac{g}{f}\sec\theta \tag{7.65}$$

Since

$$\ddot{x} = f\cos\theta \tag{7.66a}$$

$$\ddot{y} = f\sin\theta - g \tag{7.66b}$$

the line is traversed at a constant acceleration:

$$\begin{aligned} a^2 &= \ddot{x}^2 + \ddot{y}^2 \\ &= f^2 - 2gf\sin\theta + g^2 \end{aligned} \tag{7.67}$$

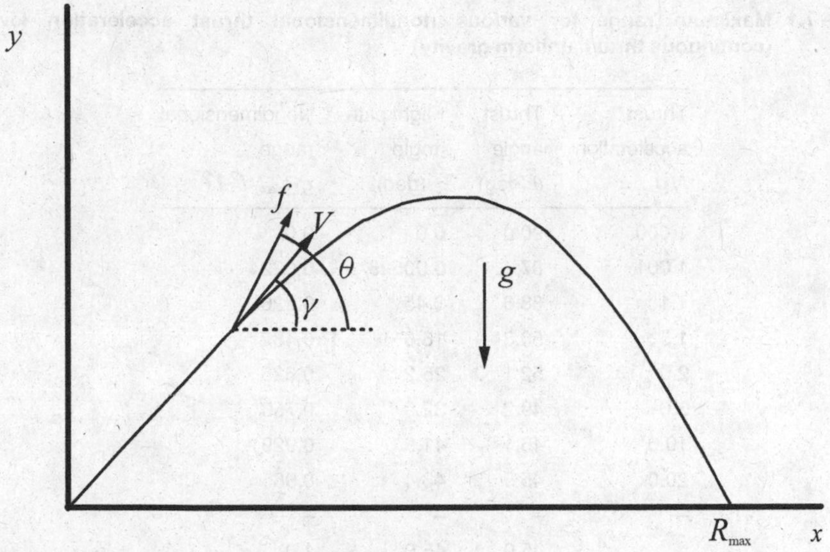

Figure 7.8. Solution for maximum rocket range.

7.2.3. Maximum Range Equation

The maximum range of the rocket is found from Eq. (7.40):

$$R = x_1 + \frac{V_{x1}}{g}\left(V_{y1} + r\right) \tag{7.68}$$

where r is given in Eq. (7.54e). Using Eq. (7.58) to eliminate r from Eq. (7.68) we obtain

$$R = x_1 + \frac{V_{x1}}{g}\left(V_{y1} + \frac{V_{x1}}{\tan\theta}\right) \tag{7.69}$$

Substituting the solutions for x_1, V_{x1} and V_{y1} from Eqs. (7.59) gives:

$$\begin{aligned} R &= \frac{1}{2}fT^2\cos\theta + \frac{1}{g}fT\cos\theta\left(fT\sin\theta - gT + \frac{\cos\theta}{\tan\theta}fT\right) \\ &= -\frac{1}{2}fT^2\cos\theta + \frac{f^2T^2}{g}\cos\theta\left(\frac{\sin^2\theta + \cos^2\theta}{\sin\theta}\right) \end{aligned} \tag{7.70}$$

Table 7.1. Maximum range for various nondimensional thrust acceleration levels (continuous thrust, uniform gravity)

Thrust acceleration f/g	Thrust angle θ (deg)	Flight path angle γ (deg)	Nondimensional range $gR_{\max}/f^2T^2$
1.000	90.0	0.0	0.0
1.001	87.4	0.00526	0.0224
1.1	68.6	3.45	0.226
1.5	56.3	16.6	0.482
2.0	52.1	25.2	0.625
3.0	49.2	32.9	0.756
10.0	46.1	41.8	0.929
20.0	45.5	43.4	0.965
–	–	–	–
∞	45.0	45.0	1.0

Thus, the maximum range of the rocket is

$$R_{\max} = fT^2\left(\frac{f}{g}\cot\theta - \frac{1}{2}\cos\theta\right) \tag{7.71}$$

where f is the constant acceleration from the thrust, $f > g$, and θ is the constant thrust angle given by Eq. (7.63).

Table 7.1 shows the result for various nondimensional thrust acceleration levels, f/g, increasing from 1 to ∞. The infinite value represents an impulsive thrust and the trajectory is the ballistic projectile path studied in elementary mechanics, for which the maximum range is achieved by the launch angle, θ, of 45 deg. In the limit as $f \to \infty$, $T \to 0$ and the product fT approaches the launch velocity (the impulsive ΔV).

In Table 7.1 the thrust angle θ is greater than the flight path angle γ for finite duration thrusts in order to counteract gravity acting downward. The nondimensional range shown is the actual $R_{\max}$ divided by the range of the ballistic path. The fact that this nondimensional range is less than one for finite duration thrust is due to gravity loss.

7.3. Time Optimal Launching of a Satellite

We now return to one of the fundamental examples of the text: the time-optimal launching of a satellite into orbit, illustrated in Fig. 7.9. We continue with the same assumptions of Example 4.5, *Launch into circular orbit from flat-Earth*.

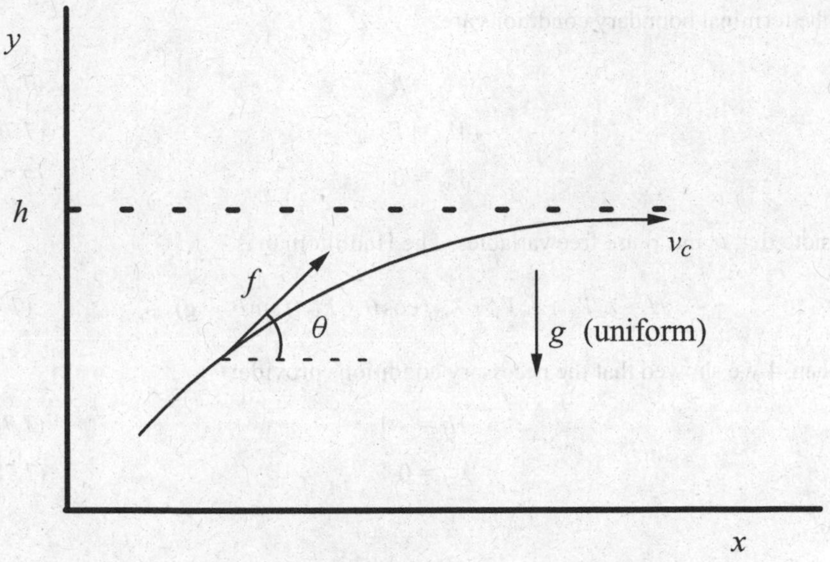

Figure 7.9. Time-optimal launching of a satellite into a circular orbit at velocity v_C and altitude h. In the flat-Earth assumption the gravity field is uniform.

Specifically, our problem is as follows.
Minimize:

$$J = t_f \tag{7.72}$$

subject to:

$$\dot{x} = V_x \tag{7.73a}$$

$$\dot{y} = V_y \tag{7.73b}$$

$$\dot{V}_x = f\cos\theta \tag{7.73c}$$

$$\dot{V}_y = f\sin\theta - g \tag{7.73d}$$

where

$$f(t) = -\dot{m}I_{spm}/m \tag{7.74}$$

is the thrust acceleration, a predetermined function of time. The initial conditions are

$$t_o = x_o = y_o = V_{xo} = V_{yo} = 0 \tag{7.75}$$

and the terminal boundary conditions are:

$$y_f = h \tag{7.76a}$$

$$V_{xf} = V_c \tag{7.76b}$$

$$V_{yf} = 0 \tag{7.76c}$$

We note that t_f and x_f are free variables. The Hamiltonian is

$$H = \lambda_x V_x + \lambda_y V_y + \lambda_{vx} f \cos\theta + \lambda_{vy}(f \sin\theta - g) \tag{7.77}$$

In Chap. 4 we showed that the necessary conditions provide:

$$H_f = -1 \tag{7.78a}$$

$$\lambda_{xf} = 0 \tag{7.78b}$$

Also:

$$\lambda_x = c_1 = 0 \tag{7.79}$$

$$\lambda_y = c_2 \tag{7.80}$$

$$\lambda_{vx} = -c_1 t + c_3 = c_3 \tag{7.81}$$

$$\lambda_{vy} = -c_2 t + c_4 \tag{7.82}$$

and

$$\begin{aligned} \tan\theta &= \frac{-\lambda_{vy}}{-\lambda_{vx}} \\ &= \frac{c_2 t - c_4}{c_1 t - c_3} \\ &= \frac{-c_2}{c_3} t + \frac{c_4}{c_3} \\ &= at + b \end{aligned} \tag{7.83}$$

which is called the linear tangent steering law. This law was provided earlier, in Eq. (4.82).

7.3.1. Integration of the EOMs

If f is a constant, the equations of motion, Eqs. (7.73a)–(7.73d), can be integrated as follows.

Differentiate the steering law [Eq. (7.83)] with respect to time:

$$\frac{d}{dt}(\tan\theta) = \frac{d}{dt}(at + b) \tag{7.84}$$

so that

$$(\sec^2\theta)\frac{d\theta}{dt} = a \tag{7.85}$$

We can now change the independent variable from t to θ by the chain rule:

$$\frac{dx}{dt} = \frac{dx}{d\theta}\frac{d\theta}{dt} = \frac{a}{\sec^2\theta}\frac{dx}{d\theta} \tag{7.86}$$

Thus, the equations of motion become the following:

$$\frac{dx}{d\theta} = \frac{V_x}{a}\sec^2\theta \tag{7.87a}$$

$$\frac{dy}{d\theta} = \frac{V_y}{a}\sec^2\theta \tag{7.87b}$$

$$\begin{aligned}\frac{dV_x}{d\theta} &= \frac{f}{a}\cos\theta\sec^2\theta \\ &= \frac{f}{a}\sec\theta \end{aligned} \tag{7.87c}$$

$$\begin{aligned}\frac{dV_y}{d\theta} &= \frac{f}{a}\sec^2\theta\sin\theta - \frac{g}{a}\sec^2\theta \\ &= \frac{f}{a}\sec\theta\tan\theta - \frac{g}{a}\sec^2\theta \end{aligned} \tag{7.87d}$$

Integrating the V_x equation, Eq. (7.87c), we have:

$$\begin{aligned}V_x(\theta) &= \frac{f}{a}\int_{\theta_o}^{\theta}\sec\theta d\theta \\ &= \frac{f}{a}[\ln(\tan\theta + \sec\theta)]|_{\theta_o}^{\theta} \\ &= \frac{f}{a}\ln\left[\frac{\tan\theta + \sec\theta}{\tan\theta_o + \sec\theta_o}\right]\end{aligned} \tag{7.88}$$

For V_y we have, by integrating Eq. (7.87d):

$$\begin{aligned}V_y(\theta) &= \frac{f}{a}\int_{\theta_o}^{\theta}\sec\theta\tan\theta d\theta - \frac{g}{a}\int_{\theta_o}^{\theta}\sec^2\theta d\theta \\ &= \frac{f}{a}(\sec\theta - \sec\theta_o) - \frac{g}{a}(\tan\theta - \tan\theta_o)\end{aligned} \tag{7.89}$$

To find $x(\theta)$ we integrate Eq. (7.87a):

$$\begin{aligned} x(\theta) &= \frac{1}{a}\int_{\theta_o}^{\theta} V_x(\theta)\sec^2\theta d\theta \\ &= \frac{f}{a^2}\int_{\theta_o}^{\theta} \ln\left(\frac{\tan\theta+\sec\theta}{\tan\theta_o+\sec\theta_o}\right)\sec^2\theta d\theta \\ &= \frac{f}{a^2}\int_{\theta_o}^{\theta} [\ln(\tan\theta+\sec\theta)-\ln(\tan\theta_o+\sec\theta_o)]\sec^2\theta d\theta \end{aligned} \tag{7.90}$$

Using integration by parts provides:

$$\begin{aligned} &\int_{\theta_o}^{\theta} [\ln(\tan\theta+\sec\theta)]\sec^2\theta d\theta \\ &\quad = [\ln(\tan\theta+\sec\theta)]\tan\theta - [\ln(\tan\theta_o+\sec\theta_o)]\tan\theta_o - \int_{\theta_o}^{\theta}\tan\theta\sec\theta d\theta \\ &\quad = [\ln(\tan\theta+\sec\theta)]\tan\theta - [\ln(\tan\theta_o+\sec\theta_o)]\tan\theta_o - (\sec\theta-\sec\theta_o) \end{aligned} \tag{7.91}$$

Also we note that,

$$\int_{\theta_o}^{\theta} -\ln(\tan\theta_o+\sec\theta_o)\sec^2\theta d\theta = -[\ln(\tan\theta_o+\sec\theta_o)](\tan\theta-\tan\theta_o) \tag{7.92}$$

Thus, using Eqs. (7.91) and (7.92) in Eq. (7.90), we obtain:

$$x(\theta) = \frac{f}{a^2}\left[\ln\left(\frac{\tan\theta+\sec\theta}{\tan\theta_o+\sec\theta_o}\right)\tan\theta + \sec\theta_o - \sec\theta\right] \tag{7.93}$$

Now, to find $y(\theta)$ from Eqs. (7.87b) and (7.89), we write:

$$\begin{aligned} y(\theta) &= \frac{1}{a}\int_{\theta_o}^{\theta} V_y\sec^2\theta d\theta \\ &= \frac{1}{a}\int_{\theta_o}^{\theta}\left[\frac{f}{a}(\sec\theta-\sec\theta_o) - \frac{g}{a}(\tan\theta-\tan\theta_o)\right]\sec^2\theta d\theta \end{aligned} \tag{7.94}$$

We evaluate

$$\begin{aligned} \int_{\theta_o}^{\theta}\sec^3\theta d\theta &= \frac{1}{2}\tan\theta\sec\theta - \frac{1}{2}\tan\theta_o\sec\theta_o + \frac{1}{2}\int_{\theta_o}^{\theta}\sec\theta d\theta \\ &= \frac{1}{2}\tan\theta\sec\theta - \frac{1}{2}\tan\theta_o\sec\theta_o + \frac{1}{2}\ln\left(\frac{\tan\theta+\sec\theta}{\tan\theta_o+\sec\theta_o}\right) \end{aligned} \tag{7.95}$$

and

$$\int_{\theta_o}^{\theta} \tan\theta \sec^2\theta d\theta = \frac{1}{2}\tan^2\theta - \frac{1}{2}\tan^2\theta_o \tag{7.96}$$

to obtain

$$\begin{aligned} y(\theta) = \frac{f}{a^2}\left[\frac{1}{2}\tan\theta\sec\theta - \frac{1}{2}\tan\theta_o\sec\theta_o + \frac{1}{2}\ln\left(\frac{\tan\theta + \sec\theta}{\tan\theta_o + \sec\theta_o}\right)\right] \\ -\frac{f}{a^2}(\tan\theta - \tan\theta_o)\sec\theta_o - \frac{g}{a^2}\left(\frac{1}{2}\tan^2\theta - \frac{1}{2}\tan^2\theta_o\right) \\ +\frac{g}{a^2}(\tan\theta - \tan\theta_o)\tan\theta_o \end{aligned} \tag{7.97}$$

The last two terms on the right hand side of Eq. (7.97) can be combined to form a perfect square so we have

$$\begin{aligned} y(\theta) = \frac{f}{2a^2}\left[(\tan\theta_o - \tan\theta)\sec\theta_o - (\sec\theta_o - \sec\theta)\tan\theta + \ln\left(\frac{\tan\theta + \sec\theta}{\tan\theta_o + \sec\theta_o}\right)\right] \\ - \frac{g}{2a^2}[\tan\theta - \tan\theta_o]^2 \end{aligned} \tag{7.98}$$

Let us now consider how to solve the problem. First of all, we have four unknowns: θ_o, θ_f, t_f, and a. Second, we have four equations. Three equations are provided by our analytical solutions that make use of the three specified final boundary conditions: $y_f = h$, $V_{xf} = V_c$, and $V_{yf} = 0$ (three knowns) obtained from Eqs. (7.98), (7.88), and (7.89), respectively. Our fourth equation is

$$a = \frac{\tan\theta_f - \tan\theta_o}{t_f} \tag{7.99}$$

which comes from evaluating the steering law, Eq. (7.83), at $t_o = 0$ and at t_f.

By setting $V_y(\theta_f) = 0$ in Eq. (7.89) we obtain:

$$\frac{f}{g} = \frac{\tan\theta_f - \tan\theta_o}{\sec\theta_f - \sec\theta_o} \tag{7.100}$$

From Eq. (7.98) we write $a^2 y(\theta_f) = a^2 h$ which gives

$$a^2 h = \frac{f}{2}\left[(T_o - T_f)S_o - (S_o - S_f)T_f + \ln\left(\frac{S_f + T_f}{S_o + T_o}\right)\right] - \frac{g}{2}(T_f - T_o)^2 \tag{7.101}$$

where we use "T" to represent tangent and "S" to represent secant and subscripts "o" and "f" to indicate initial and final values, respectively.

Using $V_x(\theta_f) = V_c$ in Eq. (7.88) provides

$$a^2 = \frac{f^2}{V_c^2} \ln^2 \left(\frac{T_f + S_f}{T_o + S_o} \right) \tag{7.102}$$

Substituting Eq. (7.102) into Eq. (7.101), dividing both sides by $\ln^2()$, and multiplying by $\frac{2}{f}$ we obtain:

$$\frac{2hf}{V_c^2} = \frac{\left[(T_o - T_f)S_o - (S_o - S_f)T_f + \ln\left(\frac{S_f + T_f}{S_o + T_o}\right) \right] - \frac{g}{f}(T_f - T_o)^2}{\ln^2\left(\frac{S_f + T_f}{S_o + T_o}\right)} \tag{7.103}$$

After substituting for $\frac{g}{f}$, using Eq. (7.100), the numerator becomes:

$$T_oS_o - T_fS_o - S_oT_f + S_fT_f + \ln\left(\frac{S_f + T_f}{S_o + T_o}\right) - S_fT_f + S_fT_o + S_oT_f - S_oT_o$$

$$= -S_oT_f + \ln\left(\frac{S_f + T_f}{S_o + T_o}\right) + S_fT_o \tag{7.104}$$

Substituting Eq. (7.104) into Eq. (7.103) gives:

$$\frac{2hf}{V_c^2} = \frac{\tan\theta_o \sec\theta_f - \tan\theta_f \sec\theta_o + \ln\left(\frac{\sec\theta_f + \tan\theta_f}{\sec\theta_o + \tan\theta_o}\right)}{\ln^2\left(\frac{\sec\theta_f + \tan\theta_f}{\sec\theta_o + \tan\theta_o}\right)} \tag{7.105}$$

The expressions for $\frac{f}{g}$ and $\frac{2hf}{V_c^2}$, Eqs. (7.100) and (7.105), are functions of two unknowns, θ_o and θ_f, so we have two equations to solve for the two unknowns. After solving for θ_o and θ_f we can use $V_x(\theta_f) = V_c$, Eq. (7.88), to solve for a:

$$a = \frac{f}{V_c} \ln\left(\frac{\sec\theta_f + \tan\theta_f}{\sec\theta_o + \tan\theta_o}\right) \tag{7.106}$$

Finally, t_f can be obtained from the steering law itself, Eq. (7.99):

$$t_f = \frac{\tan\theta_f - \tan\theta_o}{a} \tag{7.107}$$

We note that Eqs. (7.100) and (7.105) are implicit relations for the unknowns, θ_o and θ_f. They cannot be solved analytically, so we must resort to numerical methods. One approach is as follows.

1. Guess a value for θ_o and use Eq. (7.100) to find θ_f.
2. Substitute θ_o and θ_f into Eq. (7.105) and compare to $\frac{2hf}{V_c^2}$.

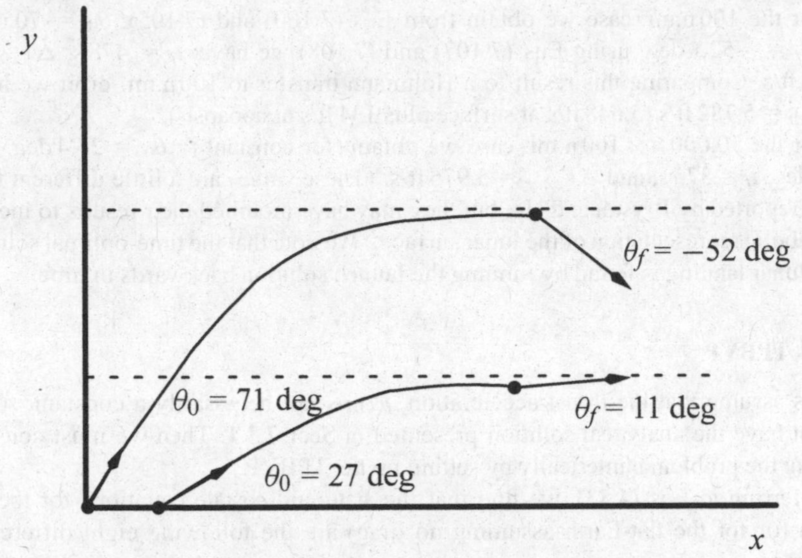

Figure 7.10. Schematic of lunar takeoff (Adapted from Bryson and Ho [1975]), with updated values that satisfy Eqs. (7.100) and (7.105). The assumptions include constant thrust, constant mass, no drag, and uniform flat-Moon gravity. The upper trajectory represents a time-optimal launch into a 100 n.mi. circular orbit; the lower into a 50,000 ft × 100 n.mi. elliptical orbit.

3. Adjust θ_o. Note that increasing θ_o increases $\frac{2hf}{V_c^2}$, which in turn increases height.

Alternatively, MATLAB provides a convenient subroutine (fsolve) for such problems.

Example 7.2 Lunar takeoff.

Bryson and Ho [1975] provide numerical results for a launch from the lunar surface. They give the values in the units historically used as follows.

Let $\frac{f}{g} = 3$; $g = 5.32$ ft/s^2 (lunar gravity), $r = 938$ nautical miles (lunar radius). We note that 1 nautical mile = 6,080 ft.

Consider two trajectories (illustrated in Fig. 7.10):

1. Launch to 100 n.mi. circular orbit.
2. Launch to periapsis of an elliptical orbit of 50,000 ft × 100 n.mi. (then inject into 100 n.mi. circular orbit using a ΔV of 464 ft/s).

The characteristic velocity for launch at constant acceleration, f, is computed from

$$\Delta V_{char} = f t_f \tag{7.108}$$

where t_f is the thrust duration.

For the 100 n.mi. case we obtain from Eqs. (7.100) and (7.105): $\theta_0 = 70.6$ deg and $\theta_f = -52.3$ deg; using Eqs. (7.107) and (7.108) we have: $t_f = 478$ s, $\Delta V_{char} =$ 7,624 ft/s. Comparing this result to a Hohmann transfer to 100 n.mi. orbit we have: $\Delta V_{char} = 5{,}782$ ft/s (5,648 ft/s at surface plus 134 ft/s at apoapsis).

For the 50,000 ft × 100 n.mi. case we obtain (for constant f): $\theta_0 = 27.4$ deg, $\theta_f =$ 11.1 deg, $t_f = 375$ s, and $\Delta V_{char} = 5{,}976$ ft/s. (These values are a little different from those reported by Bryson and Ho, but they may have modified their results to include a parabolic representation of the lunar surface.) We note that the time-optimal solution for a lunar landing is found by running the launch solution backwards in time.

7.3.2. TPBVP

Let us assume that the thrust acceleration, $f(t)$, is not necessarily a constant, so we do not have the analytical solution presented in Sect. 7.3.1. Then we must consider solving the problem numerically by setting up the TPBVP.

Referring to Eqs. (4.83), we find that the state and costate equations for the flat Moon (or for the flat-Earth assuming no drag) are the following eight differential equations:

$$\dot{x} = V_x \tag{7.109a}$$

$$\dot{y} = V_y \tag{7.109b}$$

$$\dot{V}_x = f(t)\left(\frac{-\lambda_3}{\sqrt{\lambda_3^2+\lambda_4^2}}\right) \tag{7.109c}$$

$$\dot{V}_y = f(t)\left(\frac{-\lambda_4}{\sqrt{\lambda_3^2+\lambda_4^2}}\right) - g \tag{7.109d}$$

$$\dot{\lambda}_1 = 0 \tag{7.109e}$$

$$\dot{\lambda}_2 = 0 \tag{7.109f}$$

$$\dot{\lambda}_3 = -\lambda_1 \tag{7.109g}$$

$$\dot{\lambda}_4 = -\lambda_2 \tag{7.109h}$$

with ten B.C.s:

$$t_0 = 0,\ x(0) = y(0) = 0,\ V_x(0) = V_y(0) = 0 \tag{7.110a}$$

$$y(t_f) = h,\ V_x(t_f) = V_c,\ V_y(t_f) = 0,\ \lambda_1(t_f) = 0,\ H(t_f) = -1 \tag{7.110b}$$

where λ_{1f} and H_f are found from the transversality condition. Since $\lambda_1 = 0$, we can drop it from further consideration, so now we have three costate equations

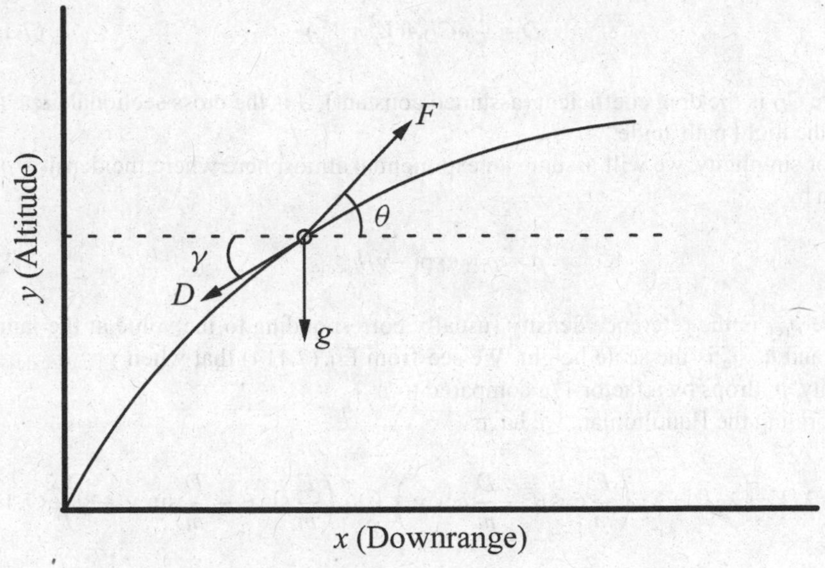

Figure 7.11. Launch vehicle subject to drag.

$$\dot{\lambda}_2 = 0 \tag{7.111a}$$

$$\dot{\lambda}_3 = 0 \tag{7.111b}$$

$$\dot{\lambda}_4 = -\lambda_2 \tag{7.111c}$$

for a total of seven differential equations with nine boundary conditions, which provides a well-defined TPBVP. Appendix A provides a numerical solution for the launch from the Moon using the TPBVP in MATLAB, bvp4c.

7.3.3. Flat-Earth Launch Including Atmospheric Drag

Now let's consider the case of launch into circular orbit in which we have atmospheric drag acting on the vehicle. See Fig. 7.11. Retaining the flat-Earth model, we have:

$$\dot{x} = V_x \tag{7.112a}$$

$$\dot{y} = V_y \tag{7.112b}$$

$$\dot{V}_x = \frac{F}{m}\cos\theta - \frac{D}{m}\cos\gamma \tag{7.112c}$$

$$\dot{V}_y = \frac{F}{m}\sin\theta - \frac{D}{m}\sin\gamma - g \tag{7.112d}$$

where the thrust, F, and the mass, m, are in general prescribed functions of time and the drag, D, is given by

$$D = \frac{1}{2}\rho C_D A(V_x^2 + V_y^2) \tag{7.113}$$

where C_D is the drag coefficient (assumed constant), A is the cross sectional area, and γ is the flight path angle.

For simplicity, we will assume an exponential atmosphere where the density, ρ, is given by

$$\rho = \rho_{\text{ref}} \exp(-y/h_{\text{scale}}) \tag{7.114}$$

where ρ_{ref} is the reference density (usually corresponding to the value at the launch site) and h_{scale} is the scale height. We see from Eq. (7.114) that when $y = h_{\text{scale}}$, the density, ρ, drops by a factor $1/e$ compared to ρ_{ref}.

Forming the Hamiltonian, we have

$$H = \lambda_1 V_x + \lambda_2 V_y + \lambda_3 \left(\frac{F}{m}\cos\theta - \frac{D}{m}\cos\gamma\right) + \lambda_4 \left(\frac{F}{m}\sin\theta - \frac{D}{m}\sin\gamma - g\right) \tag{7.115}$$

We obtain the control law from the procedure described after Eq. (4.66). The Hamiltonian in Eq. (7.115) can be written as:

$$H = \lambda_1 V_x + \lambda_2 V_y - \lambda_3 \left(\frac{D}{m}\cos\gamma\right) - \lambda_4 \left(\frac{D}{m}\sin\gamma - g\right) + \frac{F}{m}\boldsymbol{\mu}^T\boldsymbol{\sigma} \tag{7.116}$$

then

$$\cos\theta = \frac{-\lambda_3}{\sqrt{\lambda_3^2 + \lambda_4^2}} \tag{7.117a}$$

$$\sin\theta = \frac{-\lambda_4}{\sqrt{\lambda_3^2 + \lambda_4^2}} \tag{7.117b}$$

From Eqs. (7.112) to (7.115), the Hamiltonian is

$$H = \lambda_1 V_x + \lambda_2 V_y + \lambda_3 \left[\frac{F}{m}\left(\frac{-\lambda_3}{\sqrt{\lambda_3^2 + \lambda_4^2}}\right) - \frac{D}{m}\frac{V_x}{\sqrt{V_x^2 + V_y^2}}\right] + \lambda_4 \left[\frac{F}{m}\left(\frac{-\lambda_4}{\sqrt{\lambda_3^2 + \lambda_4^2}}\right) - \frac{D}{m}\frac{V_y}{\sqrt{V_x^2 + V_y^2}} - g\right] \tag{7.118}$$

or

$$H = \lambda_1 V_x + \lambda_2 V_y - \frac{F}{m}\sqrt{\lambda_3^2 + \lambda_4^2} - \left[\frac{\rho_{ref} C_D A}{2m}\exp\left(\frac{-y}{h_{\text{scale}}}\right)\sqrt{V_x^2 + V_y^2}\left(\lambda_3 V_x + \lambda_4 V_y\right)\right] - \lambda_4 g \tag{7.119}$$

Let us assume that the final conditions are

$$y(t_f) = h,\ V_x(t_f) = V_c,\ V_y(t_f) = 0 \tag{7.120}$$

The differential form of the transversality condition requires that

$$H_f dt_f - \boldsymbol{\lambda}_f^T d\mathbf{x}_f + d\phi = 0 \tag{7.121}$$

Since the final time, t_f, and the horizontal position, x_f, are free, but the other final states are specified [by Eqs. (7.120)], the transversality condition provides:

$$H_f dt_f - \lambda_{1f} dx_f + dt_f = 0 \tag{7.122}$$

From Eq. (7.122) we obtain

$$H_f = -1 \tag{7.123a}$$

$$\lambda_{1f} = 0 \tag{7.123b}$$

We note from the first of the Euler-Lagrange equations that

$$\dot{\lambda}_1 = 0 \tag{7.124}$$

so that

$$\lambda_1 = 0 \tag{7.125}$$

and we can drop λ_1 from further consideration.

For convenience let us define

$$K_1 = \frac{\rho_{ref} C_D A}{2m} \tag{7.126}$$

The remaining costate equations are:

$$\dot{\lambda}_2 = -\frac{\partial H}{\partial y} = (\lambda_3 V_x + \lambda_4 V_y) \exp\left(\frac{-y}{h_{\text{scale}}}\right)\left(\frac{-K_1 |V|}{h_{\text{scale}}}\right) \tag{7.127a}$$

$$\dot{\lambda}_3 = -\frac{\partial H}{\partial V_x} = -\lambda_1 + K_1 \exp\left(\frac{-y}{h_{\text{scale}}}\right)\left[\lambda_3\left(\frac{V_x^2}{|V|} + |V|\right) + \lambda_4 \frac{V_x V_y}{|V|}\right] \tag{7.127b}$$

$$\dot{\lambda}_4 = -\frac{\partial H}{\partial V_y} = -\lambda_2 + K_1 \exp\left(\frac{-y}{h_{\text{scale}}}\right)\left[\lambda_4\left(\frac{V_y^2}{|V|} + |V|\right) + \lambda_3 \frac{V_x V_y}{|V|}\right] \tag{7.127c}$$

where we have used $|V|$ to represent the magnitude of the velocity, $(V_x^2+V_y^2)^{1/2}$. We can specify the TPBVP by substituting the control law, Eqs. (7.117), into the equations of motion, Eqs. (7.112)–(7.114), and by using

$$\lambda_3 \cos\theta + \lambda_4 \sin\theta = -\sqrt{\lambda_3^2 + \lambda_4^2} \tag{7.128}$$

in the costate equations, Eqs. (7.127), using $\lambda_1 = 0$:

$$\dot{x} = V_x \tag{7.129a}$$

$$\dot{y} = V_y \tag{7.129b}$$

$$\dot{V}_x = \frac{F}{m}\frac{-\lambda_3}{\sqrt{\lambda_3^2+\lambda_4^2}} - K_1 \exp\left(\frac{-y}{h_{\text{scale}}}\right) V_x\sqrt{V_x^2+V_y^2} \tag{7.129c}$$

$$\dot{V}_y = \frac{F}{m}\frac{-\lambda_4}{\sqrt{\lambda_3^2+\lambda_4^2}} - K_1 \exp\left(\frac{-y}{h_{\text{scale}}}\right) V_y\sqrt{V_x^2+V_y^2} - g \tag{7.129d}$$

$$\dot{\lambda}_2 = (\lambda_3 V_x + \lambda_4 V_y)\exp\left(\frac{-y}{h_{\text{scale}}}\right)\left(\frac{-K_1\sqrt{V_x^2+V_y^2}}{h_{\text{scale}}}\right) \tag{7.129e}$$

$$\dot{\lambda}_3 = K_1 \exp\left(\frac{-y}{h_{\text{scale}}}\right)\left[\lambda_3\left(\frac{V_x^2}{\sqrt{V_x^2+V_y^2}} + \sqrt{V_x^2+V_y^2}\right) + \lambda_4\frac{V_xV_y}{\sqrt{V_x^2+V_y^2}}\right] \tag{7.129f}$$

$$\dot{\lambda}_4 = -\lambda_2 + K_1 \exp\left(\frac{-y}{h_{\text{scale}}}\right)\left[\lambda_4\left(\frac{V_y^2}{\sqrt{V_x^2+V_y^2}} + \sqrt{V_x^2+V_y^2}\right) + \lambda_3\frac{V_xV_y}{\sqrt{V_x^2+V_y^2}}\right] \tag{7.129g}$$

with B.C.s

$$t_0 = 0,\ x(0) = y(0) = 0,\ V_x(0) = V_y(0) = 0 \tag{7.130a}$$

$$y(t_f) = h,\ V_x(t_f) = V_c,\ V_y(t_f) = 0 \tag{7.130b}$$

$$H_f = -1 \tag{7.130c}$$

Let us examine the constraint on the final Hamiltonian, Eq. (7.130c), in detail:

$$H_f = \lambda_{1f}V_{xf} + \lambda_{2f}V_{yf} - \frac{F_f}{m_f}\sqrt{\lambda_{3f}^2+\lambda_{4f}^2} - \exp\left(\frac{-y_f}{h_{\text{scale}}}\right)\left[\frac{\rho_{ref}C_DA}{2m}\sqrt{V_{xf}^2+V_{yf}^2}\,(\lambda_{3f}V_{xf}+\lambda_{4f}V_{yf})\right] \tag{7.131}$$

Since $\lambda_{1f} = 0$, $V_{yf} = 0$, $y_f = h$, and $V_{xf} = V_c$, Eq. (7.131) becomes:

$$H_f = -\frac{F_f}{m_f}\sqrt{\lambda_{3f}^2 + \lambda_{4f}^2} - \rho_{\text{ref}} \exp(-h/h_{\text{scale}})\frac{C_D A}{2m_f} V_c^2 \lambda_{3f} - \lambda_{4f} g = -1 \qquad (7.132)$$

We recall that the thrust, $F(t)$, and the mass, $m(t)$, are prescribed functions of time. The seven differential equations, Eqs. (7.129), along with the nine B.C.s, Eqs. (7.130) and Eq. (7.132) provide a well-defined TPBVP.

Example 7.3 Time-optimal flat-Earth launch of a Titan II rocket including the effects of atmospheric drag and time-varying mass.

Let us consider the case of launching a Titan II, subject to atmospheric drag, into a circular LEO in minimum time, including the effect of time-varying mass (with constant thrust).

Assume the following numerical values:

$$F = 2.10 \times 10^6 \text{ N} \qquad (7.133a)$$

$$m_o = 1.1702 \times 10^5 \text{ kg} \qquad (7.133b)$$

$$A = 7.069 \text{ m}^2 \qquad (7.133c)$$

$$C_D = 0.5 \qquad (7.133d)$$

$$\rho_{ref} = 1.225 \text{ kg/m}^3 \qquad (7.133e)$$

$$h_{\text{scale}} = 8.44 \times 10^3 \text{ m} \qquad (7.133f)$$

$$g = 9.80665 \text{ m/s}^2 \qquad (7.133g)$$

We use Eqs. (7.112) for our equations of motion along with Eq. (7.113) for the drag model and an exponential atmosphere for the density, given in Eq. (7.114). The thrust of the launch vehicle is constant as given above in Eq. (7.133a) and the acceleration of gravity is the standard free-fall constant given in Eq. (7.133g).

Equations (7.129), (7.130), and (7.132) provide the TPBVP.

For the final altitude and velocity we use

$$h = 1.80 \times 10^5 \text{ m} \qquad (7.134a)$$

$$V_c = 7.796 \times 10^3 \text{ m/s} \qquad (7.134b)$$

To include the effect of time-varying mass we use a constant mass-flow rate of:

$$\dot{m} = -(1.1702 \times 10^5 \text{ kg} - 4.76 \times 10^3 \text{ kg})/139 \text{ s} = -807.6 \text{ kg/s} \qquad (7.135)$$

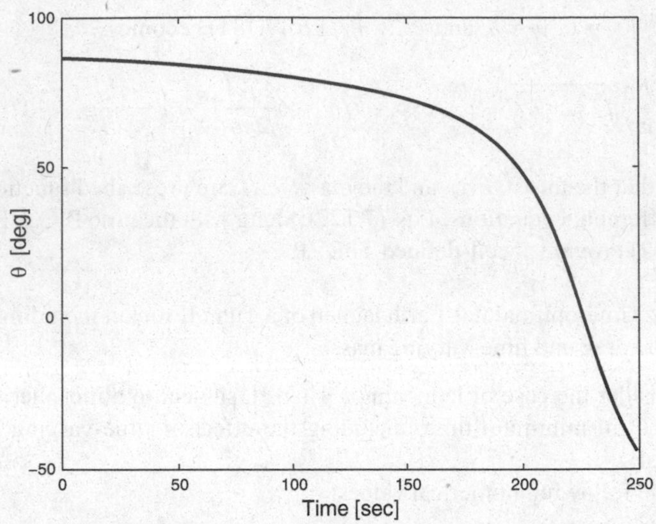

Figure 7.12. Steering angle for the time-optimal launch of a Titan II rocket into a 180 km circular orbit. Assumptions include constant thrust, time-varying mass, drag from exponential atmosphere, and uniform flat-Earth gravity.

We note that we are only considering the first stage of the Titan II, which is capable of getting into orbit (i.e. we are not including the second stage in this problem). In the real case, the initial mass of the first stage is 117,020 kg (fully fueled), the final mass is 4,760 kg (empty), and the burn time is 139 s. (Note that, because of our simplified model, we do not expect to obtain precisely these values in our TPBVP solver.) We note that the burn time we are using here (of 139 s) is merely an approximation and is not the optimal time we are seeking; it allows us to calculate a mass-flow rate. The TPBVP solver will provide the minimum time.

Figures 7.12 and 7.13 show results from the numerical solution of the TPBVP. (See Appendix B for more details on how to solve this problem in MATLAB.) In Fig. 7.12 we see that the initial steering angle is near 90 deg, in contrast to the lunar takeoff problem, Example 7.2, where the initial steering angle can be quite small. The reasons for the high steering angle in the Earth-launch case are two-fold. The first reason is that the vehicle must climb to higher altitude to reduce the effect of drag losses. The second is that the initial acceleration, F/m, is very low, about 1.8 g's, so any steering angle less than about 57 deg is too low to counter the gravitational acceleration. Near the end of the trajectory the steering angle is about −50 deg, which cancels the vertical velocity in a similar manner to that of the lunar takeoff problem. Figure 7.13 shows the trajectory profile of the altitude vs. the downrange distance.

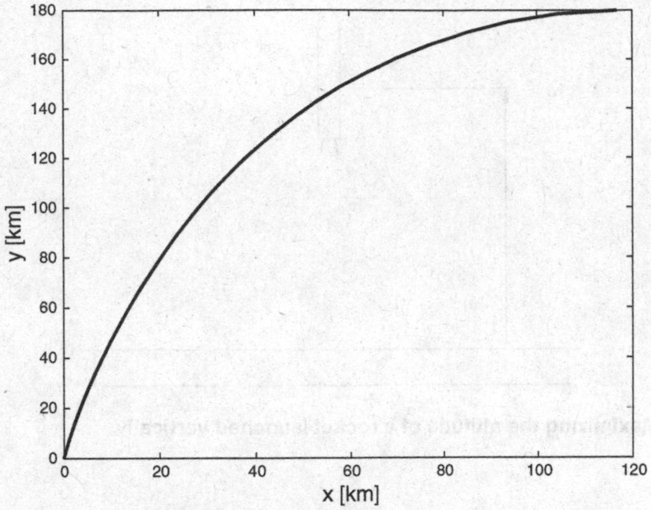

Figure 7.13. Altitude vs. range for time-optimal launch of Titan II.

7.4. Summary

We have considered three classical problems (aircraft minimum time to climb, maximum range of a rocket, and the optimal launch of a satellite), in which approximate analytical solutions are available. These solutions give special insight into some of the most important applications in trajectory optimization. It is particularly interesting to note that before optimal control theory was applied to the aircraft minimum time-to-climb problem, pilots were not aware of the strategy of diving through the sound barrier. Knowledge of the optimal solution ushered in many new time-to-climb records for existing aircraft.

7.5. Exercises

1. Solve the problem of maximizing the altitude of a rocket that is launched vertically using the following assumptions. The process equations are

$$\dot{x} = v$$

$$\dot{v} = f - g$$

$$\dot{m} = -\alpha f$$

where we assume uniform gravity, g, and no drag (Fig. 7.14). The initial conditions are

$$t_0 = 0,\ x(t_0) = 0,\ v(t_0) = 0,\ m(t_0) = m_0$$

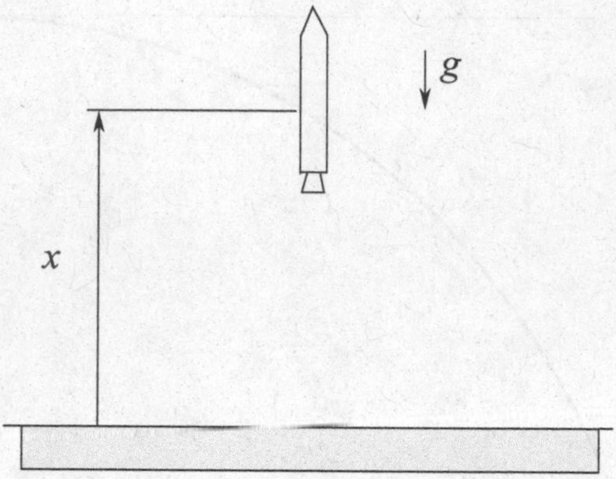

Figure 7.14. Maximizing the altitude of a rocket launched vertically.

where x is the altitude, v is the velocity, and m is the mass of the rocket which has a final value:

$$m(t_b) = m_b$$

where t_b is the burnout time. Note that the burnout time is not prescribed, but depends on the acceleration f. The acceleration is the only control and is bounded as follows:

$$0 \leq f \leq f_{\max}$$

Assume that f can be a time-varying function in general and that α, in the third process equation, is a positive constant.

Set up the optimal control problem as a Mayer problem and solve according to the following steps.

1a. Show that Max $J = x_f$ can be written as Max $J = x_b + \frac{1}{2}v_b^2/g$ where x_b and v_b are the altitude and velocity at burnout and x_f is the maximum altitude reached.
1b. Write the Hamiltonian using λ_x, λ_v, and λ_m as the costates, respectively.
1c. Indicate why we know that the Hamiltonian is a constant in this case.
1d. Give the costate differential equations and the form of their solutions.
1e. Give an expression for the switching function in terms of λ_v and λ_m.
1f. Write the transversality condition and give explicit solutions for $\lambda_x(t)$ and $\lambda_v(t)$. Also state the value of $H(t)$. Deduce that $\lambda_m = v_b/(g\alpha)$. (Hint: write the transversality condition at $t = t_b$.)
1g. Give your final expression for the switching function in terms of t and t_b and give the control law, based on the switching function.

2. Use MATLAB's bvp4c to solve the two cases of the lunar takeoff problem described in Example 7.2. (See Appendix A.)

2a. For the first case use the following numerical values:

$$g_{\text{Moon}} = 1.62 \text{ m/s}^2$$

$$y_f = 185.2 \text{ km}$$

$$F/m = 3g_{\text{Moon}}$$

$$V_c = 1.627 \text{ km/s}$$

2b. Determine the maximum ranges for initial guesses for t_f, $\lambda_2(0)$, and $\lambda_4(0)$ in order to converge on the optimal solution. (Note: to simplify this analysis you may hold two values fixed while varying the third.)

2c. Propagate the EOMs (using I.C.s and the steering law found in solving the TPBVP) to see if the satellite achieves the desired orbit. Investigate the effect of small errors in the propagated solution. (For example, consider inserting small errors in the I.C.s or in the steering law.) Comment on the behavior and accuracy of your propagated solution.

2d. Compare the results obtained in Exercise 2a with the initial and final values for the steering values given by Eqs. (7.100) and (7.105).

2e. For the second case repeat parts Exercises 2a through 2d using the same values for g_{Moon} and F/m, but use the values for y_f and V_f (in metric units) that correspond to a launch to an elliptical orbit of 50,000 ft × 100 n.mi.

3. Use bvp4c to solve the Titan launch problem described in Example 7.3. (See Appendix B.)

3a. Make plots for the zero-drag, constant-mass case.

3b. Make plots for the drag plus variable-mass case.

3c. Discuss the differences between the results for Exercise 3a and for Exercise 3b.

4. Consider the following projects.

4a. Compare the flat-Moon TPBVP solution to the complete analytical solution given in Sect. 7.3.1.

4b. Create a more historically accurate version of the lunar takeoff problem by modeling the variable mass of the Lunar Module in your TPBVP.

4c. Compute time-optimal launches from Titan, Venus, or Mars.

4d. Investigate the advantages of air launch (from balloons or aircraft).

4e. Study the effect of improved thrust efficiency as a function of altitude in a time-optimal approach.

4f. Examine the effect of ballistic coefficient, $BC = m/(C_D A)$, which tends to be large for large launch vehicles and small for small launch vehicles (i.e. it is proportional to scale length of the vehicle).

4g. Solve the launch problem assuming predetermined staging.

5. Answer the following questions based on the material presented in Chap. 7.

5a. In the aircraft performance analysis it was shown that to minimize the time to climb the following rule applies: at a given energy height pick the velocity which maximizes the excess power.

True False

5b. Sometimes a supersonic aircraft may need to dive in order to achieve the minimum time to climb to a particular altitude. (Assume the aircraft trajectory begins at zero altitude with zero velocity.)

True False

5c. In our simplified analysis of the aircraft minimum time to climb problem we assumed that the time required for zoom climbs and dives is negligible.

True False

5d. The optimal trajectory for launch from the flat moon using $f/g = 3$ requires more propellant than a Hohmann transfer.

True False

5e. In the maximization of the range of a rocket (for no drag and for uniform gravity), it was found that the steering control law is $\tan\theta = at + b$ where θ is the thrust angle with respect to the horizontal and a and b are nonzero constants.

True False

Solution: aT, bT, cT, dT, eF.

References

J.D. Anderson Jr., *Introduction to Flight*, 4th edn. (McGraw Hill, New York, 1999)
A.E. Bryson Jr., Y.C. Ho, *Applied Optimal Control* (Hemisphere Publishing, Washington, D.C., 1975)
Y.C. Ho, J.L. Speyer, In appreciation of Arthur E. Bryson, Jr. J. Guid. Control Dyn. **13**(5), 770–774 (1990)
D.F. Lawden, *Optimal Trajectories for Space Navigation* (Butterworths, London, 1963)
A. Miele, *Flight Mechanics: Theory of Flight Paths*, vol. 1 (Addison-Wesley, Reading, 1962)
N.X. Vinh, *Flight Mechanics of High-Performance Aircraft* (Cambridge University Press, New York, 1993)

Chapter 8

Weierstrass-Erdmann Corner Conditions

8.1. Statement of the Weierstrass-Erdmann Corner Conditions

So far, we have discussed four sets of necessary conditions which must be met by $\boldsymbol{x}^*(t)$ and $\boldsymbol{u}^*(t)$, over the class of admissible functions:

Necessary Condition I:
Euler-Lagrange equations: $\dot{\boldsymbol{\lambda}} = -H_{\boldsymbol{x}}$, $H_{\boldsymbol{u}} = 0$
along with the transversality condition.

Necessary Condition II:
Legendre-Clebsch condition: $H_{\boldsymbol{uu}} \geq 0$.

Necessary Condition III:
Weierstrass condition: $H(t, \boldsymbol{x}^*, \boldsymbol{\lambda}, \boldsymbol{u}^*) \leq H(t, \boldsymbol{x}^*, \boldsymbol{\lambda}, \boldsymbol{u})$.

Necessary Condition IV:
Jacobi condition: requires the non-existence of a conjugate point on (t_o, t_f).

These four conditions are also treated in the classical texts (e.g., Bliss [1968], Bolza [1961], Bryson et al. [1975], and Lawden [1963]). Now, let us consider the conditions that must be met at a *corner* of $\boldsymbol{x}^*(t)$, where $\dot{\boldsymbol{x}}^*(t_1)$ is discontinuous.

Suppose u is a scalar with $|u| \leq 1$ for the standard Bolza problem. If $[\boldsymbol{x}^*(t), u^*(t)]$ minimizes J with $u^*(t) \in P.C.[t_o, t_f]$, then:

1. $\dot{\boldsymbol{\lambda}} = -H_{\boldsymbol{x}}$ on each sub-arc between corners (i.e. points where the control is discontinuous).
2. $\boldsymbol{\lambda}(t)$ and $H(t)$ are continuous on the entire trajectory and, in particular, across corners (the Weierstrass-Erdmann corner conditions).
3. Also:

$$H_u^*(t) \begin{cases} \leq 0 & \text{if} \quad u^*(t) = +1 \\ = 0 & \text{if} \ -1 < u^*(t) < +1 \\ \geq 0 & \text{if} \quad u^*(t) = -1 \end{cases} \tag{8.1}$$

J.M. Longuski et al., *Optimal Control with Aerospace Applications*,
Space Technology Library 32, DOI 10.1007/978-1-4614-8945-0_8,

8.2. Proof Outline of Weierstrass-Erdmann Corner Conditions

In this section we prove statements 1–3 of Sect. 8.1 including statement 2, the Weierstrass-Erdmann corner conditions. (For a similar treatment see Hull [2003] and Vagners [1983]). Without loss of generality, we assume:

$$u^*(t) \begin{cases} \text{interior} & \text{on} \quad [t_o, t_1) \\ +1 & \text{on} \quad (t_1, t_f^*] \end{cases} \tag{8.2}$$

where t_1 (the time when the corner occurs) and t_f are not known beforehand. Figure 8.1 illustrates a one-sided variation of the control for $t \in (t_1, t_f] : \delta u \leq 0$.

Varying the control can alter the time required to reach the final state. Now, as in the proof of the Euler-Lagrange theorem (in Chap. 3), we make use of a one-parameter family, $u(t,\varepsilon)$, where $u^*(t) = u(t,0)$ and $\varepsilon \geq 0$. In this analysis it is convenient to use the un-adjoined approach. Let us form $\left.\frac{dJ}{d\varepsilon}\right|_{\varepsilon=0} \geq 0$ as follows:

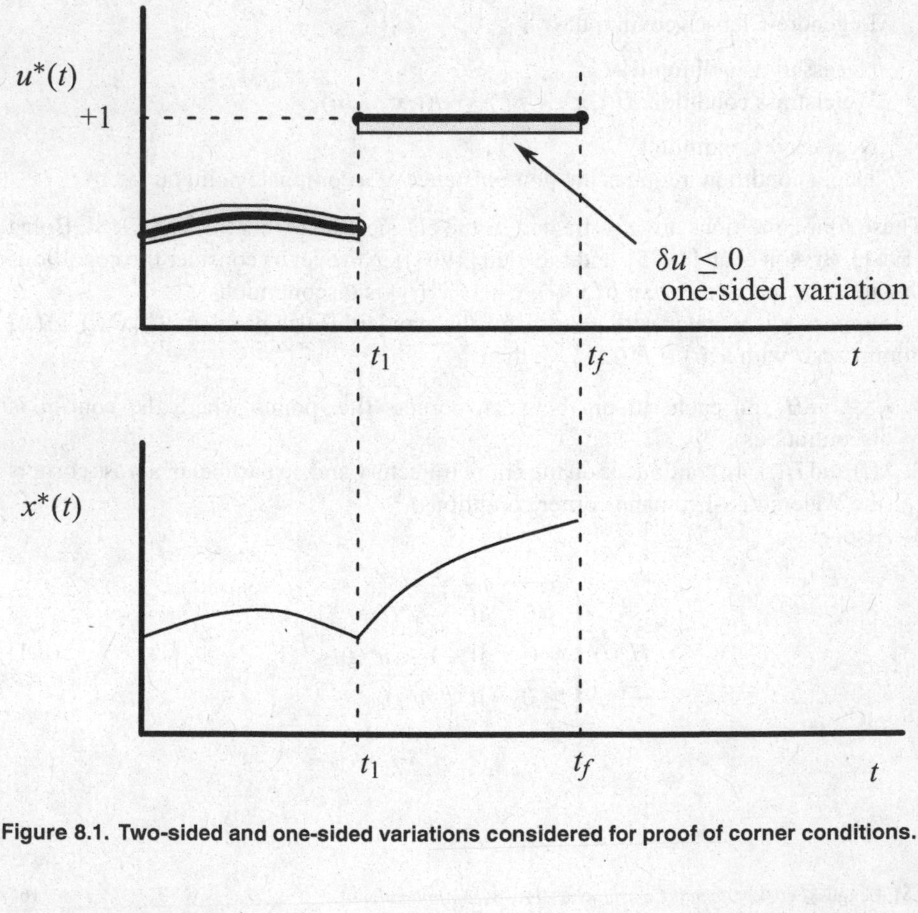

Figure 8.1. Two-sided and one-sided variations considered for proof of corner conditions.

$$J = \phi\{t_f(\varepsilon), \boldsymbol{x}[t_f(\varepsilon), \varepsilon]\} + \int_{t_o}^{t_1^-(\varepsilon)} \left\{H[t, \boldsymbol{x}(t,\varepsilon), \boldsymbol{\lambda}(t), u(t,\varepsilon)] - \boldsymbol{\lambda}^T(t)\dot{\boldsymbol{x}}(t,\varepsilon)\right\} dt$$
$$+ \int_{t_1^+(\varepsilon)}^{t_f(\varepsilon)} \left\{H[t, \boldsymbol{x}(t,\varepsilon), \boldsymbol{\lambda}(t), u(t,\varepsilon)] - \boldsymbol{\lambda}^T(t)\dot{\boldsymbol{x}}(t,\varepsilon)\right\} dt \quad (8.3)$$

where $t_1^-(\varepsilon)$ represents the limit of $t_1(\varepsilon)$ approached from the left side and $t_1^+(\varepsilon)$ represents the limit of $t_1(\varepsilon)$ approached from the right. Applying Leibniz' rule, Eq. (3.40), to Eq. (8.3) we obtain

$$\frac{dJ}{d\varepsilon} = \frac{d\phi}{d\varepsilon} + \frac{dt_1^-(\varepsilon)}{d\varepsilon}\left(H - \boldsymbol{\lambda}^T\dot{\boldsymbol{x}}\right)\Big|_{t_1^-} - \frac{dt_1^+(\varepsilon)}{d\varepsilon}\left(H - \boldsymbol{\lambda}^T\dot{\boldsymbol{x}}\right)\Big|_{t_1^+} + \frac{dt_f(\varepsilon)}{d\varepsilon}\left(H - \boldsymbol{\lambda}^T\dot{\boldsymbol{x}}\right)\Big|_{t_f}$$
$$+ \int_{t_o}^{t_1^-(\varepsilon)} \left(H_{\boldsymbol{x}}\frac{\partial \boldsymbol{x}}{\partial \varepsilon} + H_u\frac{\partial u}{\partial \varepsilon} - \boldsymbol{\lambda}^T\frac{\partial \dot{\boldsymbol{x}}}{\partial \varepsilon}\right) dt + \int_{t_1^+(\varepsilon)}^{t_f} \left(H_{\boldsymbol{x}}\frac{\partial \boldsymbol{x}}{\partial \varepsilon} + H_u\frac{\partial u}{\partial \varepsilon} - \boldsymbol{\lambda}^T\frac{\partial \dot{\boldsymbol{x}}}{\partial \varepsilon}\right) dt \quad (8.4)$$

Then, the differential change in J is:

$$dJ = \frac{dJ}{d\varepsilon}\Big|_{\varepsilon=0} d\varepsilon = d\phi^* + \left(H - \boldsymbol{\lambda}^T\dot{\boldsymbol{x}}\right)\Big|_{t_1^{-*}} dt_1^- - \left(H - \boldsymbol{\lambda}^T\dot{\boldsymbol{x}}\right)\Big|_{t_1^{+*}} dt_1^+ + \left(H - \boldsymbol{\lambda}^T\dot{\boldsymbol{x}}\right)\Big|_{t_f^*} dt_f$$
$$+ \int_{t_o}^{t_1^{-*}} \left(H_{\boldsymbol{x}}^*\delta\boldsymbol{x} + H_u^*\delta u - \boldsymbol{\lambda}^T\delta\dot{\boldsymbol{x}}\right) dt + \int_{t_1^{+*}}^{t_f^*} \left(H_{\boldsymbol{x}}^*\delta\boldsymbol{x} + H_u^*\delta u - \boldsymbol{\lambda}^T\delta\dot{\boldsymbol{x}}\right) dt \geq 0 \quad (8.5)$$

Using integration by parts we have the following:

$$\int_{t_o}^{t_1^{-*}} -\boldsymbol{\lambda}^T\delta\dot{\boldsymbol{x}}dt = (-\boldsymbol{\lambda}^T\delta\boldsymbol{x})\Big|_{t_0}^{t_1^{-*}} + \int_{t_o}^{t_1^{-*}} \dot{\boldsymbol{\lambda}}^T\delta\boldsymbol{x}dt \quad (8.6)$$

$$\int_{t_1^{+*}}^{t_f^*} -\boldsymbol{\lambda}^T\delta\dot{\boldsymbol{x}}dt = (-\boldsymbol{\lambda}^T\delta\boldsymbol{x})\Big|_{t_1^{+*}}^{t_f^*} + \int_{t_1^{+*}}^{t_f^*} \dot{\boldsymbol{\lambda}}^T\delta\boldsymbol{x}dt \quad (8.7)$$

By choosing $\boldsymbol{\lambda}(t)$ on each sub-arc to cause

$$\dot{\boldsymbol{\lambda}}^T = -H_{\boldsymbol{x}}^* \quad (8.8)$$

on $[t_o, t_1^{-*}]$ and $[t_1^{+*}, t_f^*]$, we eliminate the $\delta\boldsymbol{x}$ terms in the integrands of Eq. (8.5). Next we consider how to evaluate $\delta\boldsymbol{x}(t_1^-)$, $\delta\boldsymbol{x}(t_1^+)$, $d\boldsymbol{x}(t_1^-)$, $d\boldsymbol{x}(t_1^+)$, dt_1^-, and dt_1^+ arising in Eqs. (8.5)–(8.7). In Fig. 8.2 we see that at the corner:

$$dt_1^- = t_1^-|_{\text{varied path}} - t_1^-|_{\text{optimal path}} \quad (8.9)$$

$$dt_1^+ = t_1^+|_{\text{varied path}} - t_1^+|_{\text{optimal path}} \quad (8.10)$$

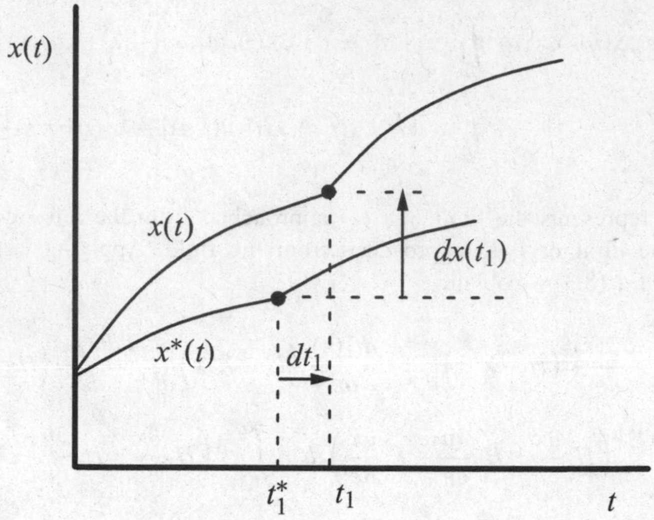

Figure 8.2. Variations near a corner at t_1^*.

Thus,

$$dt_1^+ = dt_1^- \equiv dt_1 \tag{8.11}$$

Also, we have:

$$d\boldsymbol{x}(t_1^-) = \boldsymbol{x}(t_1^-) - \boldsymbol{x}^*(t_1^{-*}) \equiv d\boldsymbol{x}(t_1) \tag{8.12}$$

$$d\boldsymbol{x}(t_1^+) = \boldsymbol{x}(t_1^+) - \boldsymbol{x}^*(t_1^{+*}) \equiv d\boldsymbol{x}(t_1) \tag{8.13}$$

because $\boldsymbol{x}$ is continuous across the corner.

In Fig. 8.3 we consider each sub-arc separately, to conclude that

$$d\boldsymbol{x}(t_1^-) = \delta\boldsymbol{x}(t_1^{-*}) + \dot{\boldsymbol{x}}^*(t_1^{-*})dt_1 \tag{8.14}$$

$$d\boldsymbol{x}(t_1^+) = \delta\boldsymbol{x}(t_1^{+*}) + \dot{\boldsymbol{x}}^*(t_1^{+*})dt_1 \tag{8.15}$$

Figure 8.4 shows a combined picture of the sub-arcs illustrated in Fig. 8.3 to emphasize the difference between $\delta\boldsymbol{x}(t_1^{-*})$ and $\delta\boldsymbol{x}(t_1^{+*})$.

Rearranging Eqs. (8.14) and (8.15) and generalizing Eq. (3.29) to the case of a vector (as we did in the proof of the Euler-Lagrange theorem), we can substitute the relations:

$$\delta\boldsymbol{x}(t_1^{-*}) = d\boldsymbol{x}_1 - \dot{\boldsymbol{x}}^*(t_1^{-*})dt_1 \tag{8.16}$$

$$\delta\boldsymbol{x}(t_1^{+*}) = d\boldsymbol{x}_1 - \dot{\boldsymbol{x}}^*(t_1^{+*})dt_1 \tag{8.17}$$

$$\delta\boldsymbol{x}(t_f^*) = d\boldsymbol{x}_f - \dot{\boldsymbol{x}}^*(t_f^*)dt_f \tag{8.18}$$

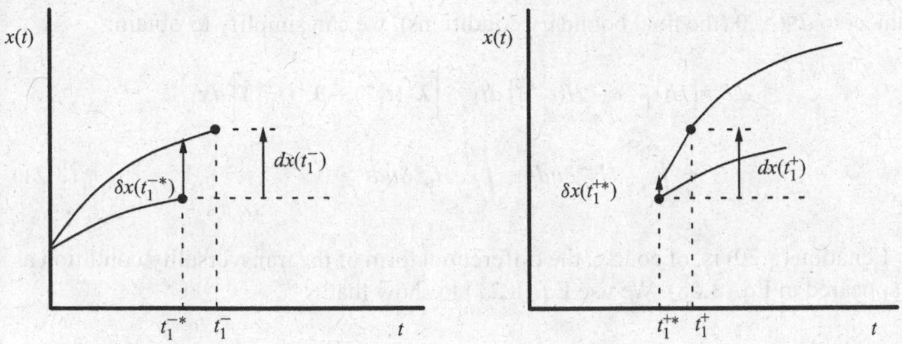

Figure 8.3. **Sub-arcs to the left and right of a corner in which $dx(t_1^-) = dx(t_1^+)$.**

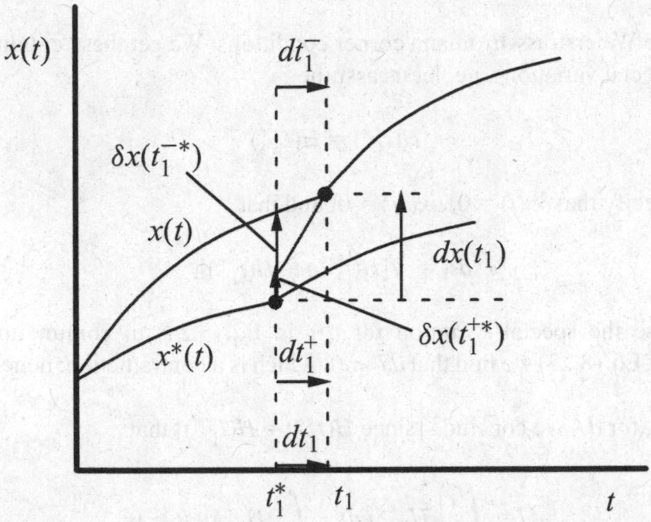

Figure 8.4. **Combined picture of sub-arcs on either side of a corner from Figs. 8.2 and 8.3.**

into the expression for dJ, Eqs. (8.5)–(8.8), to obtain:

$$
\begin{aligned}
dJ = d\phi^* &+ (H - \boldsymbol{\lambda}^T\dot{\boldsymbol{x}})|_{t_1^{-*}} dt_1^- - (H - \boldsymbol{\lambda}^T\dot{\boldsymbol{x}})|_{t_1^{+*}} dt_1^+ + (H - \boldsymbol{\lambda}^T\dot{\boldsymbol{x}})|_{t_f^*} dt_f \\
&- \boldsymbol{\lambda}^T(t_1^{-*})\left[d\boldsymbol{x}_1 - \dot{\boldsymbol{x}}^*(t_1^{-*})dt_1\right] + \boldsymbol{\lambda}^T(t_1^{+*})\left[d\boldsymbol{x}_1 - \dot{\boldsymbol{x}}^*(t_1^{+*})dt_1\right] \\
&- \boldsymbol{\lambda}^T(t_f^*)\left[d\boldsymbol{x}_f - \dot{\boldsymbol{x}}^*(t_f^*)dt_f\right] + \int_{t_o}^{t_1^{-*}} H_u{}^*\delta u\, dt + \int_{t_1^{+*}}^{t_f^*} H_u{}^*\delta u\, dt \geq 0
\end{aligned} \tag{8.19}
$$

Now, if we choose $\boldsymbol{\lambda}_f$ so that

$$
d\phi^* + H_f^* dt_f - \boldsymbol{\lambda}_f^T d\boldsymbol{x}_f = 0 \tag{8.20}
$$

subject to $d\boldsymbol{\Psi} = 0$ (the final boundary conditions), we can simplify to obtain:

$$dJ = \left[H(t_1^{-*}) - H(t_1^{+*})\right] dt_1 + \left[\boldsymbol{\lambda}^T(t_1^{+*}) - \boldsymbol{\lambda}^T(t_1^{-*})\right] d\boldsymbol{x}_1$$
$$+ \int_{t_o}^{t_1^{-*}} H_u{}^* \delta u dt + \int_{t_1^{+*}}^{t_f^*} H_u{}^* \delta u dt \geq 0 \quad (8.21)$$

Equation (8.20) is, of course, the differential form of the transversality condition as it appeared in Eq. (3.68). We use Eq. (8.21) to show that:

$$H(t_1^{+*}) = H(t_1^{-*}) \quad (8.22a)$$
$$\boldsymbol{\lambda}^T(t_1^{+*}) = \boldsymbol{\lambda}^T(t_1^{-*}) \quad (8.22b)$$

which are the Weierstrass-Erdmann corner conditions. We get these conditions by the following special variations, i.e. let us assume

$$H(t_1^{+*}) \neq H(t_1^{-*}) \quad (8.23)$$

and let us specify that $\delta u(t) = 0$, $d\boldsymbol{x}(t_1) = 0$, and that

$$dt_1 = -[H(t_1^{-*}) - H(t_1^{+*})] \quad (8.24)$$

When we use the special variation for dt_1 in Eq. (8.24) in conjunction with the hypothesis of Eq. (8.23) we find that $dJ < 0$, which is a contradiction; hence Eq. (8.23) is not true.

Therefore, for dJ, we conclude [since $H(t_1^{+*}) = H(t_1^{-*})$] that:

$$dJ = \int_{t_o}^{t_1^{-*}} H_u{}^* \delta u dt + \int_{t_1^{+*}}^{t_f^*} H_u{}^* \delta u dt \geq 0 \quad (8.25)$$

Next let us assume $\delta u(t) = 0$ on (t_1, t_f) and let us choose $\delta u(t)$ to be a nonzero variation on (t_0, t_1), as shown in Fig. 8.5. Since any variation (δu) on this interval could be reversed in sign $(-\delta u)$, it is necessary that $H_u^* = 0$ on (t_o, t_1). Suppose $H_u^* \neq 0$ at some $t \in (t_o, t_1)$. Without loss of generality assume $H_u^* > 0$. Because of continuity, $H_u^* > 0$ on some interval $(t - \delta, t + \delta)$ where $\delta > 0$. Let us choose:

$$\delta u = \begin{cases} 0 & \text{on} \quad (t_o, t - \delta) \cup (t + \delta, t_1) \\ -H_u^*(t) & \text{on} \quad (t - \delta, t + \delta) \end{cases} \quad (8.26)$$

which implies that

$$dJ = \int_{t-\delta}^{t+\delta} -H_u^{*2}(s) ds < 0 \quad (8.27)$$

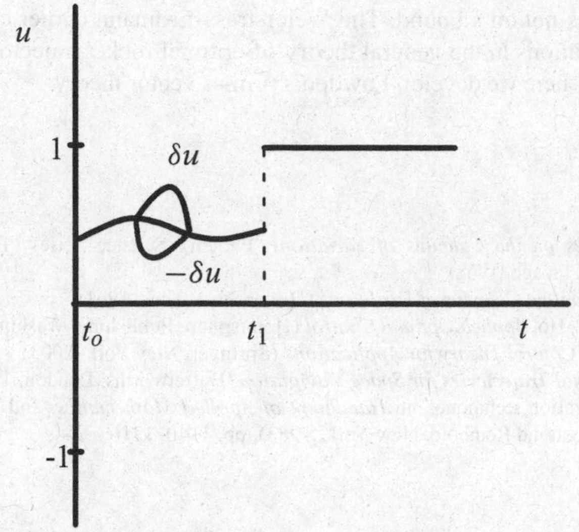

Figure 8.5. Positive and negative variations on (t_0, t_1).

which is a contradiction; hence H_u^* must be zero on (t_0, t_1). Next, we consider the time interval (t_1^+, t_f). Our remaining expression for the cost variation is

$$dJ = \int_{t_1^+}^{t_f} H_u{}^* \delta u dt \geq 0 \tag{8.28}$$

Recall our one-sided variation illustrated in Fig. 8.1, where $\delta u \leq 0$. If δu could be a two-sided variation then we would have $H_u(t) = 0$, as we just demonstrated for the interval (t_0, t_1). Thus, we see that when $u^*(t) = +1$, $\delta u \leq 0$, so we must have $H_u \leq 0$ to satisfy Eq (8.28). On the other hand, when $u^*(t) = -1$, $\delta u \geq 0$, and we must have $H_u \geq 0$. So, in conclusion:

$$H_u^*(t) \begin{cases} \leq 0 & \text{if} \quad u^*(t) = +1 \\ = 0 & \text{if} \; -1 < u^*(t) < +1 \\ \geq 0 & \text{if} \quad u^*(t) = -1 \end{cases} \tag{8.29}$$

8.3. Summary

For the standard Bolza problem, the Hamiltonian and the costates are continuous across discontinuous jumps in the control that result in a corner (discontinuity at $\dot{\boldsymbol{x}}^*$). The costate equation, $\dot{\boldsymbol{\lambda}} = -H_{\boldsymbol{x}}$, holds on either side of the corner. Also, $H_u \equiv 0$ means that the control is not on a bound. The Weierstrass-Erdmann corner conditions have important applications in the general theory of optimal rocket trajectories as we will see in Chap. 10 where we develop Lawden's primer vector theory.

References

G.A. Bliss, *Lectures on the Calculus of Variations*. Phoenix Science Series (The University of Chicago Press, Chicago, 1968)

O. Bolza, *Lectures on the Calculus of Variations* (Dover, New York, 1961)

A.E. Bryson Jr., Y.C. Ho, *Applied Optimal Control* (Hemisphere Publishing, Washington, D.C., 1975)

D.G. Hull, *Optimal Control Theory for Applications* (Springer, New York, 2003)

D.F. Lawden, *Optimal Trajectories for Space Navigation* (Butterworths, London, 1963)

J. Vagners, Optimization techniques, in *Handbook of Applied Mathematics*, 2nd edn., ed. by C.E. Pearson (Van Nostrand Reinhold, New York, 1983), pp. 1140–1216

Chapter 9

Bounded Control Problems

9.1. Optimal Control Problems with Constraints

We recall that the general form of the minimization problem can be stated as

Minimize:

$$J = \phi(t_o, \boldsymbol{x}_o, t_f, \boldsymbol{x}_f) + \int_{t_o}^{t_f} L(t, \boldsymbol{x}, \boldsymbol{u})dt \tag{9.1}$$

subject to:

$$\dot{\boldsymbol{x}} = \boldsymbol{f}(t, \boldsymbol{x}, \boldsymbol{u}) \tag{9.2}$$

$$\boldsymbol{x}(t_o) = \boldsymbol{x}_o \tag{9.3}$$

$$\boldsymbol{u} \in U \tag{9.4}$$

$$\boldsymbol{\Psi}(t_o, \boldsymbol{x}_o, t_f, \boldsymbol{x}_f) = \boldsymbol{0} \tag{9.5}$$

So far we have discussed optimization problems where endpoint constraints may appear. In this chapter we discuss the problem of bounded control where u is a scalar. The form of the constraint can be written as

$$G(u) \leq 0 \tag{9.6}$$

e.g., $G = u^2 - 1 \leq 0$ or $|u| \leq 1$. In this case, the Minimum Principle applies without modification.

Before addressing the bounded control problem, we briefly mention that other constraints often appear (in the literature and in applications) involving not only the control but the state as well. There may be equality constraints of the form

$$C(\boldsymbol{x}, \boldsymbol{u}, t) = 0 \tag{9.7}$$

and inequality constraints of the form

$$C(\boldsymbol{x}, \boldsymbol{u}, t) \leq 0 \tag{9.8}$$

J.M. Longuski et al., *Optimal Control with Aerospace Applications*,
Space Technology Library 32, DOI 10.1007/978-1-4614-8945-0_9,

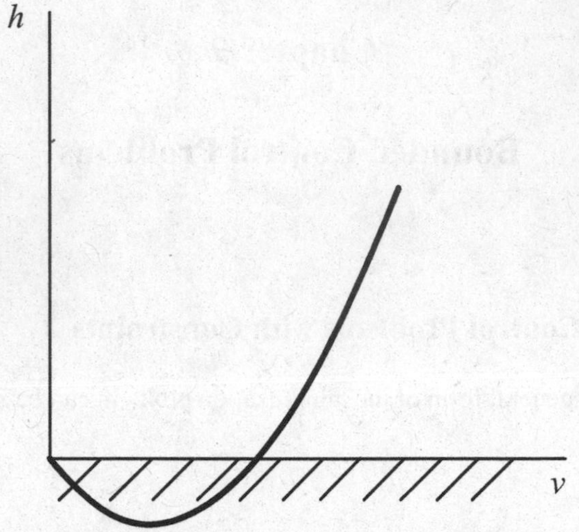

Figure 9.1. Aircraft state inequality constraint example: $S(h) = -h \leq 0$.

When the control does not appear in Eq. (9.8) we have the *state variable inequality constraint*:

$$S(\boldsymbol{x}, t) \leq 0 \tag{9.9}$$

An example of Eq. (9.9) occurs in the aircraft performance problem (Fig. 9.1) where the aircraft may not fly below the ground level ($-h \leq 0$).

In addition there is the isoperimetric (or integral) constraint:

$$G(t_f) = \int_{t_0}^{t_f} \gamma(\boldsymbol{x}, \boldsymbol{u}, t)\, dt = 0 \tag{9.10}$$

which applies to the ancient problem presented to Carthaginian Princess Dido: find the largest land area enclosed between a shoreline and a closed curve made from a bull's hide.

These problems involving constraints on the state (while important and fascinating in their own right) require special techniques beyond the scope of this introductory text. We refer the reader to the treatment of state constraints in Bryson and Ho [1975], in Vagners [1983], and in Hull [2003].

9.2. Examples of Bounded Control Problems

Example 9.1 Single-axis spacecraft attitude control problem.

Let us assume that a spacecraft (illustrated in Fig. 9.2) has a principal moment of inertia given by I. Our problem is to find the minimum-time control law to achieve the

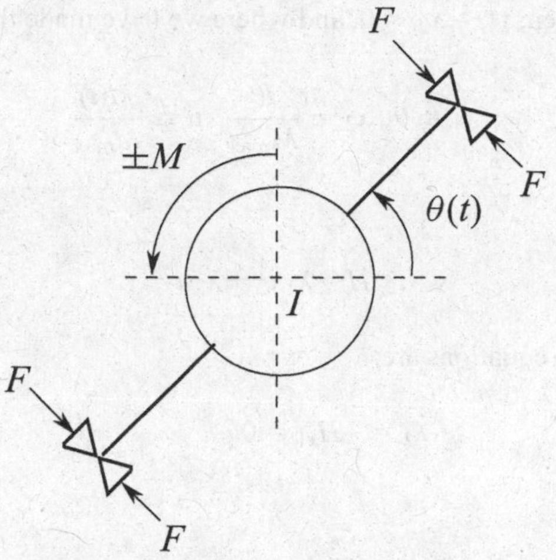

Figure 9.2. Single-axis spacecraft attitude control problem in which the initial state is driven to a final state, $\theta(t_f) = \dot{\theta}(t_f) = 0$.

final state $\theta(t_f) = \dot{\theta}(t_f) = 0$, starting from the initial state, $\theta(0)$ and $\dot{\theta}(0)$, where $\theta(t)$ is the orientation angle measured from an inertial reference direction.

We assume that the thrusters on the spacecraft can produce a variable moment, $M(t)$, with a maximum magnitude:

$$|M(t)| \leq M_{max} \tag{9.11}$$

From Euler's equations of motion we have:

$$\ddot{\theta} = \frac{M(t)}{I} \tag{9.12}$$

for this single degree-of-freedom problem.

More formally, we can state our bounded control problem as follows:

Minimize:

$$J = t_f \tag{9.13}$$

subject to:

$$\dot{x}_1 = x_2 \tag{9.14a}$$

$$\dot{x}_2 = u \tag{9.14b}$$

$$|u| \leq 1 \tag{9.14c}$$

with x_{1o} and x_{2o} given, $x_{1f} = x_{2f} = 0$, and where we have made the variable changes:

$$x_1 \equiv \theta,\ x_2 \equiv \frac{I\dot{\theta}}{M_{\max}},\ u \equiv \frac{M(t)}{M_{\max}} \tag{9.15}$$

The Hamiltonian is:

$$H = \lambda_1 x_2 + \lambda_2 u \tag{9.16}$$

The Euler-Lagrange equations are:

$$\dot{\lambda}_1 = -H_{x1} = 0 \tag{9.17}$$

so

$$\lambda_1 = c_1 \tag{9.18}$$

and

$$\dot{\lambda}_2 = -H_{x2} = -\lambda_1 = -c_1 \tag{9.19}$$

so that

$$\lambda_2 = -c_1 t + c_2 \tag{9.20}$$

where c_1 and c_2 are constants.

Here we note that we cannot set $H_u = 0$, which is only good for *interior arcs*. We wish to minimize H with respect to u, but H_u is not necessarily equal to zero. In fact, we see that $H_u = \lambda_2$ provides the switching function, λ_2, which we introduced in Chap. 4 (Example 4.7). Thus, to minimize H with respect to u:

$$u = \begin{cases} +1 & \text{if } \lambda_2 < 0 \\ -1 & \text{if } \lambda_2 > 0 \end{cases} \tag{9.21}$$

It turns out that "singular arcs" are non-optimal in this problem, i.e., $\lambda_2 \not\equiv 0$ on a non-zero time interval. We will define and discuss singular arcs later in this chapter. For the present, we merely wish to show that it is not possible in this problem to have λ_2 identically equal to zero for a nonzero time interval.

Assume the contrary that $\lambda_2 \equiv 0$. Then:

$$c_1 = 0,\ c_2 = 0,\ \lambda_1 = 0 \tag{9.22}$$

From the differential form of the transversality condition

$$H_f dt_f - \boldsymbol{\lambda}_f^T d\boldsymbol{x}_f + d\phi = 0 \tag{9.23}$$

and since $\boldsymbol{\lambda} \equiv 0$

$$H_f dt_f + dt_f = 0 \tag{9.24}$$

so that

$$H_f = -1 \tag{9.25}$$

But evaluating the Hamiltonian from Eq. (9.16) at t_f and using $\lambda_1 = \lambda_2 = 0$, we find that:

$$H_f = 0 \tag{9.26}$$

Equations (9.25) and (9.26) contradict each other. Therefore, it is impossible to have $\lambda_2 \equiv 0$ in the optimal solution. Thus, according to Eq. (9.21) the control u must always be either $+1$ or -1 depending on the sign of λ_2; u never takes on intermediate values ($-1 < u < 1$) since λ_2 is never identically zero.

The next question to be answered is: How many switches are possible? This question can be answered by plotting the switching function, λ_2, as shown in Fig. 9.3. Recalling that $\lambda_2 = -c_1 t + c_2$, we have the following possibilities:

1. If $-c_1 > 0$, $c_2 > 0$, then λ_2 is always positive and there is no switch.
2. If $-c_1 < 0$, $c_2 > 0$, then λ_2 is positive initially and may become negative so that one switch is possible.
3. If $-c_1 > 0$, $c_2 < 0$, then λ_2 is negative initially and may become positive so that one switch is possible.
4. If $-c_1 < 0$, $c_2 < 0$, then λ_2 is always negative and there is no switch.

Since this problem is two dimensional, a phase plane analysis can be applied (in which we plot x_2 as a function of x_1). We have a *bang-bang* controller where:

$$u = \pm 1 = k \tag{9.27}$$

The term "bang-bang" is used throughout the literature for controls that suddenly change from one extreme to the other. (In early lab experiments with servomechanisms, loud banging noises were heard when the servos switched control direction. Similarly, Shuttle astronauts have noted the banging of their attitude control thrusters.)

Using Eq. (9.27) in the equations of motion, Eqs. (9.14a) and (9.14b), we have

$$\dot{x}_1 = x_2 \tag{9.28}$$

$$\dot{x}_2 = k \tag{9.29}$$

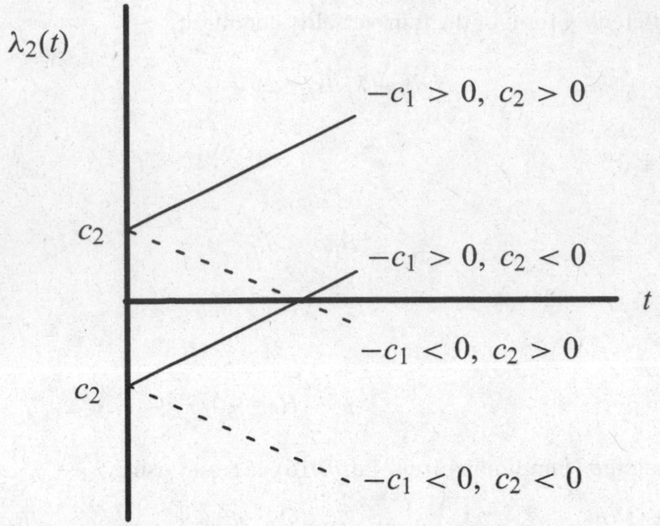

Figure 9.3. Switches for the attitude control problem occur when λ_2 in Eq. (9.20) changes sign.

Dividing Eq. (9.28) by Eq. (9.29) we have $\dot{x}_1/\dot{x}_2 = dx_1/dx_2$ to obtain:

$$dx_1 = \frac{1}{k}x_2 dx_2 \tag{9.30}$$

Integrating Eq. (9.30) we get

$$x_1 - x_{1o} = \frac{1}{2k}(x_2^2 - x_{2o}^2) \tag{9.31}$$

which is the equation of a parabola. In Fig. 9.4 we show two contours corresponding to a parabola opening to the right (where $u = +1$) and to a parabola opening to the left (where $u = -1$). By its very nature, the trajectory flow in this phase plane plot is to the right when $x_2 > 0$ (since this implies that $\dot{x}_1$ is positive and therefore x_1 is increasing) and to the left when $x_2 < 0$ (when $\dot{x}_1$ is negative).

In Fig. 9.5, we show several examples of these parabolas for $u = \pm 1$, including the special case where two parabolas (one with $u = +1$ and one with $u = -1$) intersect at the origin.

This special case is called the *switching curve*. We note that any point on the switching curve will be driven to the origin and requires no switches in the direction of the control. (Of course, once the trajectory reaches the origin, the control is turned off—but this is not considered to be a switch.)

In Fig. 9.6, we show the switching curve and consider the initial state to begin at point *a*. If we start at *a* with $u = -1$, then only one switch is required at *b* (i.e. to $u = +1$) to get to the origin in minimum time. If we start at *a* with $u = +1$, then at

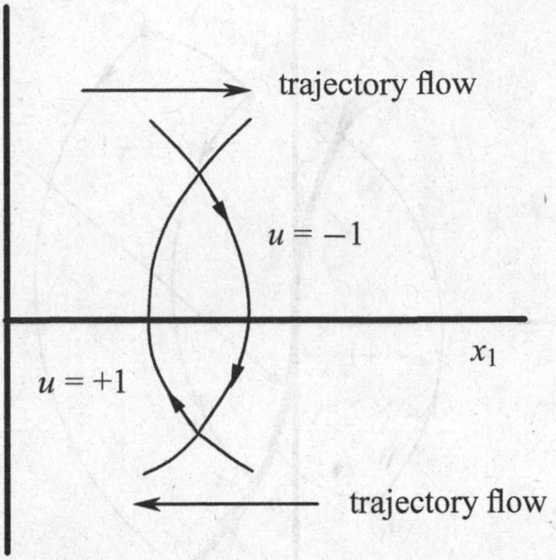

Figure 9.4. Trajectory flow: when $x_2 > 0$ the flow is from *left* to *right*; when $x_2 < 0$ the flow is from *right* to *left*.

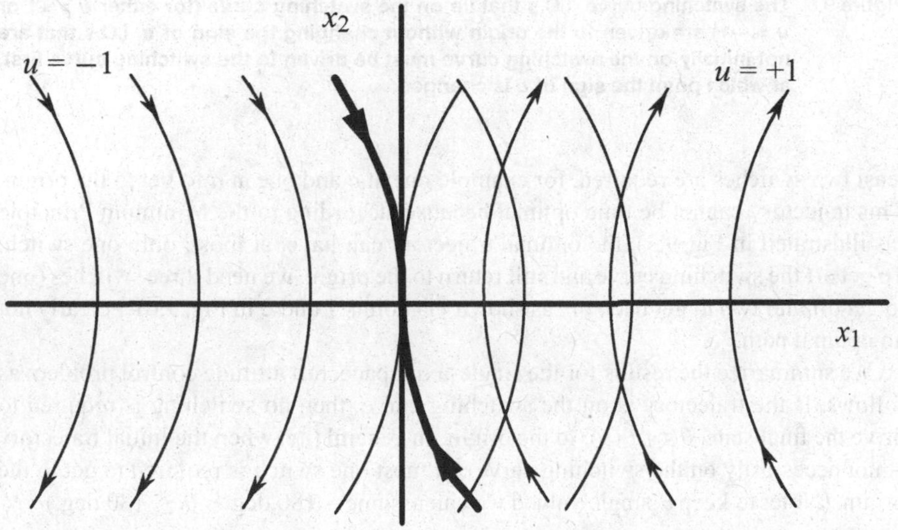

Figure 9.5. Sample trajectories for the spacecraft attitude control problem consist of parabolas that open to the *right* for $u = +1$ and to the *left* for $u = -1$. A special case occurs when both parabolas meet at the origin.

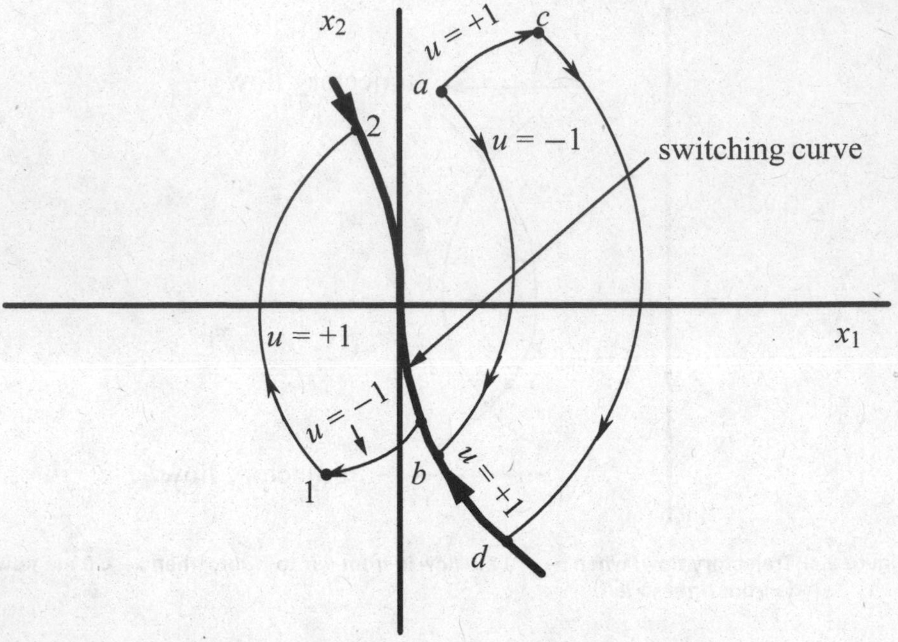

Figure 9.6. **The switching curve. I.C.s that lie on the switching curve (for either $u = +1$ or $u = -1$) are driven to the origin without changing the sign of u. I.C.s that are not initially on the switching curve must be driven to the switching curve first, at which point the sign of u is changed.**

least two switches are required, for example one at c and one at d to get to the origin. This trajectory cannot be time optimal because, according to the Minimum Principle (as illustrated in Fig. 9.3), the optimal trajectory can have, at most, only one switch. To get off the switching curve and still return to the origin, we need three switches (one to get off and two to get back on, as shown via points 1 and 2 in Fig. 9.6)—clearly not an optimal path.

We summarize the results for the single-axis spacecraft attitude control problem as follows. If the trajectory is on the switching curve, then no switching is required to drive the final state, $\theta(t_f)$, $\dot{\theta}(t_f)$, to the origin. In general (i.e. when the initial trajectory is not necessarily on the switching curve), at most one switch is required to get to the origin. (Note: to keep θ single valued we can assume -180 deg $< \theta \leq 180$ deg.)

Example 9.2 Oscillator with bounded control.

Let us consider the following problem:

Minimize:

$$J = t_f \tag{9.32}$$

subject to:

$$\dot{x}_1 = \omega x_2 \tag{9.33a}$$

$$\dot{x}_2 = -\omega x_1 + u \tag{9.33b}$$

$$|u| \leq 1 \tag{9.33c}$$

with initial conditions $x_1(0) = x_2(0) = 1$ and final conditions $x_1(t_f) = x_2(t_f) = 0$.

For this problem we will show the following:

1. The singular arc is non-optimal,
2. The control must switch at least every $\frac{\pi}{\omega}$ time units,
3. A sketch of the optimal trajectory in the phase plane, and
4. The solution in the sketch (3) satisfies the Minimum Principle in terms of the switching time.

The Hamiltonian is:

$$H = \lambda_1 \omega x_2 - \lambda_2 \omega x_1 + \lambda_2 u \tag{9.34}$$

and the Euler-Lagrange equations are:

$$\dot{\lambda}_1 = -H_{x1} = \lambda_2 \omega \tag{9.35a}$$

$$\dot{\lambda}_2 = -H_{x2} = -\lambda_1 \omega \tag{9.35b}$$

Differentiating the expression for $\dot{\lambda}_1$ in Eq. (9.35a), we obtain

$$\ddot{\lambda}_1 = \omega \dot{\lambda}_2 \tag{9.36}$$

Substituting Eq. (9.35b) into Eq. (9.36) provides

$$\ddot{\lambda}_1 + \omega^2 \lambda_1 = 0 \tag{9.37}$$

The solutions for λ_1 and λ_2 can be written as

$$\lambda_1 = A\cos(\omega t) + B\sin(\omega t) \tag{9.38a}$$

$$\lambda_2 = \frac{\dot{\lambda}_1}{\omega} = -A\sin(\omega t) + B\cos(\omega t) \tag{9.38b}$$

Since H is a linear function of u, the switching function is:

$$H_u = \lambda_2 \tag{9.39}$$

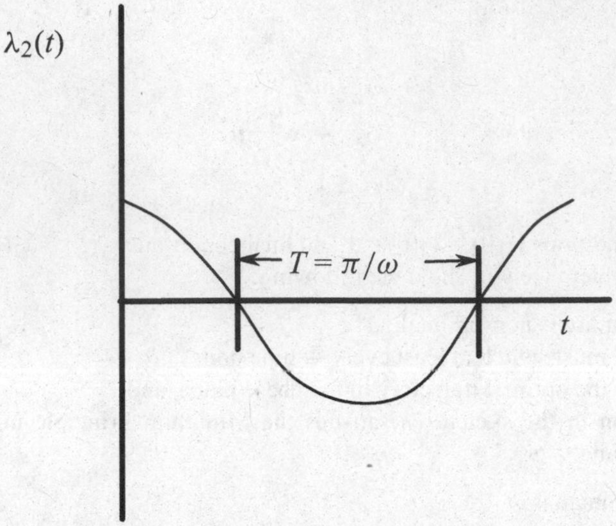

Figure 9.7. Switches occur every half period for the oscillator with bounded control.

1. We can show that a singular arc is non-optimal, that is, $\lambda_2 \not\equiv 0$ on a nonzero time interval. Assuming the contrary, $\lambda_2 \equiv 0$ implies that $A = B = 0$ and that $\lambda_1 \equiv 0$. Then from the transversality condition

$$H_f dt_f - \boldsymbol{\lambda}_f^T d\boldsymbol{x}_f + d\phi = 0 \tag{9.40}$$

we obtain

$$H_f = -1 \tag{9.41}$$

On the other hand, evaluating the Hamiltonian using Eq. (9.34) at t_f, with $\lambda_1 = \lambda_2 = 0$, gives $H_f = 0$. Thus, we get a contradiction and λ_2 cannot be identically zero for an optimal solution.

2. We now show that the control must switch at least every π/ω time units. From the switching function:

$$u = \begin{cases} +1 & \text{if } \lambda_2 < 0 \\ -1 & \text{if } \lambda_2 > 0 \end{cases} \tag{9.42}$$

Since

$$\lambda_2 = -A\sin(\omega t) + B\cos(\omega t) = C\cos(\omega t + \phi) \tag{9.43}$$

the half period is π/ω, when a sign change will occur in λ_2, as illustrated in Fig. 9.7.

3. The equations of motion can be integrated for phase plane analysis. Thus, dividing Eq. (9.33a) by Eq. (9.33b), where u is a constant (+1 or −1) we have

$$\frac{dx_1}{dx_2} = \frac{\omega x_2}{-\omega x_1 + u} \tag{9.44}$$

which implies that

$$(-\omega x_1 + u)dx_1 = (\omega x_2)dx_2 \tag{9.45}$$

Integrating Eq. (9.45) with initial conditions $x_1(0) = x_2(0) = 1$ provides

$$-\frac{1}{2}\omega x_1^2 + \frac{1}{2}\omega + ux_1 - u = \frac{1}{2}\omega x_2^2 - \frac{1}{2}\omega \tag{9.46}$$

Rearranging Eq. (9.46) in perfect squares we obtain

$$\left(x_1 - \frac{u}{\omega}\right)^2 + x_2^2 = \left(\frac{u}{\omega} - 1\right)^2 + 1 \tag{9.47}$$

where we note that the key parameter is u/ω. Thus, the trajectory is a circle with center at $(u/\omega, 0)$ and radius

$$r = \sqrt{\left(\frac{u}{\omega} - 1\right)^2 + 1} \tag{9.48}$$

Without loss of generality, let us consider the two cases: $u/\omega = 1$ and $u/\omega = -1$. We have two circles: one with center at $(1, 0)$ and radius of 1 and the other with center at $(-1, 0)$ and radius of $\sqrt{5}$, as shown in Fig. 9.8. In the figure we see that there are two paths that drive the state to the origin. The first path remains on the small circle (of radius 1), starts from the coordinates $(1, 1)$ and ends at $(0, 0)$ with $u = +1$. Clearly this first path uses the fixed value of $u/\omega = +1$ for more than a half revolution, so we expect it is not optimal (since for this example the Minimum Principle requires a switch to occur every half period, π/ω). The second path starts at $(1, 1)$ with $u/\omega = -1$ and travels along the large circle (of radius $\sqrt{5}$) until it reaches the point $(1, -1)$. At that point the control switches to $u/\omega = +1$ and the trajectory continues on the small circle until it reaches the origin $(0, 0)$. The second path is our candidate for the time-optimal trajectory because the control switches before a half revolution is completed.
4. We can show that the solution in (3) satisfies the Minimum Principle in terms of the switching time. Solving the equations of motion for $u/\omega = -1$, $x_1(0) = x_2(0) = 1$, and $\omega = 1$ provides

$$x_1(t) = 2\cos(t) + \sin(t) - 1 \tag{9.49a}$$

$$x_2(t) = -2\sin(t) + \cos(t) \tag{9.49b}$$

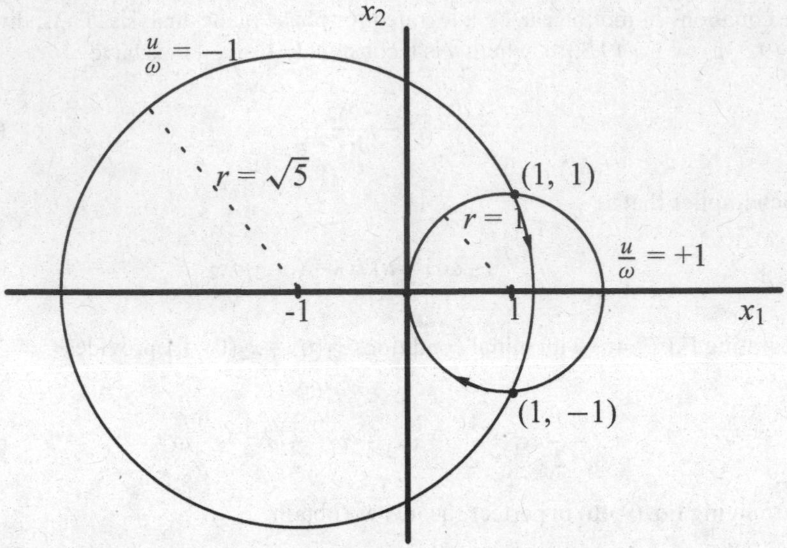

Figure 9.8. **Phase plane analysis for the oscillator problem. The I.C. is (1,1) when $u/\omega = -1$ drives the state on the large radius circle ($r = \sqrt{5}$) to intersect (1, −1) at which point $u/\omega = +1$ and the system follows the switching curve to the origin.**

Now, with $x_1(t) = 1$ and $x_2(t) = -1$, we see that the time required on this path is

$$t = \cos^{-1}\left(\frac{3}{5}\right) = 0.927 \leq \pi \text{ seconds} \tag{9.50}$$

The time is less than the half period, π, as required by the Minimum Principle. Thus we have found the time-optimal solution that drives the initial state to the origin.

9.3. Singular Arcs

In some optimization problems, extremal arcs, $H_{\boldsymbol{u}} = 0$, occur on which the matrix $H_{\boldsymbol{uu}}$ is singular. Such arcs are called *singular arcs* (as discussed in Bryson and Ho [1975]). In the singular problem, neither the Minimum Principle nor classical calculus of variations provides adequate tests for optimality. We will consider cases in which H is linear in a control variable.

Assume that the Hamiltonian has the form

$$H = H_o(t, \boldsymbol{x}, \boldsymbol{\lambda}) + H_1(t, \boldsymbol{x}, \boldsymbol{\lambda})u \tag{9.51}$$

where $|u| \leq 1$ and u is a scalar. From the Minimum Principle we have

$$u = \begin{cases} +1 & \text{if } H_1(t, \boldsymbol{x}, \boldsymbol{\lambda}) < 0 \\ -1 & \text{if } H_1(t, \boldsymbol{x}, \boldsymbol{\lambda}) > 0 \end{cases} \tag{9.52}$$

We now introduce the *Generalized Legendre-Clebsch condition* which may be regarded as an extension and generalization of the work of Contensou [1962] and Kelley [1964].

If $u^*(t)$ is a scalar singular optimal control, then the Generalized Legendre-Clebsch condition states that

$$(-1)^q \frac{\partial}{\partial u}\left(\frac{d^{2q} H_u^*}{dt^{2q}}\right) \geq 0 \tag{9.53}$$

where the $2q$-th derivative is the first even derivative of H_u that contains u explicitly. (See Kelley et al. [1967].)

To determine candidate singular controls, we differentiate $H_u = 0$ with respect to time. For example:

$$\frac{d}{dt}(H_u) = 0 = \frac{\partial H_u}{\partial t} + \frac{\partial H_u}{\partial x}\dot{x} + \frac{\partial H_u}{\partial \lambda}\dot{\lambda} + \frac{\partial H_u}{\partial u}\dot{u} \tag{9.54}$$

where the last term vanishes. Next we substitute $\dot{x} = f(t, x, u)$ and $\dot{\lambda} = -H_x(t, x, \lambda, u)$ into Eq. (9.54). We then form $\frac{d^2}{dt^2}H_u \equiv 0$, and if u appears explicitly we apply Eq. (9.53). If u does not appear we continue taking even time derivatives until Eq. (9.53) can be applied.

Equation (9.53) is referred to as the Kelley-Contensou condition (Marec 1979). When $q = 0$, Eq. (9.53) reduces to the Legendre-Clebsch condition, $H_{uu} \geq 0$.

Several investigators have contributed to the generalization of the Legendre-Clebsch condition and its application to different types of problems. These investigators include Miele [1958], Tait [1965], Kopp and Moyer [1965], Goh [1966], Robbins [1967], and Bryson and Ho [1975]. Goh [1966] provides a necessary condition that applies to systems with multiple controls.

Example 9.3 Application of the Generalized Legendre-Clebsch condition.

Consider the case of a boat moving at constant speed, V, traveling from the origin to (x_f, y_f) in minimum time, as illustrated in Fig. 9.9. We assume that the control is the turn rate, $u = \dot{\alpha}$, which is bounded so that $|u| \leq k$. Also, we assume that the turn angle, α, is initially zero. We state our problem formally as follows.

Minimize:

$$J = t_f \tag{9.55}$$

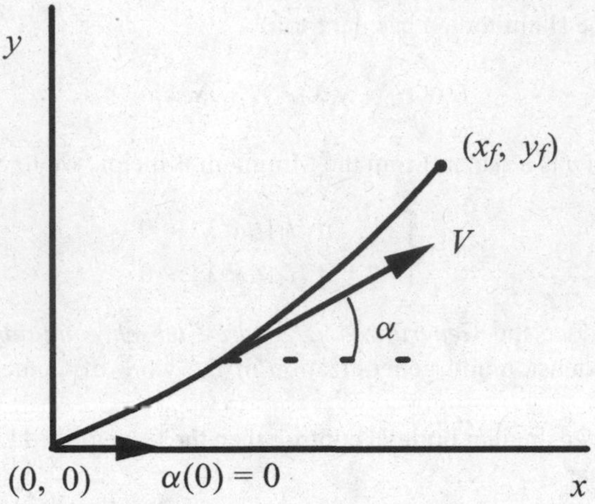

Figure 9.9. Example of a boat crossing a river requiring the Generalized Legendre-Clebsch condition. In this example the maximum turn rate is bounded.

subject to:

$$\dot{x} = V \cos\alpha \tag{9.56a}$$

$$\dot{y} = V \sin\alpha \tag{9.56b}$$

$$\dot{\alpha} = u \tag{9.56c}$$

$$|u| \leq k \tag{9.56d}$$

where V is a constant and we have I.C.s: $\alpha(0) = x(0) = y(0) = 0$.

Intuitively, we expect that the boat should turn at the maximum rate until the turn angle aligns with the terminal point (x_f, y_f) and that the turn angle would remain constant for the remainder of the trajectory, as shown in Fig. 9.10.

For the Hamiltonian we have

$$H = \lambda_1 V \cos\alpha + \lambda_2 V \sin\alpha + \lambda_3 u \tag{9.57}$$

so that from the Minimum Principle we have

$$u = \begin{cases} +k & \text{if } \lambda_3 < 0 \\ -k & \text{if } \lambda_3 > 0 \end{cases} \tag{9.58}$$

We expect the switching function, λ_3, to behave as illustrated in Fig. 9.11 where λ_3 becomes zero for the interval (t_1, t_f). We now apply the Generalized Legendre-Clebsch condition over the interval (t_1, t_f).

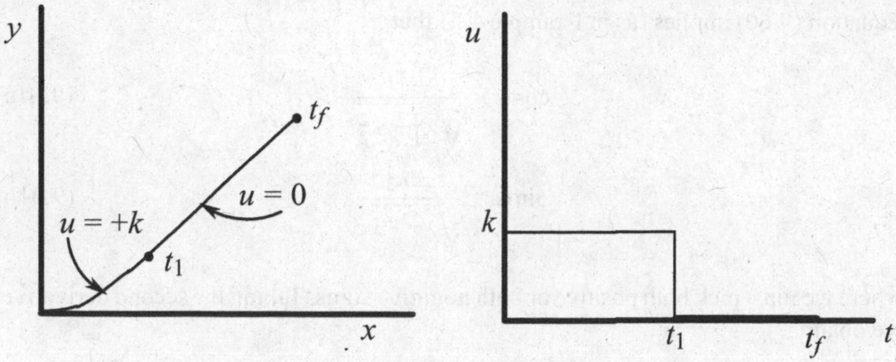

Figure 9.10. Expected solution for the river crossing problem. The maximum turn rate is used until the boat is pointed at the destination at which point the turn rate, $\dot{\alpha} = k$, is switched to zero. This solution is verified by the Minimum Principle and the Generalized Legendre-Clebsch condition.

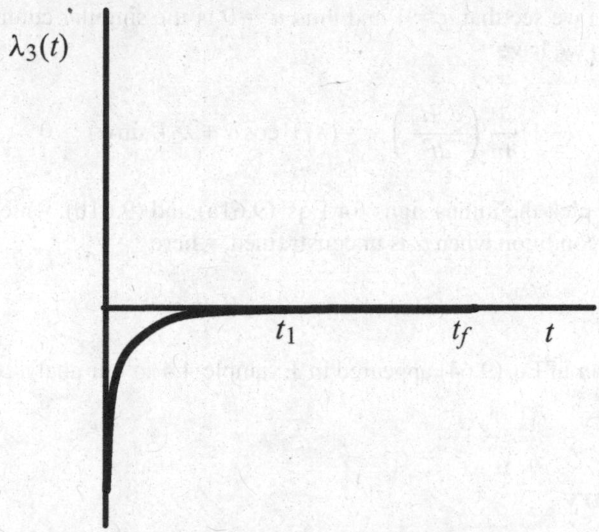

Figure 9.11. Switching function λ_3 for the river crossing problem.

The first derivative of H_u with respect to time is given by

$$\frac{d}{dt}(H_u) = \dot{\lambda}_3 = -H_\alpha = \lambda_1 V \sin\alpha - \lambda_2 V \cos\alpha = 0 \tag{9.59}$$

where we have made use of the Euler-Lagrange equation, $\dot{\lambda}_3 = -H_\alpha$. We observe that, on the singular arc

$$\tan\alpha = \frac{\lambda_2}{\lambda_1} = \frac{-\lambda_2}{-\lambda_1} \tag{9.60}$$

Equation (9.60) implies (as in Example 4.5) that

$$\cos\alpha = \frac{\pm\lambda_1}{\sqrt{\lambda_1^2+\lambda_2^2}} \tag{9.61a}$$

$$\sin\alpha = \frac{\pm\lambda_2}{\sqrt{\lambda_1^2+\lambda_2^2}} \tag{9.61b}$$

where we must pick both positive or both negative signs. Taking the second derivative, we obtain

$$\begin{aligned}\frac{d^2}{dt^2}(H_u) &= (\lambda_1 V\cos\alpha)\dot{\alpha} + (\lambda_2 V\sin\alpha)\dot{\alpha} = 0\\ &= (\lambda_1 V\cos\alpha + \lambda_2 V\sin\alpha)u = 0\end{aligned} \tag{9.62}$$

From Eq. (9.62) we see that $q = 1$ and thus $u = 0$ is the singular control. Therefore, using Eq. (9.53) we have

$$(-1)\frac{\partial}{\partial u}\left(\frac{d^2H_u}{dt^2}\right) = -(\lambda_1 V\cos\alpha + \lambda_2 V\sin\alpha) \geq 0 \tag{9.63}$$

As a result, we pick the minus signs for Eqs. (9.61a) and (9.61b), which corresponds to the $\tilde{H}_{uu} \geq 0$ condition when $\dot{\alpha}$ is unconstrained, where

$$\tilde{H} = \lambda_1 V\cos\alpha + \lambda_2 V\sin\alpha \tag{9.64}$$

The Hamiltonian in Eq. (9.64) appeared in Example 4.4 in our analysis of Zermelo's problem.

9.4. Summary

In this chapter we have addressed the problem of bounded control where the control is scalar. In some cases the Minimum Principle provides the control law, so the problem can be solved. There are, however, cases in which the matrix, $H_{\boldsymbol{uu}}$, is singular on the extremal arc (where $H_{\boldsymbol{u}} = 0$) so that the Minimum Principle does not provide an adequate test for optimality. In these singular arc problems an additional condition, the Generalized Legendre-Clebsch condition, can provide the control law.

9.5. Exercises

1. Consider the following bounded control problem:

$$\text{Min } J = t_f$$

subject to

$$\dot{x}_1 = x_2$$

$$\dot{x}_2 = -x_1 + u$$

$$x_1(0) = 4.5,\ x_1(t_f) = 0$$

$$x_2(0) = 0,\ x_2(t_f) = 0$$

$$|u| \leq 1$$

1a. Show that the control must switch at least every π time units.
1b. Sketch the optimal trajectory in the phase plane. (Include a sketch of the time optimal switching curve.)

2. It's raining outside. An aerospace engineering student must run from point A to point B a distance L. Assume that the fastest the student can run is V_{max}, and that the control is the student's velocity, V. The rain is falling straight down at a velocity, V_{rain}, and the density of the rain is ρ_{rain}. The student's goal is to run at a velocity that results in minimum wetness. Make appropriate assumptions and state the problem as an optimization problem. Determine the student's velocity to achieve minimum wetness.

References

A.E. Bryson Jr., Y.C. Ho, *Applied Optimal Control* (Hemisphere Publishing, Washington, D.C., 1975)
P. Contensou, Etude théorique des trajectoires optimales dans un champ de gravitation. Application au cas d'un centre d'attraction unique. Astronaut. Acta **8**, 134–150 (1962)
B.S. Goh, Necessary conditions for singular extremals involving multiple control variables. SIAM J. Control **4**(4), 716–731 (1966)
D.G. Hull, *Optimal Control Theory for Applications* (Springer, New York, 2003)
H.J. Kelley, A second variation test for singular extremals. AIAA J. **2**, 1380–1382 (1964)
H.J. Kelley, R.E. Kopp, A.G. Moyer, Singular extremals, Chapter 3, in *Topics in Optimization*, ed. by G. Leitmann (Academic, New York, 1967), p. 63
R.E. Kopp, A.G. Moyer, Necessary conditions for singular extremals. AIAA J. **3**, 1439–1444 (1965)
J.P. Marec, *Optimal Space Trajectories* (Elsevier Scientific, New York, 1979)
A. Miele, Flight mechanics and variational problems of a linear type. J. Aerosp. Sci. **25**(9), 581–590 (1958)
H.M. Robbins, A generalized Legendre-Clebsch condition for the singular cases of optimal control. IBM J. Res. Dev. **11**(4), 361–372 (1967)
K.S. Tait, Singular problems in optimal control. PhD thesis, Harvard University, Cambridge (1965)
J. Vagners, Optimization techniques, in *Handbook of Applied Mathematics*, 2nd edn., ed. by C.E. Pearson (Van Nostrand Reinhold, New York, 1983), pp. 1140–1216

Chapter 10

General Theory of Optimal Rocket Trajectories

10.1. Introduction

In this chapter we develop a general theory of optimal spacecraft trajectories based on two pioneering works: Breakwell [1959] and Lawden [1963]. Lawden introduced the concept of the primer vector, which plays a dominant role in minimum-propellant trajectories and also in other types of optimal trajectories. A more complete discussion of the topics in this chapter, including several example trajectories, is in Prussing [2010].

10.2. Equations of Motion

The equation of motion of a spacecraft which is thrusting in a gravitational field can be expressed in terms of the orbital radius vector **r** as:

$$\ddot{\boldsymbol{r}}(t) = \boldsymbol{g}(\boldsymbol{r}) + \boldsymbol{\Gamma}(t), \quad \boldsymbol{\Gamma}(t) = \Gamma(t)\boldsymbol{u}(t) \tag{10.1}$$

The variable $\boldsymbol{\Gamma}$ is the thrust acceleration vector, whose magnitude Γ is defined as the thrust (force), T, divided by the mass of the vehicle, m. The variable $\boldsymbol{u}$ is a unit vector in the thrust direction, and $\boldsymbol{g}(\boldsymbol{r})$ is the gravitational acceleration vector. A careful derivation of Eq. (10.1) requires deriving the *rocket equation* by equating the net external force (such as gravity) to the time rate of change of the linear momentum of the vehicle/exhaust particle system (see Sects. 6.1–6.4 of Prussing and Conway [2013]).

An additional equation expresses the change in mass of the spacecraft due to the generation of thrust:

$$\dot{m} = -b, \quad b \geq 0 \tag{10.2}$$

In Eq. (10.2) b is the (nonnegative) mass flow rate. The thrust magnitude, T, is given by $T = bc$, where c is the *effective* exhaust velocity of the engine. The word "effective" applies to high-thrust chemical engines where the exhaust gases may not be fully expanded at the nozzle exit. In this case there exists an additional contribution to the thrust which is incorporated into the effective exhaust velocity by defining

J.M. Longuski et al., *Optimal Control with Aerospace Applications*,
Space Technology Library 32, DOI 10.1007/978-1-4614-8945-0_10,

$$c \equiv c_a + (p_e - p_\infty)\frac{A_e}{b} \tag{10.3}$$

In Eq. (10.3) the subscript "e" refers to the pressure and area at the nozzle exit, c_a is the actual exhaust velocity at the exit, and p_∞ is the ambient pressure. If the gases are exhausted into the vacuum of space, $p_\infty = 0$.

An alternative to specifying the effective exhaust velocity is to describe the engine in terms of its *specific impulse*, defined to be:

$$I_{sp} \equiv \frac{(bc)\Delta t}{(b\Delta t)g_o} = \frac{c}{g_o} \tag{10.4}$$

where g_o is the standard acceleration of free fall on Earth, equal to 9.80665 m/s^2. As shown in Eq. (10.4) the specific impulse is obtained by dividing the mechanical impulse delivered to the vehicle by the weight of propellant consumed. The mechanical impulse provided by the thrust force over a time Δt is simply $(bc)\Delta t$ and, in the absence of other forces acting on the vehicle, is equal to the change in its linear momentum. The weight (on Earth) of propellant consumed during that same time interval is $(b\Delta t)g_o$. We note that if instead one divides by the *mass* of the propellant (which, of course, is the fundamental measure of the amount of substance), the specific impulse would be identical to the exhaust velocity. However, the definition in Eq. (10.4) is in standard use with the value typically expressed in units of seconds.

10.3. High and Low-Thrust Engines

We can distinguish between high- and low-thrust engines based on the value of the nondimensional ratio Γ_{max}/g_o. For high-thrust devices this ratio is greater than unity and thus these engines can be used to launch vehicles from the surface of the Earth. This ratio may extend to as high as 100. The corresponding range of specific impulse values is between 200 and approximately 850 s, with the lower values corresponding to chemical rockets, both solid and liquid, and the higher values corresponding to nuclear thermal rockets.

For low-thrust devices the ratio Γ_{max}/g_o is quite small, ranging from approximately 10^{-2} down to 10^{-5}. These values are typical of electric rocket engines such as magnetohydrodynamic (MHD), plasma arc, and ion engines, and also for solar sails. The ratio for solar sails is of the order of 10^{-5}. An electric engine requires a separate power generator such as a radioisotope thermoelectric generator (RTG) or solar cells.

Other engine designs such as the magnetoplasmadynamic (MPD) thrusters and the variable specific impulse magnetoplasma rocket (VASIMR) seek to achieve higher levels of thrust while maintaining the high specific impulse typical of low-thrust engines.

10.4. Cost Functionals for Rocket Engines

Two basic types of rocket engines exist: *constant specific impulse* (CSI) and *variable specific impulse* (VSI), also called *power limited* (PL) engines. The CSI category includes both high- and low-thrust devices. The mass flow rate b in some cases can be continuously varied, but is limited by a maximum value b_{max}. For this reason a CSI engine is also described as a thrust-limited engine, with $0 \leq \Gamma \leq \Gamma_{max}$.

The VSI category includes those low-thrust engines, such as electric engines, which need a separate power source to run the engine. For these engines, the power is limited by a maximum value P_{max}, but the specific impulse can be varied over a range of values. The propellant expenditure for the CSI and VSI categories is handled separately.

The equation of motion, Eq. (10.1), can be expressed as:

$$\dot{\mathbf{v}} = \frac{cb}{m}\boldsymbol{u} + \boldsymbol{g}(\boldsymbol{r}), \quad \frac{cb}{m} \equiv \Gamma \tag{10.5}$$

For the CSI case we solve Eq. (10.5) using the fact that c is constant as follows:

$$d\mathbf{v} = \frac{cb}{m}\boldsymbol{u}dt + \boldsymbol{g}(\boldsymbol{r})dt \tag{10.6}$$

Using Eq. (10.2),

$$d\mathbf{v} = -c\boldsymbol{u}\frac{dm}{m} + \boldsymbol{g}(\boldsymbol{r})dt \tag{10.7}$$

which can be integrated (assuming constant $\boldsymbol{u}$) to yield:

$$\Delta\mathbf{v} = \mathbf{v}(t_f) - \mathbf{v}(t_o) = -c\boldsymbol{u}\left(\ln m_f - \ln m_o\right) + \int_{t_o}^{t_f} \boldsymbol{g}(\boldsymbol{r})dt \tag{10.8}$$

or

$$\Delta\mathbf{v} = c\boldsymbol{u}\ln\left(\frac{m_o}{m_f}\right) + \int_{t_o}^{t_f} \boldsymbol{g}(\boldsymbol{r})dt \tag{10.9}$$

which correctly indicates that, in the absence of gravity, the velocity change would be in the thrust direction $\boldsymbol{u}$. The actual velocity change achieved also depends on the gravitational acceleration $\boldsymbol{g}(\boldsymbol{r})$ which is acting during the thrust period. The term in Eq. (10.9) involving the gravitational acceleration $\boldsymbol{g}(\boldsymbol{r})$ is called the *gravity loss*. We note that there is no gravity loss due to an (instantaneous) impulsive thrust, described in Sect. 10.5.2.

If we ignore the gravity loss term for the time being, a cost functional representing propellant consumed can be formulated. As will be seen, minimizing this cost functional is equivalent to maximizing the final mass of the vehicle. Since the thrust is equal to the product of the mass flow rate b and the exhaust velocity c, we can write:

$$\dot{m} = -b = \frac{-m\Gamma}{c} \tag{10.10}$$

$$\frac{dm}{m} = -\frac{\Gamma}{c}dt \tag{10.11}$$

For the CSI case the exhaust velocity, c, is constant and Eq. (10.11) can be integrated to yield

$$\ln\left(\frac{m_f}{m_o}\right) = -\frac{1}{c}\int_{t_o}^{t_f} \Gamma dt \tag{10.12}$$

or

$$c\ln\left(\frac{m_o}{m_f}\right) = \int_{t_o}^{t_f} \Gamma dt \equiv J_{\text{CSI}} \tag{10.13}$$

J_{CSI} is referred to as the *characteristic velocity* of the maneuver or the $\Delta\mathbf{V}$ (pronounced "delta vee") and it is clear from Eq. (10.13) that minimizing J_{CSI} is equivalent to maximizing the final mass m_f. This form for J_{CSI} is also derived in Marec [1979].

In the impulsive thrust approximation for the unbounded thrust case ($\Gamma_{max} \to \infty$) the vector thrust acceleration is represented by

$$\boldsymbol{\Gamma}(t) = \sum_{k=1}^{n} \Delta\boldsymbol{v}_k \delta(t - t_k) \tag{10.14}$$

with $t_o \leq t_1 < t_2 \ldots < t_n \leq t_f$ representing the times of the n thrust impulses. (See Sects. 6.1–6.3 of Prussing and Conway [2013].) Using the definition of a unit impulse,

$$\int_{t_k^-}^{t_k^+} \delta(t - t_k)\, dt = 1 \tag{10.15}$$

where $t_k^{\pm} \equiv \lim_{\epsilon\to 0}(t_k \pm \epsilon), \ \epsilon > 0$.

Using Eq. (10.14) in Eq. (10.13) we obtain:

$$J_{\text{CSI}} = \int_{t_o}^{t_f} \Gamma\, dt = \sum_{k=1}^{n} \Delta v_k \tag{10.16}$$

and the total propellant cost is given by the sum of the magnitudes of the velocity changes.

The corresponding cost functional for the VSI case is obtained differently. The exhaust power (stream or beam power) is half of the product of the thrust and the exhaust velocity:

$$P = \frac{1}{2}Tc = \frac{1}{2}m\Gamma c = \frac{1}{2}bc^2 \tag{10.17}$$

Using Eq. (10.17) along with

$$\frac{b}{m^2} = \frac{-\dot{m}}{m^2} = \frac{d}{dt}\left(\frac{1}{m}\right) \tag{10.18}$$

results in

$$\frac{d}{dt}\left(\frac{1}{m}\right) = \frac{\Gamma^2}{2P} \tag{10.19}$$

which integrates to

$$\frac{1}{m_f} - \frac{1}{m_o} = \frac{1}{2}\int_{t_o}^{t_f} \frac{\Gamma^2}{P} dt \tag{10.20}$$

Maximizing m_f for a given value of m_o regardless of whether it is optimal or not is obtained by running the engine at maximum power $P = P_{max}$. This conclusion is not as obvious as it looks in Eq. (10.20), because the value of Γ might be different for different values of P. To see that the engine should be run at maximum power we note that for a specified trajectory, $\boldsymbol{r}(t)$, the required vector thrust acceleration is given by Eq. (10.1) as

$$\boldsymbol{\Gamma}(t) = \ddot{\boldsymbol{r}}(t) - \boldsymbol{g}[\boldsymbol{r}(t)] \tag{10.21}$$

Thus, for a given trajectory $\boldsymbol{r}(t)$ (optimal or not), the final mass in Eq. (10.20) is maximized by running the engine at maximum power.

For this reason the VSI cost functional can be taken to be

$$J_{\text{VSI}} = \frac{1}{2}\int_{t_o}^{t_f} \Gamma^2 dt \tag{10.22}$$

This form for J_{VSI} is also derived in Marec [1979].

To summarize, the cost functionals representing minimum-propellant expenditure are given by

$$J_{\text{CSI}} = \int_{t_o}^{t_f} \Gamma dt \tag{10.23}$$

and

$$J_{\text{VSI}} = \frac{1}{2}\int_{t_o}^{t_f} \Gamma^2 dt \tag{10.24}$$

We see from Eqs. (10.23) and (10.24) that the minimum-propellant cost can be written in terms of the control magnitude $\Gamma(t)$ rather than introducing the mass as an additional state variable whose final value is to be maximized.

10.5. First-Order Necessary Conditions

10.5.1. Optimal Constant Specific Impulse Trajectory

For a constant specific impulse (CSI) engine the thrust is bounded by $0 \leq T \leq T_{max}$ (where T_{max} is a constant), corresponding to bounds on the mass flow rate: $0 \leq b \leq b_{max}$ (where b_{max} is a constant). Note that we can also prescribe bounds on the thrust acceleration (thrust per unit mass) $\Gamma \equiv T/m$ as $0 \leq \Gamma \leq \Gamma_{max}$, where Γ_{max} is achieved by running the engine at T_{max}. However, Γ_{max} is not constant, but increases due to the decreasing mass. One must keep track of the changing mass in order to compute Γ for a given thrust level, but this is easy to do, especially if the thrust is held constant, e.g., at its maximum value. However, if the propellant mass required is a small fraction of the total mass, a constant Γ_{max} approximation can be made.

The cost functional representing minimum-propellant consumption for the CSI case is given in Eq. (10.13) as

$$J = \int_{t_o}^{t_f} \Gamma(t)dt \tag{10.25}$$

The state vector is defined as

$$\boldsymbol{x}(t) = \begin{bmatrix} \boldsymbol{r}(t) \\ \boldsymbol{v}(t) \end{bmatrix} \tag{10.26}$$

where $\boldsymbol{r}(t)$ is the spacecraft position vector and $\boldsymbol{v}(t)$ is its velocity vector. The mass m can be kept track of without defining it to be a state variable by noting that

$$m(t) = m_o e^{-F(t)/c} \tag{10.27}$$

where

$$F(t) = \int_{t_o}^{t} \Gamma(\xi)d\xi \tag{10.28}$$

We note from Eq. (10.28) that $F(t_f)$ is equal to the cost J_{CSI}. In the constant thrust case Γ varies according to $\dot{\Gamma} = \frac{1}{c}\Gamma^2$, which is consistent with the mass decreasing linearly with time. (See Exercise 1.)

The equation of motion is

$$\dot{\boldsymbol{x}} = \begin{bmatrix} \dot{\boldsymbol{r}} \\ \dot{\boldsymbol{v}} \end{bmatrix} = \begin{bmatrix} \boldsymbol{v} \\ \boldsymbol{g}(\boldsymbol{r}) + \Gamma \boldsymbol{u} \end{bmatrix} \tag{10.29}$$

with the initial state $\boldsymbol{x}(t_o)$ specified.

The first-order necessary conditions for an optimal CSI trajectory were first derived by Lawden [1963] using classical calculus of variations. In the derivation that follows, an optimal control theory formulation is used, but the derivations and examples are analogous to those of Lawden and Breakwell. One significant difference is that the mass is not considered a state variable, but is kept track of separately which simplifies the state equations and the adjoint equations by having fewer variables.

In order to minimize the cost in Eq. (10.25) we form the Hamiltonian using Eq. (10.29) as

$$H = \Gamma + \boldsymbol{\lambda}_r^T \boldsymbol{v} + \boldsymbol{\lambda}_v^T[\boldsymbol{g}(\boldsymbol{r}) + \Gamma \boldsymbol{u}] \tag{10.30}$$

The adjoint (costate) equations are then

$$\dot{\boldsymbol{\lambda}}_r^T = -\frac{\partial H}{\partial \boldsymbol{r}} = -\boldsymbol{\lambda}_v^T \boldsymbol{G}(\boldsymbol{r}) \tag{10.31}$$

$$\dot{\boldsymbol{\lambda}}_v^T = -\frac{\partial H}{\partial \boldsymbol{v}} = -\boldsymbol{\lambda}_r^T \tag{10.32}$$

where

$$\boldsymbol{G}(\boldsymbol{r}) \equiv \frac{\partial \boldsymbol{g}(\boldsymbol{r})}{\partial \boldsymbol{r}} \tag{10.33}$$

is the symmetric 3×3 *gravity gradient matrix*. (See Exercise 2.)

Example 10.1 Derivatives of the gravity gradient matrix.

For the inverse-square gravitational field: $\boldsymbol{g}(\boldsymbol{r}) = -(\mu/r^2)\boldsymbol{r}/r = -(\mu/r^3)\boldsymbol{r}$, show that the gravity gradient matrix $\boldsymbol{G}(\boldsymbol{r})$ of Eq. (10.33) is equal to $\boldsymbol{G}(\boldsymbol{r}) = \frac{\mu}{r^5}(3\boldsymbol{r}\boldsymbol{r}^T - r^2\boldsymbol{I}_3)$, where $\boldsymbol{I}_3$ is the 3×3 identity matrix.

For $\mathrm{g} = -\frac{\mu \boldsymbol{r}}{r^3}$ we have:

$$\boldsymbol{G} = \frac{\partial \boldsymbol{g}}{\partial \boldsymbol{r}} = \left[r^3(-\mu \frac{\partial \boldsymbol{r}}{\partial \boldsymbol{r}}) + \mu \boldsymbol{r}(3r^2 \frac{\partial r}{\partial \boldsymbol{r}}) \right] / r^6 \tag{10.34}$$

Using $\partial \boldsymbol{r}/\partial \boldsymbol{r} = \boldsymbol{I}_3$ and differentiating $r^2 = \boldsymbol{r}^T\boldsymbol{r}$ yields $2r(\partial r/\partial \boldsymbol{r}) = 2\boldsymbol{r}^T$, so $\partial r/\partial \boldsymbol{r} = \boldsymbol{r}^T/r$ and we obtain:

$$\boldsymbol{G} = \frac{3\mu r^2 \boldsymbol{r}\boldsymbol{r}^T/r - \mu r^3 \boldsymbol{I}_3}{r^6} = \frac{\mu}{r^5}\left(3\boldsymbol{r}\boldsymbol{r}^T - r^2\boldsymbol{I}_3\right) \tag{10.35}$$

Following the notation of Sect. 3.3.1 the terminal constraints for optimal rocket trajectories are of the form

$$\boldsymbol{\Psi}[t_f, \boldsymbol{r}(t_f), \boldsymbol{v}(t_f)] = \boldsymbol{0} \tag{10.36}$$

which may describe an orbital intercept, rendezvous, etc. The boundary conditions on Eqs. (10.31) and (10.32) are given in terms of

$$\Phi \equiv \boldsymbol{\nu}^T\boldsymbol{\Psi}[\boldsymbol{r}(t_f),\ \boldsymbol{v}(t_f),\ t_f] \tag{10.37}$$

as

$$\boldsymbol{\lambda}_r^T(t_f) = \frac{\partial \Phi}{\partial \boldsymbol{r}(t_f)} = \boldsymbol{\nu}^T\frac{\partial \boldsymbol{\Psi}}{\partial \boldsymbol{r}(t_f)} \tag{10.38}$$

$$\boldsymbol{\lambda}_{\mathrm{v}}^T(t_f) = \frac{\partial \Phi}{\partial \boldsymbol{v}(t_f)} = \boldsymbol{\nu}^T\frac{\partial \boldsymbol{\Psi}}{\partial \boldsymbol{v}(t_f)} \tag{10.39}$$

where $\boldsymbol{\nu}$ is a constant Lagrange multiplier vector.

There are two control variables, the thrust direction $\boldsymbol{u}$ and the thrust acceleration magnitude Γ, that must be chosen to satisfy the Minimum Principle, i.e., to minimize the instantaneous value of the Hamiltonian H, as discussed in Chap. 6. By inspection, the Hamiltonian of Eq. (10.30) is minimized over the choice of thrust direction by aligning the unit vector $\boldsymbol{u}(t)$ opposite to the adjoint vector $\boldsymbol{\lambda}_{\mathrm{v}}(t)$. Because of the significance of the vector $-\boldsymbol{\lambda}_{\mathrm{v}}(t)$ Lawden [1963] termed it the *primer vector* $\boldsymbol{p}(t)$:

$$\boldsymbol{p}(t) \equiv -\boldsymbol{\lambda}_{\mathrm{v}}(t) \tag{10.40}$$

The optimal thrust unit vector is then in the direction of the primer vector, specifically:

$$\boldsymbol{u}(t) = \frac{\boldsymbol{p}(t)}{p(t)} \tag{10.41}$$

and

$$\boldsymbol{\lambda}_{\mathrm{v}}^T\boldsymbol{u} = -\lambda_{\mathrm{v}} = -p \tag{10.42}$$

in the Hamiltonian of Eq. (10.30).

From Eqs. (10.32) and (10.38) we see that

$$\dot{\boldsymbol{p}}(t) = \boldsymbol{\lambda}_r(t) \tag{10.43}$$

Equations (10.31), (10.32), (10.38), and (10.41) combine to yield the *primer vector equation*

$$\ddot{\boldsymbol{p}} = \boldsymbol{G}(\boldsymbol{r})\boldsymbol{p} \tag{10.44}$$

The boundary conditions on the solution to Eq. (10.42) are obtained from Eqs. (10.36) and (10.37):

$$\boldsymbol{p}^T(t_f) = -\,\boldsymbol{\nu}^T \frac{\partial \boldsymbol{\Psi}}{\partial \boldsymbol{v}(t_f)} \tag{10.45}$$

$$\dot{\boldsymbol{p}}^T(t_f) = \boldsymbol{\nu}^T \frac{\partial \boldsymbol{\Psi}}{\partial \boldsymbol{r}(t_f)} \tag{10.46}$$

We note that in Eq. (10.43) the final value of the primer vector for an optimal intercept (only final position specified) is the zero vector, because the terminal constraint $\boldsymbol{\Psi}$ does not depend on the final velocity $\boldsymbol{v}(t_f)$.

Using Eqs. (10.38) and (10.41) the Hamiltonian of Eq. (10.30) can be rewritten as

$$H = -\,(p-1)\,\Gamma + \dot{\boldsymbol{p}}^T\boldsymbol{v} - \boldsymbol{p}^T\boldsymbol{g} \tag{10.47}$$

To minimize the Hamiltonian over the choice of the thrust acceleration magnitude, Γ, we note that the Hamiltonian is a linear function of Γ, and thus the minimizing value for $0 \leq \Gamma \leq \Gamma_{max}$ will depend on the algebraic sign of the coefficient of Γ in Eq. (10.45). It is convenient to define the *switching function*

$$S(t) \equiv p(t) - 1 \tag{10.48}$$

The choice of the thrust acceleration magnitude, Γ, that minimizes H is then given by the bang-bang control law:

$$\Gamma = \begin{cases} \Gamma_{\max} \text{ for } S > 0 \; (p > 1) \\ 0 \;\; \text{for } S < 0 \; (p < 1) \end{cases} \tag{10.49}$$

That is, the thrust magnitude switches between its limiting values of 0 (an NT, null-thrust, arc) and $T_{\max}$ (an MT, maximum-thrust, arc) each time $S(t)$ passes through 0 [i.e. $p(t)$ passes through 1] according to Eq. (10.47). Figure 10.1 shows an example switching function for a three-burn sequence.

The possibility also exists that $S(t) \equiv 0$ [$p(t) \equiv 1$] on an interval of finite duration. From Eq. (10.45) it is evident that in this case the thrust acceleration magnitude is not determined by the Minimum Principle and may take on intermediate values between 0 and Γ_{max}. This IT (intermediate thrust arc) in Lawden [1963] is called a *singular arc* in optimal control.

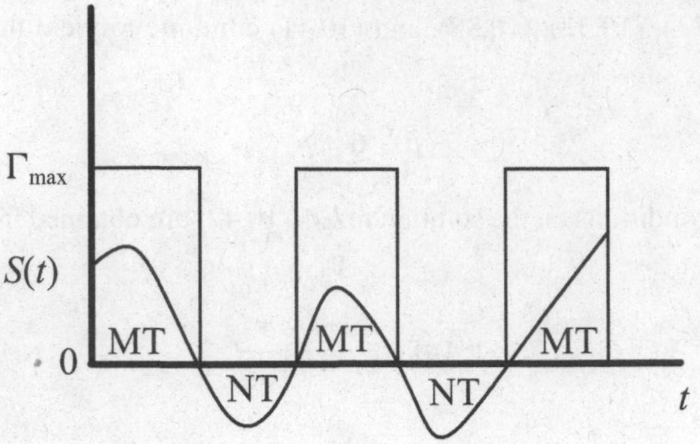

Figure 10.1. Three-burn CSI switching function and thrust profile.

From the Weierstrass-Erdmann corner conditions of Chap. 8, we know that, at a corner separating thrust arcs of different types, $\boldsymbol{\lambda}_r$ and $\boldsymbol{\lambda}_v$, and hence $\boldsymbol{p}$ and $\dot{\boldsymbol{p}}$, are continuous. Equation (10.46) then indicates that the switching function $S(t)$ is also continuous.

Lawden explained to co-author Prussing the origin of the term *primer vector* in a personal letter in 1990: "In regard to the term 'primer vector' you are quite correct in your supposition. I served in the artillery during the war [World War II] and became familiar with the initiation of the burning of cordite by means of a primer charge. Thus, p = 1 is the signal for the rocket motor to be ignited."

It follows then from Eq. (10.27) that, if $T = T_{max}$ and the engine is on for a total of Δt time units,

$$\Gamma_{max}(t) = e^{F(t)/c}\, T_{max}/m_o = T_{max}/\,(m_o - b_{max}\Delta t) \tag{10.50}$$

Even though the gravitational field is time-invariant, the Hamiltonian in this formulation does not provide a first integral (constant of the motion) on an MT arc, because Γ_{max} is an explicit function of time as shown in Eq. (10.48). On NT and IT arcs, however, the Hamiltonian is constant. (See Example 10.2.) From the corner conditions we note that the Hamiltonian is continuous at a corner between arcs of different types, which is also evident from Eq. (10.45) because $S = 0$ (i.e. $p = 1$) at those instants when Γ is discontinuous.

10.5.2. Optimal Impulsive Trajectory

For a high-thrust CSI engine the thrust durations are very small compared with the times between thrusts. Because of these short durations we can approximate each MT arc as an impulse (a Dirac delta function) having unbounded magnitude ($\Gamma_{max} \to \infty$) and zero duration. The primer vector then determines both the optimal times and directions of the thrust impulses with $p \leq 1$ corresponding to $S \leq 0$. The impulses can occur only at those instants at which $S = 0$ ($p = 1$). These impulses are separated

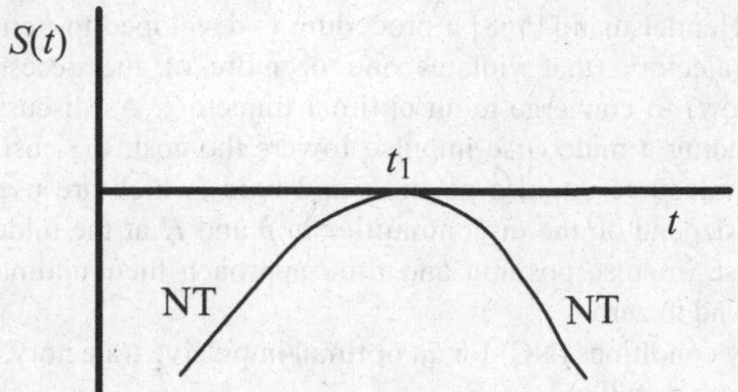

Figure 10.2. Switching function for an impulsive thrust.

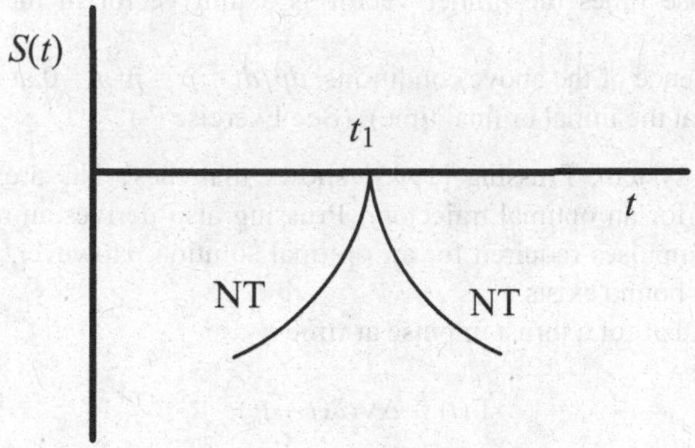

Figure 10.3. Hypothetical cusp for which $\dot{S} \neq 0$ when $S = 0$.

by NT arcs along which $S < 0$ ($p < 1$). At the impulse times, the primer vector is a unit vector in the optimal thrust direction. Figure 10.2 shows a switching function for an impulsive thrust at time t_1.

The primer vector $\boldsymbol{p}$ is defined at an impulse time and $\boldsymbol{r}$ is continuous at an impulse (recall that it is $\dot{\boldsymbol{r}} = \boldsymbol{v}$ that is discontinuous). Combined with the primer vector Eq. (10.42) it follows that $\boldsymbol{p}$, $\dot{\boldsymbol{p}}$, and $\ddot{\boldsymbol{p}}$ are continuous at an impulse.

Figure 10.3 shows a hypothetical cusp for which $\dot{S} \neq 0$ when $S = 0$. However, as we will show, a cusp is not possible. From the Hamiltonian of Eq. (10.45) and the Weierstrass-Erdmann corner condition

$$H^+ - H^- = 0 = \dot{\boldsymbol{p}}^T(\boldsymbol{v}^+ - \boldsymbol{v}^-) = \dot{\boldsymbol{p}}^T \Delta \boldsymbol{v} = \Delta v \dot{\boldsymbol{p}}^T \boldsymbol{p} \tag{10.51}$$

because $\boldsymbol{p}$, $\dot{\boldsymbol{p}}$, and $\boldsymbol{g}$ are continuous, $S = 0$ ($p = 1$), and $\Delta \boldsymbol{v} = \Delta v \boldsymbol{p}$. Thus $\dot{\boldsymbol{p}}^T \boldsymbol{p} = \dot{p} = 0$ (see Exercise 7), which implies that $\dot{S} = 0$. So the hypothetical case shown in Fig. 10.3 does not exist and there can be no cusp. This argument does not apply at the terminals of the trajectory because the corner conditions do not apply at the terminals. So in general $\dot{p} \neq 0$ at the terminals.

In Lion and Handelsman [1968] a procedure is developed to iteratively improve a nonoptimal trajectory (that violates one or more of the necessary conditions summarized below) to converge to an optimal trajectory. As discussed in Prussing [2010], when adding a midcourse impulse lowers the cost, the cost gradients with respect to the midcourse impulse position and time (which are used to iterate on these variables) depend on the discontinuities in $\dot{\boldsymbol{p}}$ and H at the midcourse impulse. As the midcourse impulse position and time approach their optimal values, these discontinuities tend to zero.

The necessary conditions (NC) for an optimal impulsive trajectory, first derived by Lawden [1963], are as follows

1. The primer vector and its first derivative are continuous everywhere.
2. The magnitude of the primer vector satisfies $p(t) \leq 1$ with the impulses occurring at those instants at which $p = 1$.
3. At the impulse times the primer vector is a unit vector in the optimal thrust direction.
4. As a consequence of the above conditions, $dp/dt = \dot{p} = \dot{\boldsymbol{p}}^T \boldsymbol{p} = 0$ at an intermediate impulse (not at the initial or final time). (See Exercise 7.)

For a linear system, Prussing [1995] shows that these NC are also sufficient conditions (SC) for an optimal trajectory. Prussing also derives an upper bound on the number of impulses required for an optimal solution. However, for a nonlinear system no upper bound exists.

We note also that for a thrust impulse at time t_k

$$\Gamma(t) = \Delta v_k \delta(t - t_k) \tag{10.52}$$

and, from Eq. (10.28), the Δv_k can be expressed as

$$\Delta v_k = \int_{t_k^-}^{t_k^+} \Gamma(t)\, dt = F(t_k^+) - F(t_k^-) \tag{10.53}$$

where t_k^+ and t_k^- are times immediately after and before the impulse time, respectively. Equation (10.27) then becomes the familiar solution to the rocket equation, Eq. (10.9), with the interval term equal to zero.

$$m(t_k^+) = m(t_k^-)\, e^{-\Delta v_k / c} \tag{10.54}$$

Figure 10.4 illustrates the primer vector magnitude for a three-impulse trajectory.

10.5.3. Optimal Variable Specific Impulse Trajectory

A variable specific impulse (VSI) engine is also known as a power-limited (PL) engine, because the power source is separate from the engine itself, e.g., solar panels, radioisotope thermoelectric generator, etc. The power delivered to the engine is bounded between 0 and a maximum value P_{max}, with the optimal value being constant

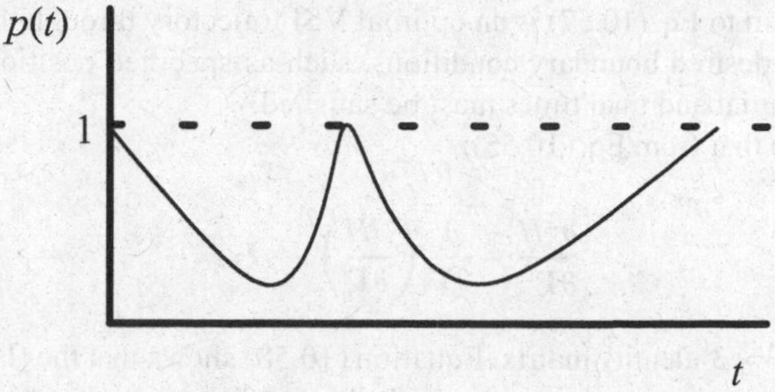

Figure 10.4. Sample primer vector history for a three-impulse trajectory. The magnitude of the primer vector satisfies $p(t) \leq 1$ with the impulses occurring at those instants at which $p = 1$.

and equal to the maximum, as discussed after Eq. (10.20). The cost functional representing minimum-propellant consumption for the VSI case is given by Eq. (10.22) as

$$J = \frac{1}{2}\int_{t_o}^{t_f} \Gamma^2(t)\, dt \tag{10.55}$$

Writing Γ^2 as $\boldsymbol{\Gamma}^T\boldsymbol{\Gamma}$ we obtain the corresponding Hamiltonian function:

$$H = \frac{1}{2}\boldsymbol{\Gamma}^T\boldsymbol{\Gamma} + \boldsymbol{\lambda}_r^T\boldsymbol{v} + \boldsymbol{\lambda}_{\text{v}}^T[\boldsymbol{g}(\boldsymbol{r}) + \boldsymbol{\Gamma}] \tag{10.56}$$

For the VSI case there is no need to consider the thrust acceleration magnitude and direction separately, so the vector $\boldsymbol{\Gamma}$ is used in place of the term $\Gamma\boldsymbol{u}$ that appears in Eq. (10.30).

Because H is a nonlinear function of $\boldsymbol{\Gamma}$, the Minimum Principle is applied by setting

$$\frac{\partial H}{\partial \boldsymbol{\Gamma}} = \boldsymbol{\Gamma}^T + \boldsymbol{\lambda}_{\text{v}}^T = \boldsymbol{0}^T \tag{10.57}$$

or

$$\boldsymbol{\Gamma}(t) = -\boldsymbol{\lambda}_{\text{v}}(t) = \boldsymbol{p}(t) \tag{10.58}$$

using the definition of the primer vector in Eq. (10.38). *Thus for a VSI engine the optimal thrust acceleration vector is equal to the primer vector*: $\boldsymbol{\Gamma}(t) = \boldsymbol{p}(t)$. Now Eq. (10.1) $\ddot{\boldsymbol{r}} = \boldsymbol{g}(\boldsymbol{r}) + \boldsymbol{\Gamma}$, can be combined with Eq. (10.42) to yield a fourth-order differential equation in $\boldsymbol{r}$:

$$\boldsymbol{r}^{iv} - \dot{\boldsymbol{G}}\dot{\boldsymbol{r}} + \boldsymbol{G}(\boldsymbol{g} - 2\ddot{\boldsymbol{r}}) = \boldsymbol{0} \tag{10.59}$$

Every solution to Eq. (10.57) is an optimal VSI trajectory through the gravity field $\boldsymbol{g}(\boldsymbol{r})$. However, desired boundary conditions, such as specified position and velocity vectors at the initial and final times must be satisfied.

We also note that from Eq. (10.55):

$$\frac{\partial^2 H}{\partial \boldsymbol{\Gamma}^2} = \frac{\partial}{\partial \boldsymbol{\Gamma}}\left(\frac{\partial H}{\partial \boldsymbol{\Gamma}}\right)^T = \boldsymbol{I}_3 \tag{10.60}$$

where $\mathbf{I}_3$ is the 3×3 identity matrix. Equation (10.58) shows that the (Hessian) matrix of second partial derivatives is positive definite, verifying that H is minimized.

Because the optimal VSI thrust acceleration is continuous, the procedure in Prussing and Sandrik [2005] to test whether second-order NC and SC are satisfied can be applied. Equation (10.57) shows that an NC for minimum cost (Hessian matrix positive semidefinite) and part of the SC (Hessian matrix positive definite) are satisfied. The other condition that is both an NC and SC is the Jacobi no-conjugate-point condition discussed in Sect. 6.6. Prussing and Sandrik [2005] provide the details of the no-conjugate-point test.

Example 10.2 The Hamiltonian for optimal rocket trajectories.

1. For an optimal CSI trajectory show that the Hamiltonian in Eq. (10.45) is constant on IT and NT arcs for a static gravitational field. Hint: calculate $\dot{H}$.
2. For an optimal VSI trajectory show that the Hamiltonian in Eq. (10.54) is constant.

Solution:

1. On an IT arc we have $p = 1$ and on an NT arc $\Gamma = 0$, so the first term in Eq. (10.45) is zero in both cases. Then $\dot{H} = \ddot{\boldsymbol{p}}^T\boldsymbol{v} + \dot{\boldsymbol{p}}^T\dot{\boldsymbol{v}} - \dot{\boldsymbol{p}}^T\boldsymbol{g} - \boldsymbol{p}^T\dot{\boldsymbol{g}}$. Using $\ddot{\boldsymbol{p}} = \boldsymbol{G}\boldsymbol{p}$, $\dot{\boldsymbol{v}} = \boldsymbol{g} + \boldsymbol{\Gamma}$ and $\dot{\boldsymbol{g}} = \boldsymbol{G}\boldsymbol{v}$ (see Exercise 3), $\dot{H} = \boldsymbol{p}^T\boldsymbol{G}\boldsymbol{v} + \dot{\boldsymbol{p}}^T(\boldsymbol{g} + \boldsymbol{\Gamma}) - \dot{\boldsymbol{p}}^T\boldsymbol{g} - \boldsymbol{p}^T\boldsymbol{G}\boldsymbol{v} = \dot{\boldsymbol{p}}^T\boldsymbol{\Gamma}$. On an NT arc $\boldsymbol{\Gamma} = \boldsymbol{0}$ and on IT arc $p = 1$, $\boldsymbol{\Gamma} = \Gamma\boldsymbol{p}$, and $\dot{\boldsymbol{p}}^T\boldsymbol{p} = \dot{p} = 0$.
2. Substituting $\boldsymbol{\Gamma} = \boldsymbol{p}$, $\boldsymbol{\lambda}_r = \dot{\boldsymbol{p}}$, and $\boldsymbol{\lambda}_v = -\boldsymbol{p}$ into the expression for $\dot{H}$ we have $\dot{H} = \boldsymbol{p}^T\dot{\boldsymbol{p}} + \ddot{\boldsymbol{p}}^T\boldsymbol{v} + \dot{\boldsymbol{p}}^T(\boldsymbol{g} + \boldsymbol{p}) - \dot{\boldsymbol{p}}^T(\boldsymbol{g} + \boldsymbol{p}) - \boldsymbol{p}^T(\boldsymbol{G}\boldsymbol{v} + \dot{\boldsymbol{p}})$. Using solution (a) and the fact that $\boldsymbol{G}$ is symmetric we find that the Hamiltonian is constant.

10.6. Optimal Trajectories in a Uniform Field

For a uniform gravitational field, we have $\boldsymbol{g}$ = constant which implies from Eqs. (10.33) and (10.44) that $\ddot{\boldsymbol{p}} = 0$. Thus, the general solution for the primer vector is

$$\boldsymbol{p} = \boldsymbol{a}t + \boldsymbol{b} \tag{10.61}$$

where $\boldsymbol{a}$ and $\boldsymbol{b}$ are constant vectors. Equation (10.61) is, obviously, the equation of a straight line. In the special case where $\boldsymbol{a} = 0$, the thrust direction never varies throughout the maneuver. In the general case, the thrust direction may vary in a plane, determined by the vectors $\boldsymbol{a}$ and $\boldsymbol{b}$, as illustrated in Fig. 10.5. Components of the primer vector (in general) are of the form:

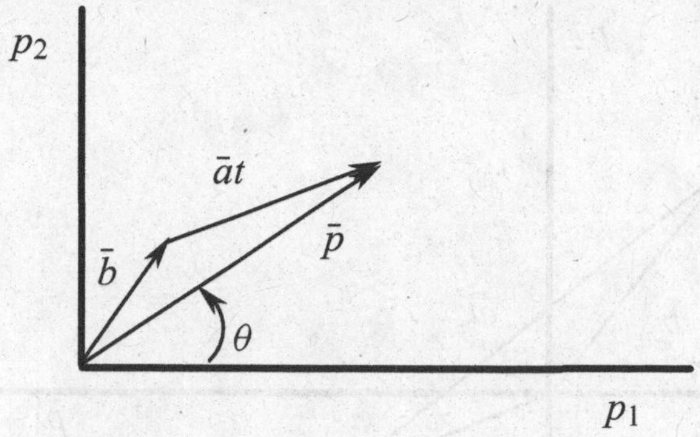

Figure 10.5. In a uniform gravity field the primer vector is restricted to a plane which means that the thrust direction may only vary in a plane.

$$p_1 = a_1 t + b_1 \tag{10.62a}$$

$$p_2 = a_2 t + b_2 \tag{10.62b}$$

$$p_3 = 0 \tag{10.62c}$$

where we have chosen a convenient set of coordinates so that $\boldsymbol{p}$ is in the xy plane.

The angle made by the primer vector (which provides us with the thrust direction) is given by

$$\tan\theta = \frac{p_2}{p_1} = \frac{a_2 t + b_2}{a_1 t + b_1} \tag{10.63}$$

which we recognize as the bilinear tangent steering law encountered in Chap. 4. [See Eq. (4.61)].

The optimal trajectory in a uniform field can include an IT arc only in special circumstances.

Lawden [1963] shows (for CSI) that for an IT arc $\boldsymbol{a} = 0$ and $\boldsymbol{p} = \boldsymbol{b}$. Thus the thrust direction is constant throughout the maneuver. Under these conditions the equations of motion can be integrated, leading to highly constrained end conditions which rule out most trajectories, in which case no IT arcs are optimal. On the other hand, if these end conditions are satisfied, it is possible to have a number of IT arcs. In the case where the end conditions rule out an IT arc, the optimal trajectory will consist of no more than three NT and MT subarcs (as shown by Leitmann [1959]). In Fig. 10.6 we illustrate possible behavior of the primer vector. In the figure we note that p is the distance from the origin. In the general case, we can have three phases:

1. The magnitude of the primer vector, p, starts out at some initial value,
2. p decreases to a minimum (or to zero), then
3. p increases to the final value.

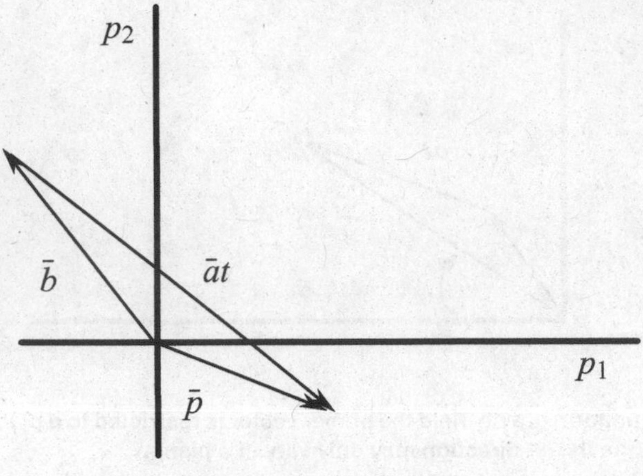

Figure 10.6. The primer vector in a uniform field may decrease in magnitude initially, reach a minimum, and then increase.

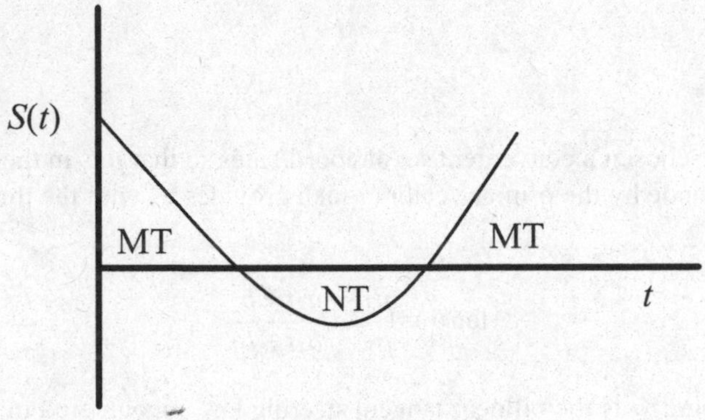

Figure 10.7. At most only three subarcs can exist MT, NT, MT according to the switching function for the primer vector in a uniform field.

For cost functional, J_{CSI}, if the initial value of p is greater than unity, the minimum less than unity, and the final greater than unity, then from Eq. (10.46) the switching function will have values of $S > 1$, $S < 1$, and $S > 1$, respectively corresponding to Eq. (10.47) and Fig. 10.7. We conclude that we can have at most three subarcs: MT, NT, MT in that order. We also note that for impulsive thrusts there are at most two impulses. These simple examples demonstrate the power of Lawden's primer vector theory.

10.7. Summary

For optimal rocket trajectories in space, we assume that the only forces acting on the spacecraft are due to gravity and rocket thrust. The thrusters can be pointed in arbitrary directions and the magnitude of the thrust may be limited. The cost to be minimized,

J, is the very important one of minimum-propellant consumption (which is equivalent to maximizing final mass). For a space mission, every kilogram of propellant saved by optimization provides an additional kilogram of payload at the final destination.

For this problem, Lawden found, using the calculus of variations, that the adjoint (costate) vector corresponding to the velocity vector provides the optimal thrust direction. Analysis of this vector, the *primer vector*, indicates when null thrust (NT), maximum thrust (MT), or intermediate thrust (IT) arcs are optimal for a constant specific impulse (CSI) engine. For a variable specific impulse (VSI) engine the magnitude of the optimal thrust acceleration is equal to the magnitude of the primer vector.

If the thrust magnitude in the CSI case is unbounded, we have the impulsive thrust case and the change in velocity is instantaneous. In reality, impulsive thrust does not exist, but the short duration of deep space thrust maneuvers compared with the time between maneuvers can be modeled as impulsive.

Lawden neatly summarizes the necessary conditions for optimal impulsive thrust trajectories in four statements involving the primer vector. These conditions also form the basis for primer vector theory, in which the primer vector evaluated on a nonoptimal impulsive trajectory can be used to determine an optimal trajectory for the same terminal conditions.

When Lawden's primer vector is applied to the problem of optimal trajectories in a uniform gravitational field with bounded thrust, we rediscover the bilinear tangent steering law from our launch from flat-Earth problem.

10.8. Exercises

1. Show that for a constant thrust, $\dot{\Gamma} = \frac{1}{c}\Gamma^2$ corresponds to a constant mass flow rate b. Hint: $m(t) = m_o - bt$.
2. Show that the gravity gradient matrix in Eq. (10.33) is symmetric for a conservative gravitational field. Hint: write $g_i = -\frac{\partial V}{\partial r_i}$, where V is a (scalar) gravitational potential function, then calculate $G_{ij} = \frac{\partial g_i}{\partial r_j}$.
3. Assuming a static gravitational field where $\boldsymbol{g}(\boldsymbol{r})$ is not an explicit function of time, determine under what circumstances $\dot{\boldsymbol{g}} \equiv \frac{d\boldsymbol{g}}{dt}$ is nonzero.
4. Consider the gravity gradient matrix for the case of a central gravitational field:

   4a. Determine the gravity gradient matrix $\boldsymbol{G}(\boldsymbol{r})$ for the general central gravitational field: $\boldsymbol{g}(\boldsymbol{r}) = g(r)\boldsymbol{r}$. Hint: the answer depends on the variables $\boldsymbol{r}$, $g(r)$, and $g'(r) \equiv dg/dr$.

   4b. Verify that your answer specializes to the result in Example 10.1 for $g(r) = -\frac{\mu}{r^3}$.

5. Consider the vector $\boldsymbol{A} = \boldsymbol{p} \times \boldsymbol{v} - \dot{\boldsymbol{p}} \times \boldsymbol{r}$ referenced by Marec [1979]:

   5a. Determine whether $\boldsymbol{A}$ is constant on an optimal continuous-thrust trajectory in the general central gravitational field of Exercise 4a. Discuss both the CSI and VSI cases, including the discontinuous thrust in the CSI case.

   5b. Determine whether the vector $\boldsymbol{A}$ is continuous at an optimal impulsive thrust.

6. Consider an optimal impulsive trajectory in an inverse-square gravitational field:
   6a. Show that the scalar a defined by Lion and Handelsman [1968] is constant on an NT arc between impulses, where $a = 2\dot{\boldsymbol{p}}^T\boldsymbol{r} + \boldsymbol{p}^T\boldsymbol{v} - 3Ht$.
   6b. Demonstrate that the discontinuity in the variable a at an optimal impulse is equal to the magnitude of the velocity change Δv.
   6c. Using the result of Exercise 6b determine a new quantity $\hat{a}$ that is both constant between impulses and continuous at an optimal impulse. Hint: form $\hat{a}$ by adding a term to a that depends on the mass.
7. Show that when $p = 1$, $\dot{p} = \dot{\boldsymbol{p}}^T\boldsymbol{p}$, where $\dot{p} \equiv dp/dt \neq |\dot{\boldsymbol{p}}|$ in general. Hint: differentiate $p^2 = \boldsymbol{p}^T\boldsymbol{p}$.
8. Consider the optimal VSI trajectory equation:
   8a. Derive Eq. (10.59).
   8b. Specialize your result for a uniform gravitational field, where $\boldsymbol{g}(\boldsymbol{r})$ is a constant vector.
   8c. Write the general solution $\boldsymbol{r}(t)$ to the equation obtained in Exercise 8b.
   8d. Based on Exercise 8c write an expression for the optimal thrust acceleration $\boldsymbol{\Gamma}(t)$.
9. Consider the problem of optimal intercept:
   9a. Based on the sentence following Eq. (10.46), specialize the form of the primer vector in Eq. (10.59) for an optimal intercept.
   9b. Qualitatively describe the behavior of the thrust direction.
   9c. Describe the possible subarc sequences involving MT, NT, and IT arcs.
   9d. For impulsive thrusts, what is the maximum number of impulses?

10.9. True or False Quiz for Chaps. 6–10

Answer the following questions in the context of the material presented in Chaps. 6–10.

1. According to the general theory of optimal rocket trajectories (and assuming $p \neq 0$) whenever the thruster is operating, the thrust must act in the direction of the primer vector, $\boldsymbol{p}$.

   True False

2. In the theory of optimal rocket trajectories, when impulsive thrusts are considered, then the switching function, $S = p - 1$, is never greater than zero.

   True False

3. Derivation of the rules for the primer vector is based mainly on the Minimum Principle and the Weierstrass-Erdmann corner conditions.

   True False

4. In the derivation of the Weierstrass-Erdmann corner conditions, it is shown that $\lambda(t)$ and $H(t)$ are continuous at corners.

True False

5. The Generalized Legendre-Clebsch condition was shown to be directly derivable from the Minimum Principle.

True False

6. If $T_{max} \rightarrow \infty$ then all optimal maneuvers must be impulsive.

True False

7. In primer theory we always have $\boldsymbol{p} = \boldsymbol{v}$ and $\dot{\boldsymbol{p}} = \boldsymbol{g}$.

True False

8. No modifications of the Minimum Principle are required for the State Variable Inequality Constraint (SVIC) Problem.

True False

9. If $x^*(t)$ is a weak extremal, then it may be possible to find a control which provides a lower cost.

True False

10. For a uniform gravity field, if the end conditions rule out an IT arc, then the optimal trajectory will consist of not more than three NT and MT subarcs.

True False

11. The Minimum Principle always tells us everything we need to know to find optimal trajectories.

True False

12. The main difference between the Pontryagin Minimum Principle and the Weierstrass condition is the class of functions of the admissible controls.

True False

13. In general, the Legendre-Clebsch condition applies to a broader class of controls than does the Pontryagin Minimum Principle.

True False

14. If $x^*(t)$ is a weak extremal, then it is also a strong extremal.

True False

15. On a corner $x(t_c^-) \neq x(t_c^+)$.

True False

Solution: 1T, 2T, 3T, 4T, 5F, 6F, 7F, 8F, 9T, 10T, 11F, 12T, 13F, 14F, 15F.

References

J.V. Breakwell, The optimization of trajectories. J. Soc. Ind. Appl. Math. **7**(2), 215–247 (1959)

D.F. Lawden, *Optimal Trajectories for Space Navigation* (Butterworths, London, 1963)

G. Leitmann, On a class of variational problems in rocket flight. J. Aerosp. Sci. **26**(9), 586–591 (1959)

P.M. Lion, M. Handelsman, Primer vector on fixed-time impulsive trajectories. AIAA J. **6**(1), 127–132 (1968)

J.P. Marec, *Optimal Space Trajectories* (Elsevier Scientific, New York, 1979)

J.E. Prussing, Optimal impulsive linear systems: sufficient conditions and maximum number of impulses. J. Astronaut. Sci. **43**(2), 195–206 (1995)

J.E. Prussing, Chapter 2: primer vector theory and applications, in *Spacecraft Trajectory Optimization*, ed. by B.A. Conway (Cambridge University Press, New York, 2010)

J.E. Prussing, B.A. Conway, *Orbital Mechanics*, 2nd edn. (Oxford University Press, New York, 2013)

J.E. Prussing, S.L. Sandrik, Second-order necessary conditions and sufficient conditions applied to continuous-thrust trajectories. J. Guid. Control Dyn. (Engineering Note) **28**(4), 812–816 (2005)

Appendices

A. Time-Optimal Lunar Ascent

Contributor: George Pollock

This appendix includes a brief introduction to MATLAB's two-point boundary-value solver and its application to the lunar takeoff problem. For the equations of motion and the derivation of the associated TPBVP, see Example 4.5 and Sect. 7.3.2.

A.1. MATLAB's Two-Point Boundary-Value Solver

MATLAB now includes a two-point boundary-value problem (TPBVP) solver, known as bvp4c, in its regular package (not in a separate toolbox). This feature allows the user to quickly setup and numerically solve TPBVPs. Here we introduce the basic use of bvp4c (for more detailed information, refer to the MATLAB help file for bvp4c).

The function bvp4c is a finite difference code that implements the three-stage Lobatto IIIa formula. This method uses a collocation formula and the collocation polynomial provides a C^1-continuous solution that is fourth-order accurate uniformly in $[t_o, t_f]$. Mesh selection and error control are based on the residual of the continuous solution.

The function integrates a system of ordinary differential equations of the form

$$\dot{\boldsymbol{y}} = \boldsymbol{f}(t, \boldsymbol{y}) \tag{A1}$$

on the known interval $[t_o, t_f]$ subject to the two-point boundary value conditions,

$$\boldsymbol{\Psi}[\boldsymbol{y}(t_o), \boldsymbol{y}(t_f)] = \boldsymbol{0} \tag{A2}$$

where in the context of trajectory optimization $\boldsymbol{y}$ is a vector that includes the state, costate (adjoint), and control variables, i.e.,

$$\boldsymbol{y}(t) = \begin{bmatrix} \boldsymbol{x}(t) \\ \boldsymbol{\lambda}(t) \\ \boldsymbol{u}(t) \end{bmatrix} \tag{A3}$$

The function bvp4c can also handle unknown parameters in the differential equations. For example, suppose that $\boldsymbol{p}$ is a vector of unknown parameters. The problem becomes

$$\dot{\boldsymbol{y}} = \boldsymbol{f}(t, \boldsymbol{y}, \boldsymbol{p}) \tag{A4a}$$

$$\boldsymbol{\Psi}[\boldsymbol{y}(t_o), \boldsymbol{y}(t_f), \boldsymbol{p}] = \boldsymbol{0} \tag{A4b}$$

In trajectory optimization, a classic example of having an unknown parameter is when the final time is unknown. We will use this feature of bvp4c in solving the optimal launch from the flat-Moon problem.

A.2. Solution Method

Follow the example code given to use MATLAB's boundary-value problem solver, bvp4c, to numerically solve the flat-Moon problem. Users can write their own versions of the script file (OptimalAscent.m) and the two function files (ascent_odes_tf.m and ascent_bcs_tf.m) that follow. For this problem, use the following initial guesses:

$$\begin{bmatrix} x_o \\ y_o \\ v_{yo} \\ \bar{\lambda}_2 \\ \bar{\lambda}_4 \end{bmatrix} = \mathbf{0} \tag{A5}$$

Note that we must provide an initial guess for the final time, even though it is a free variable (recall that we seek to minimize t_f).

In many boundary-value problems, finding a suitable initial guess is a major difficulty. No general procedure exists to develop initial guesses, and in many cases the engineer must attain some insight into the problem through numerical experimentation. Approximate solutions or expansions may also provide a suitable starting point for the numerical TPBVP solver.

To solve this problem (with a free final time) with bvp4c we must introduce the final time as an unknown parameter and remove explicit dependencies on the final time from the state and costate equations. We will accomplish this goal by nondimensionalizing the time according to the relation

$$\tau = \frac{t}{t_f} \tag{A6a}$$

$$\Rightarrow d\tau = \frac{dt}{t_f} \tag{A6b}$$

and then

$$\frac{d}{d\tau} = t_f \frac{d}{dt} \tag{A7a}$$

with $\tau \in [0, 1]$. This parameterization of the time allows bvp4c to solve the problem with assumed initial and final nondimensional times ($\tau_o = 0$ and $\tau_f = 1$) and to treat the actual final time, t_f, as an unknown parameter to be found in the optimization routine. See the following code and plots.

A.3. MATLAB Code

```
% Optimal Ascent Problem with MATLAB's bvp4c
%
% by Jose J. Guzman and George E. Pollock
%
% This script uses MATLAB's bvp4c to solve the problem of finding the
% optimal ascent trajectory for launch from a flat Moon to a 100 nautical
% mile circular orbit.  In addition to this script, we use two functions:
% one to provide the differential equations and another that gives the
% boundary conditions:
%
% This file: OptimalAscent.m
% State and Costate Equations: ascent_odes_tf.m
% Boundary Conditions:          ascent_bcs_tf.m
%
close all; clear all; clc;

% Define parameters of the problem
global g_accel Thrust2Weight     % pass these parameters to the DE function
h = 185.2e3;            % meters, final altitude (100 nmi circular orbit)
Vc = 1.627e3;           % m/s, Circular speed at 100 nmi
g_accel = 1.62;         % m/s^2, gravitational acceleration of Moon
Thrust2Weight = 3;  % Thrust to Weight ratio for Ascent Vehicle, in lunar G's
rad2deg = 180/pi;

%------------------------------------------------------------------------------
%% Boundary Conditions
%------------------------------------------------------------------------------

global x0 y0 Vx0 Vy0 yf Vxf Vyf     % pass these BCs to the BC function

% Initial conditions
% Launch from zero altitude with zero initial velocity
x0 = 0;          % meters, initial x-position
y0 = 0;          % meters, initial y-position
Vx0 = 0;         % m/s, initial downrange velocity
Vy0 = 0;         % m/s, initial vertical velocity

% Final conditions
yf = h;          % meters, final altitude
Vxf = Vc;        % m/s, final downrange velocity
Vyf = 0;         % m/s, final vertical velocity

%------------------------------------------------------------------------------
%% Initial Guesses
%------------------------------------------------------------------------------

% initial time
t0 = 0;

% list initial conditions in yinit, use zero if unknown
yinit = [x0 y0 Vx0 Vy0 0 0]; % guess for initial state and costate variables
tf_guess = 700;              % sec, initial guess for final time
```

```
% Because the final time is free, we must parameterize the problem by
% the unknown final time, tf.  Create a nondimensional time vector,
% tau, with Nt linearly spaced elements.  (tau = time/tf)  We will pass the
% unknown final time to bvp4c as an unknown parameter and the code will
% attempt to solve for the actual final time as it solves our TPBVP.
Nt = 41;
tau = linspace(0,1,Nt)';    % nondimensional time vector

% Create an initial guess of the solution using the MATLAB function
% bvpinit, which requires as inputs the (nondimensional) time vector,
% initial states (or guesses, as applicable), and the guess for the final
% time.  The initial guess of the solution is stored in the structure
% solinit.
solinit = bvpinit(tau,yinit,tf_guess);

%--------------------------------------------------------------------------
%% Solution
%--------------------------------------------------------------------------

% Call bvp4c to solve the TPBVP.  Point the solver to the functions
% containing the differential equations and the boundary conditions and
% provide it with the initial guess of the solution.
sol = bvp4c(@ascent_odes_tf, @ascent_bcs_tf, solinit);

% Extract the final time from the solution:
tf = sol.parameters(1);

% Evaluate the solution at all times in the nondimensional time vector tau
% and store the state variables in the matrix Z.
Z = deval(sol,tau);

% Convert back to dimensional time for plotting
time = t0 + tau.*(tf-t0);

% Extract the solution for each state variable from the matrix Z:
x_sol = Z(1,:);
y_sol = Z(2,:);
vx_sol = Z(3,:);
vy_sol = Z(4,:);
lambda2_bar_sol = Z(5,:);
lambda4_bar_sol = Z(6,:);

%% Plots

figure;
subplot(3,2,1);plot(time,x_sol/1000);
ylabel('x, km','fontsize',14);
xlabel('Time, sec','fontsize',14);
hold on; grid on;

title('Optimal Ascent from Flat Moon')
subplot(3,2,2);plot(time,y_sol/1000);
ylabel('y, km','fontsize',14);
xlabel('Time, sec','fontsize',14); hold on; grid on;
```

```
subplot(3,2,3);plot(time,vx_sol/1000);
ylabel('V_x, km/s','fontsize',14);
xlabel('Time, sec','fontsize',14);
hold on; grid on;

subplot(3,2,4);plot(time,vy_sol/1000);
ylabel('V_y, km/s','fontsize',14);
xlabel('Time, sec','fontsize',14);
hold on; grid on;

subplot(3,2,5);plot(time,rad2deg*atan(lambda4_bar_sol));
ylabel('\alpha, deg','fontsize',14); xlabel('Time,
sec','fontsize',14); grid on; hold on;

subplot(3,2,6);plot(time,lambda4_bar_sol);
ylabel('\lambda_4','fontsize',14);
xlabel('Time, sec','fontsize',14);
grid on; hold on;

% Plot of ascent trajectory in y versus x coordinates [km]
figure;
plot(x_sol/1000, y_sol/1000); grid on; axis equal
xlabel('Downrange Position, x, km','fontsize',14)
ylabel('Altitude, y, km','fontsize',14)
```

The boundary conditions are given in the function file, ascent_bcs_tf.m:.

```
function PSI = ascent_bcs_tf(Y0,Yf,tf)
%
% Boundary Condition Function for the Flat-Moon Optimal Ascent Problem
%
%
% pass in values of boundary conditions as global variables
global x0 y0 Vx0 Vy0 yf Vxf Vyf

% Create a column vector with the 7 boundary conditions on the state &
% costate variables (the 8th boundary condition, t0 = 0, is not used).
PSI  =   [  Y0(1) - x0        % Initial Condition
            Y0(2) - y0        % Initial Condition
            Y0(3) - Vx0       % Initial Condition
            Y0(4) - Vy0       % Initial Condition
            Yf(2) - yf        % Final Condition
            Yf(3) - Vxf       % Final Condition
            Yf(4) - Vyf       % Final Condition
        ];
return
```

The state and costate differential equations are in the function file, ascent_odes_tf.m:

```
function dX_dtau = ascent_odes_tf(tau,X,tf)
%
% State and Costate Differential Equation Function for the Flat-Moon
% Optimal Ascent Problem
%
%
% The independent variable here is the nondimensional time, tau, the state
% vector is X, and the final time, tf, is an unknown parameter that must
% also be passed to the DE function.

% Note that the state vector X has components
% X(1) = x, horizontal component of position
% X(2) = y, vertical component of position
% X(3) = Vx, horizontal component of velocity
% X(4) = Vy, vertical component of velocity
% X(5) = lambda2_bar
% X(6) = lambda4_bar

global g_accel Thrust2Weight
% Acceleration (F/m) of the Ascent Vehicle, m/s^2
Acc = Thrust2Weight*g_accel;

% State and Costate differential equations in terms of d/dt:
xdot = X(3);
ydot = X(4);
Vxdot = Acc*(1/sqrt(1+X(6)^2));
Vydot = Acc*(X(6)/sqrt(1+X(6)^2)) - g_accel;
lambda2_bar_dot = 0;
lambda4_bar_dot = -X(5);

% Nondimensionalize time (with tau = t/tf and d/dtau = tf*d/dt).  We must
% multiply each differential equation by tf to convert our derivatives from
% d/dt to d/dtau.

dX_dtau = tf*[xdot; ydot; Vxdot; Vydot; lambda2_bar_dot; lambda4_bar_dot];
return
```

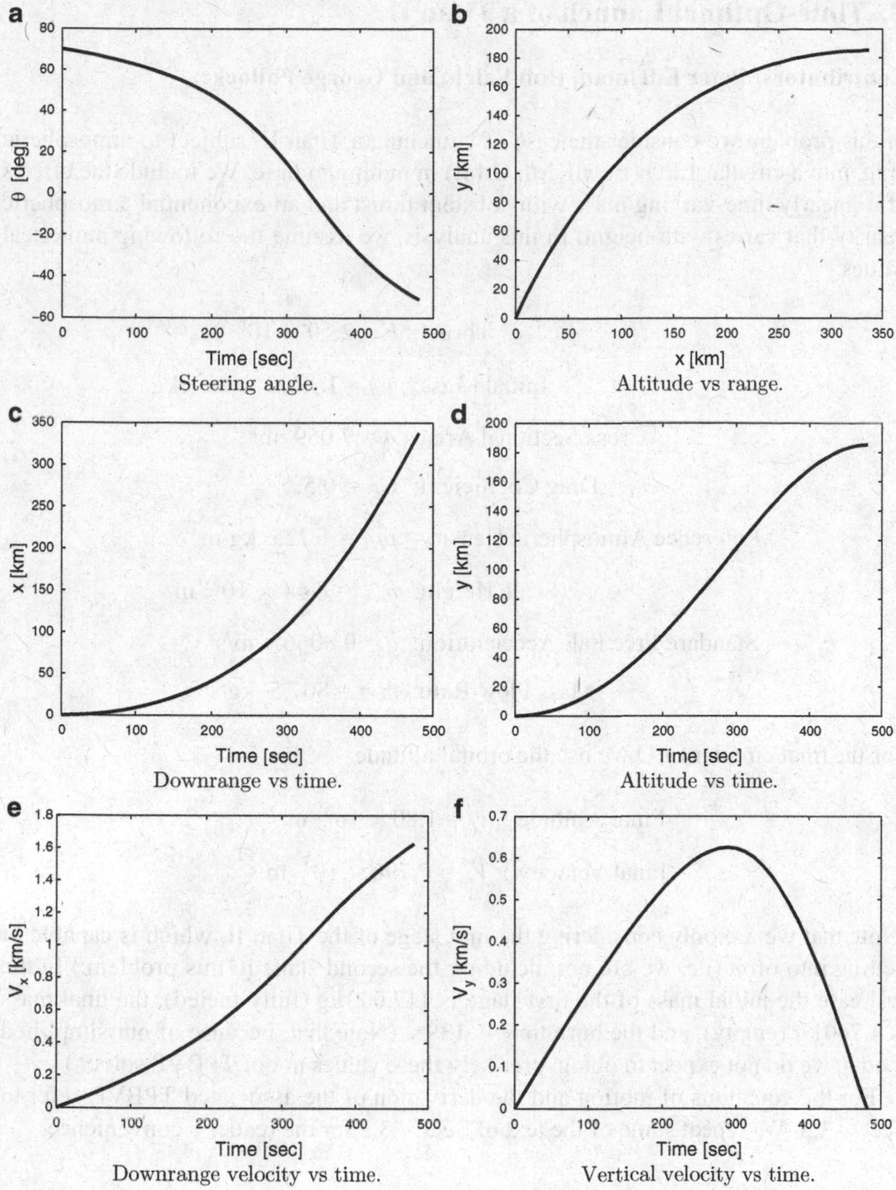

Figure A1. Plots for the time-optimal launch from flat Moon. Assumptions include constant thrust, constant mass, no drag, and uniform flat-Moon gravity

B. Time-Optimal Launch of a Titan II

Contributors: Peter Edelman, Rob Falck, and George Pollock

In this problem we consider the case of launching a Titan II, subject to atmospheric drag, into a circular LEO (Low Earth Orbit) in minimum time. We include the effects of a linearly time-varying mass with constant thrust and an exponential atmospheric density that varies with height. In this analysis, we assume the following numerical values:

$$\text{Thrust: } F = 2.10 \times 10^6 \text{ N}$$

$$\text{Initial Mass: } m_o = 1.1702 \times 10^5 \text{ kg}$$

$$\text{Cross Sectional Area: } A = 7.069 \text{ m}^2$$

$$\text{Drag Coefficient: } C_D = 0.5$$

$$\text{Reference Atmospheric Density: } \rho_{\text{ref}} = 1.225 \text{ kg/m}^3$$

$$\text{Scale Height: } h_{\text{scale}} = 8.44 \times 10^3 \text{ m}$$

$$\text{Standard Free Fall Acceleration: } g = 9.80665 \text{ m/s}^2$$

$$\text{Mass Flow Rate: } \dot{m} = -807.5 \text{ kg/s}$$

For the final circular LEO we use the orbital altitude:

$$\text{Final Altitude : } y_f = 1.80 \times 10^5 \text{ m}$$

$$\text{Final Velocity : } V_c = 7.796 \times 10^3 \text{ m/s}$$

Note that we are only considering the first stage of the Titan II, which is capable of getting into orbit (i.e. we are not including the second stage in this problem.) In the real case the initial mass of the first stage is 117,000 kg (fully fueled), the final mass is 4,760 kg (empty), and the burn time is 139 s. (Note that, because of our simplified model, we do not expect to obtain precisely these values in our TPBVP solver.)

For the equations of motion and the derivation of the associated TPBVP, refer to Sect. 7.3.3. We repeat some of the text of Sect. 7.3.3 for the reader's convenience.

B.1. Scaling the TPBVP

We scale the state variables to remain close to unity to aid in computational convergence. Using the final orbit altitude and velocity as reference values, and normalizing the time by the (unknown) final time, we define new state variables:

$$\bar{x} = \frac{x}{h}, \quad \bar{y} = \frac{y}{h}, \quad \bar{V}_x = \frac{V_x}{h}, \quad \bar{V}_y = \frac{V_y}{h} \tag{B1}$$

Applying the chain rule to change variables results in the following nondimensional EOMs:

For $\bar{x}$:

$$\frac{d\bar{x}}{d\tau} = \frac{d(x/h)}{dt}\frac{dt}{d\tau} = \frac{dx}{dt}\frac{t_f}{h} = V_x\frac{t_f}{h} \tag{B2}$$

or,

$$\frac{d\bar{x}}{d\tau} = \bar{V}_x\frac{V_c t_f}{h} \tag{B3}$$

For $\bar{y}$:

$$\frac{d\bar{y}}{d\tau} = \frac{d(y/h)}{dt}\frac{dt}{d\tau} = \frac{dy}{dt}\frac{t_f}{h} = V_y\frac{t_f}{h} \tag{B4}$$

or,

$$\frac{d\bar{y}}{d\tau} = \bar{V}_x\frac{V_c t_f}{h} \tag{B5}$$

For $\bar{V}_x$

$$\begin{aligned}\frac{d\bar{V}_x}{d\tau} &= \frac{d(V_x/V_c)}{dt}\frac{dt}{d\tau} = \frac{dV_x}{dt}\frac{t_f}{V_c}\\ &= \left[F\cos\theta - K_1\exp\left(\frac{-y}{h_{\text{scale}}}\right)V_x\sqrt{V_x^2+V_y^2}\right]\frac{t_f}{V_c(m_o-|\dot{m}|t)}\end{aligned} \tag{B6}$$

or,

$$\frac{d\bar{V}_x}{d\tau} = \left[\frac{F}{V_c}\cos\theta - K_1\exp\left(\frac{-\bar{y}h}{h_{\text{scale}}}\right)\bar{V}_x\sqrt{\bar{V}_x^2+\bar{V}_y^2}\;V_c\right]\frac{t_f}{m_o-|\dot{m}|\tau\, t_f} \tag{B7}$$

For $\bar{V}_y$

$$\begin{aligned}\frac{d\bar{V}_y}{d\tau} &= \frac{d(V_y/V_c)}{dt}\frac{dt}{d\tau} = \frac{dV_y}{dt}\frac{t_f}{V_c}\\ &= \left[F\sin\theta - K_1\exp\left(\frac{-y}{h_{\text{scale}}}\right)V_y\sqrt{V_x^2+V_y^2}\right]\frac{t_f}{V_c(m_o-|\dot{m}|t)} - \frac{gt_f}{V_c}\end{aligned} \tag{B8}$$

or,

$$\frac{d\bar{V}_y}{d\tau} = \left[\frac{F}{V_c}\sin\theta - K_1\exp\left(\frac{-\bar{y}h}{h_{\text{scale}}}\right)\bar{V}_y\sqrt{\bar{V}_x^2+\bar{V}_y^2}\;V_c\right]\frac{t_f}{m_o-|\dot{m}|\tau\, t_f} - \frac{gt_f}{V_c} \tag{B9}$$

The Hamiltonian for our problem is:

$$H = \lambda_1(\bar{V}_x\eta) + \lambda_2(\bar{V}_y\eta) + \lambda_3\left\{\left[\frac{F}{V_c}\cos\theta - K_1\exp(-\bar{y}\beta)\,\bar{V}_x\sqrt{\bar{V}_x^2 + \bar{V}_y^2}\;V_c\right]\frac{t_f}{m_o - |\dot{m}|\tau\; t_f}\right\}$$

$$+\lambda_4\left\{\left[\frac{F}{V_c}\sin\theta - K_1\exp(-\bar{y}\beta)\,\bar{V}_y\sqrt{\bar{V}_x^2 + \bar{V}_y^2}\;V_c\right]\frac{t_f}{m_o - |\dot{m}|\tau\; t_f} - \bar{g}\right\} \tag{B10}$$

where K_1 is defined in Sect. 7.3.3 and where the following substitutions have been made for certain constants:

$$\eta = \frac{V_c t_f}{h} \tag{B11a}$$

$$\beta = \frac{h}{h_{\text{scale}}} \tag{B11b}$$

$$\bar{g} = g\frac{t_f}{V_c} \tag{B11c}$$

Applying the Euler-Lagrange Equations to Eq. (B10) exactly as performed in Sect. 7.3.3 gives:

$$\frac{d\lambda_1}{d\tau} = -\frac{\partial H}{\partial \bar{x}} = 0 \tag{B12a}$$

$$\frac{d\lambda_2}{d\tau} = -\frac{\partial H}{\partial \bar{y}} = -\left(\lambda_3\bar{V}_x + \lambda_4\bar{V}_y\right)K_1\beta\exp(-\bar{y}\beta)\bar{V}_c\sqrt{\bar{V}_x^2 + \bar{V}_y^2}\;V_c\frac{t_f}{m_o - |\dot{m}|\tau\; t_f} \tag{B12b}$$

$$\frac{d\lambda_3}{d\tau} = -\frac{\partial H}{\partial \bar{V}_x} = -\lambda_1\eta + K_1\exp(-\bar{y}\beta)V_c\left[\frac{\lambda_3(2\bar{V}_x^2 + \bar{V}_x^2) + \lambda_4\bar{V}_x\bar{V}_y)}{\sqrt{\bar{V}_x^2 + \bar{V}_y^2}}\right]\frac{t_f}{m_o - |\dot{m}|\tau\; t_f} \tag{B12c}$$

$$\frac{d\lambda_4}{d\tau} = -\frac{\partial H}{\partial \bar{V}_y} = -\lambda_2\eta + K_1\exp(-\bar{y}\beta)V_c\left[\frac{\lambda_3\bar{V}_x\bar{V}_y + \lambda_4(\bar{V}_x^2 + 2\bar{V}_y^2)}{\sqrt{\bar{V}_x^2 + \bar{V}_y^2}}\right]\frac{t_f}{m_o - |\dot{m}|\tau\; t_f} \tag{B12d}$$

with the same control law as in Sect. 7.3.3 derived from the Minimum Principle:

$$\cos\theta = \frac{-\lambda_3}{\sqrt{\lambda_3^2 + \lambda_4^2}} \tag{B13a}$$

$$\sin\theta = \frac{-\lambda_4}{\sqrt{\lambda_3^2 + \lambda_4^2}} \tag{B13b}$$

All boundary conditions are the same as presented in Sect. 7.3.3, we only need to scale them to the appropriate value. Upon doing so, we obtain:

$$\tau_o = 0, \;\; \bar{x}(0) = \bar{y}(0) = 0, \;\; \bar{V}_x(0) = \bar{V}_y(0) = 0 \tag{B14a}$$

$$\bar{y}(\tau_f) = 1, \;\; \bar{V}_x(\tau_f) = 1, \;\; \bar{V}_y(\tau_f) = 0 \;\; H(\tau_f) = -1, \;\; \lambda_1(\tau_f) = 0 \tag{B14b}$$

We immediately see that the last boundary condition in Eq. (B14b) along with Eq. (B12a) implies that $\lambda_1 = 0$ for the entire trajectory, meaning we can omit λ_1 from the problem entirely. After substituting Eqs. (B13a) and (B13b) into Eqs. (B7) and (B9) respectively and reiterating all pertinent differential equations and boundary conditions gives the newly scaled well-defined TPBVP:

$$\frac{d\bar{x}}{d\tau} = \bar{V}_x \eta \tag{B15a}$$

$$\frac{d\bar{y}}{d\tau} = \bar{V}_y \eta \tag{B15b}$$

$$\frac{d\bar{V}_x}{d\tau} = \left[\frac{F}{V_c} \left(\frac{-\lambda_3}{\sqrt{\lambda_3^2 + \lambda_4^2}} \right) - K_1 \exp\left(-\bar{y}\beta\right) \bar{V}_x \sqrt{\bar{V}_x^2 + \bar{V}_y^2} \; V_c \right] \frac{t_f}{m_o - |\dot{m}|\tau \; t_f} \tag{B15c}$$

$$\frac{d\bar{V}_y}{d\tau} = \left[\frac{F}{V_c} \left(\frac{-\lambda_4}{\sqrt{\lambda_3^2 + \lambda_4^2}} \right) - K_1 \exp\left(-\bar{y}\beta\right) \bar{V}_y \sqrt{\bar{V}_x^2 + \bar{V}_y^2} \; V_c \right] \frac{t_f}{m_o - |\dot{m}|\tau \; t_f} - \bar{g} \tag{B15d}$$

$$\frac{d\lambda_2}{d\tau} = -\left(\lambda_3 \bar{V}_x + \lambda_4 \bar{V}_y\right) K_1 \beta \exp\left(-\bar{y}\beta\right) \sqrt{\bar{V}_x^2 + \bar{V}_y^2} \; V_c \frac{t_f}{m_o - |\dot{m}|\tau \; t_f} \tag{B15e}$$

$$\frac{d\lambda_3}{d\tau} = K_1 \exp\left(-\bar{y}\beta\right) V_c \left[\frac{\lambda_3 (2\bar{V}_x^2 + \bar{V}_y^2) + \lambda_4 \bar{V}_x \bar{V}_y)}{\sqrt{\bar{V}_x^2 + \bar{V}_y^2}} \right] \frac{t_f}{m_o - |\dot{m}|\tau \; t_f} \tag{B15f}$$

$$\frac{d\lambda_4}{d\tau} = -\lambda_2 \eta + K_1 \exp\left(-\bar{y}\beta\right) V_c \left[\frac{\lambda_3 \bar{V}_x \bar{V}_y + \lambda_4 (\bar{V}_x^2 + 2\bar{V}_y^2)}{\sqrt{\bar{V}_x^2 + \bar{V}_y^2}} \right] \frac{t_f}{m_o - |\dot{m}|\tau \; t_f} \tag{B15g}$$

with boundary conditions:

$$\tau_o = 0, \;\; \bar{x}(0) = \bar{y}(0) = 0, \;\; \bar{V}_x(0) = \bar{V}_y(0) = 0 \tag{B16a}$$

$$\bar{y}(\tau_f) = 1, \;\; \bar{V}_x(\tau_f) = 1, \;\; \bar{V}_y(\tau_f) = 0 \;\; H(\tau_f) = -1 \tag{B16b}$$

B.2. Solution Method

First we solve the optimal ascent problem using no drag and constant mass. The constant mass used is the average between the wet and dry mass of the Titan II rocket and provides us with a good initial guess for when we add time-varying mass and drag. Next the time-varying mass case is solved using the constant mass case as the initial guess, however since the acceleration will be higher near the end of the trajectory, the time to orbit will be shorter. Thus we also subtract some time from the guess for the final time achieved, t_f, obtained from the no drag and constant mass solution, otherwise the solution will not converge. Lastly, the effects of drag are added and the TPBVP is solved, using the varying mass, no drag solution as the initial guess.

B.3. Results

For a discussion of the results, see Sect. 7.3.3 (Fig. B1).

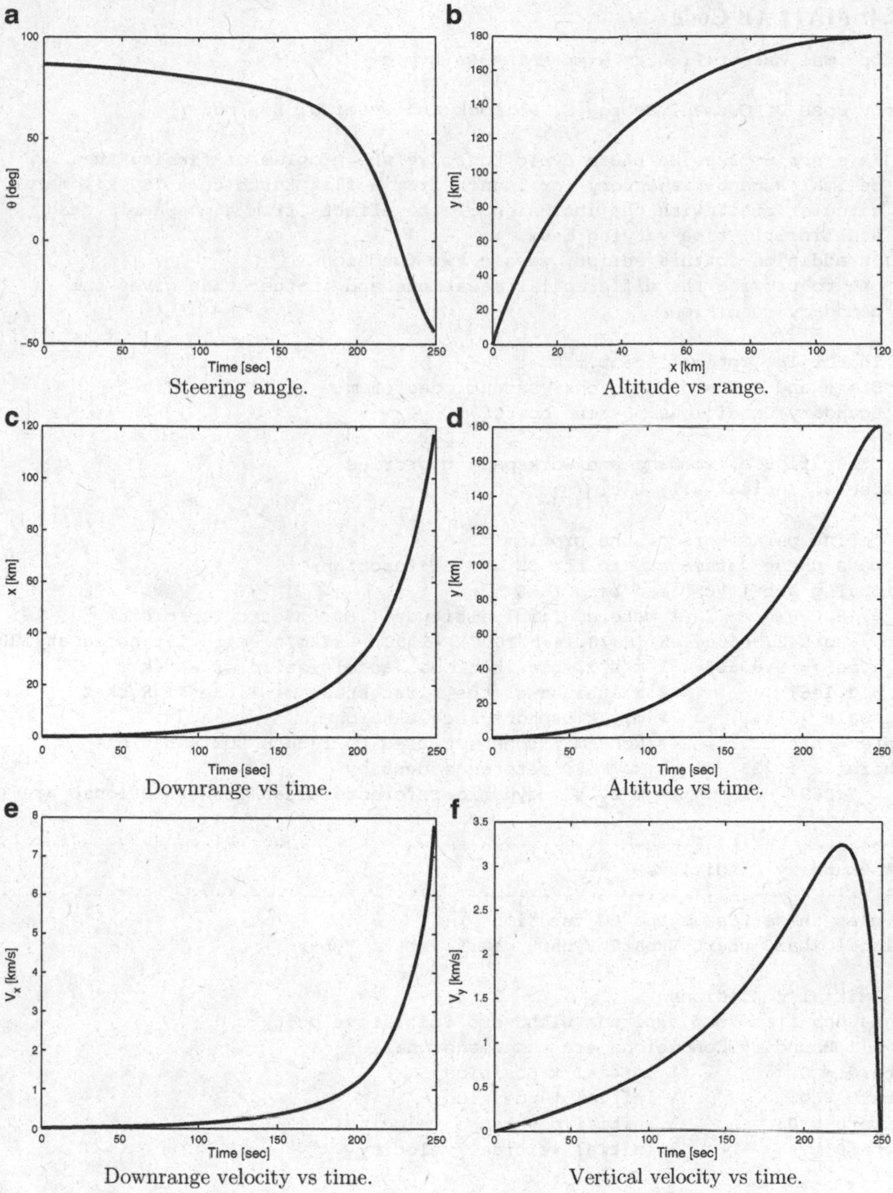

Steering angle. Altitude vs range.

Downrange vs time. Altitude vs time.

Downrange velocity vs time. Vertical velocity vs time.

Figure B1. Plots for the time-optimal launch of a Titan II rocket into a 180-km circular orbit. Assumptions include constant thrust, time-varying mass, drag from an exponential atmospheric model, and uniform flat-Earth gravity

B.4. MATLAB Code

```
% Optimal Ascent Problem with MATLAB's bvp4c
%
% by Jose J. Guzman, George E. Pollock and Peter J. Edelman
%
% This script uses MATLAB's bvp4c to solve the problem of finding the
% optimal ascent trajectory for launch from a flat Earth to a 180 kilometer
% circular orbit with the inclusion of the effects from atmospheric drag
% and linearly time-varying mass.
% In addition to this script, we use two functions:
% one to provide the differential equations and another that gives the
% boundary conditions:
%
% This file: OptimalAscent.m
% State and Costate Equations: ascent_odes_tf.m
% Boundary Conditions: ascent_bcs_tf.m
%
% Clear figure, command and workspace histories
close all; clear all; clc;

% Define parameters of the problem
% pass these parameters to the DE and BC functions
global g_accel Vc h eta beta f m0 mdot
h = 180000;          % meters, final altitude (180 km circular orbit)
Vc = sqrt(3.9860044e5/(6378.14+h/1000))*1000;    % m/s, Circular speed at 180 km
g_accel = 9.80665;  % m/s^2, gravitational acceleration of Earth
f = 2.1e6;          % N, thrust of the first stage of Titan II Rocket
h_scale = 8440;     % m, atmospheric scale-height
beta = h/h_scale;   %[Nondim], constant used to reduce EOM equations
rhoref = 1.225;     % kg/m^3, reference density
A = 7.069;          % m^2, aerodynamic reference area (cross-sectional area)

%-----------------------------------------------------------------------
%% Boundary Conditions
%-----------------------------------------------------------------------
% pass these BCs to the BC function
global xbar0 ybar0 Vxbar0 Vybar0 ybarf Vxbarf Vybarf

% Initial conditions
% Launch from zero altitude with zero initial velocity
% All Boundary Conditions are nondimensional
xbar0 = 0;          % initial x-position
ybar0 = 0;          % initial y-position
Vxbar0 = 0;         % initial downrange velocity
Vybar0 = 0;         % initial vertical velocity

% Final conditions
ybarf = h/h;        % final altitude
Vxbarf = Vc/Vc;     % final downrange velocity
Vybarf = 0;         % final vertical velocity

%% Parameters for the NO DRAG and CONSTANT MASS case
% Solve TPBVP without drag (C_D = 0) and constant mass (mdot = 0)
m0 = 60880;         % kg, average mass of a Titan II Rocket
CD = 0;             % Drag coefficient set to zero for no drag cases
mdot = 0;           % kg/s, mass flow rate for the constant mass case
% eta - a constant composed of the reference density, coefficient
```

```
% of drag and the aerodynamic reference area. It is only used to simplify
% the drag expressions in the EOMs.
eta = rhoref*CD*A/2;

%--------------------------------------------------------------------------
%% Initial Guesses
%--------------------------------------------------------------------------

% initial time
t0 = 0;
% list initial conditions in yinit, use zero if unknown
% guess for initial state and costate variables
yinit = [xbar0 ybar0 Vxbar0 Vybar0 0 -1 0];
tf_guess = 700;                  % sec, initial guess for final time
% Because the final time is free, we must parameterize the problem by
% the unknown final time, tf. Create a nondimensional time vector,
% tau, with Nt linearly spaced elements. (tau = time/tf) We will pass the
% unknown final time to bvp4c as an unknown parameter and the code will
% attempt to solve for the actual final time as it solves our TPBVP.
% Nt is the number of points that the TPBVP will be discretized into. The
% larger Nt is, the more accurate your solution. However be aware that
% if Nt is too large, the solver may not be able to compute a solution
% using its algorithm.
Nt = 80;
tau = linspace(0,1,Nt)';       % nondimensional time vector
% Create an initial guess of the solution using the MATLAB function
% bvpinit, which requires as inputs the (nondimensional) time vector,
% initial states (or guesses, as applicable), and the guess for the final
% time. The initial guess of the solution is stored in the structure
% solinit.
solinit = bvpinit(tau,yinit,tf_guess);

%--------------------------------------------------------------------------
%% Solution for the NO DRAG and CONSTANT MASS case
%--------------------------------------------------------------------------

% Call bvp4c to solve the TPBVP. Point the solver to the functions
% containing the differential equations and the boundary conditions and
% provide it with the initial guess of the solution.
sol = bvp4c(@ascent_odes_tf, @ascent_bcs_tf, solinit);

% Extract the final time from the solution:
tf = sol.parameters(1);

% Evaluate the solution at all times in the nondimensional time vector tau
% and store the state variables in the matrix Z.
Z = deval(sol,tau);

% Convert back to dimensional time for plotting
time = t0 + tau.*(tf-t0);

% Extract the solution for each state variable from the matrix Z and
% convert them back into dimensional units by multiplying each by their
% respective scaling constants.
x_sol = Z(1,:)*h/1000;
y_sol = Z(2,:)*h/1000;
vx_sol = Z(3,:)*Vc/1000;
```

```
vy_sol = Z(4,:)*Vc/1000;
lambda2_bar_sol = Z(5,:);
lambda3_bar_sol = Z(6,:);
lambda4_bar_sol = Z(7,:);

%% Parameters for VARYING MASS and NO DRAG case
m0 = 117020;           % Initial mass of Titan II rocket (kg)
mdot = (117020-4760)/139;        % Mass flow rate (kg/s)
delta_tf = 115;        % Amount subtracted from final time of constant mass
                       % case

%--------------------------------------------------------------------------
%% Solution for the VARYING MASS and NO DRAG case
%--------------------------------------------------------------------------

% Copy initial guess for the drag solution into a new structure named
% solinit_mass
solinit_mass = solinit;
% Save the time histories of the 7 state and costate variables from the NO
% DRAG, CONSTANT MASS solution in the structure of the initial guess for
% the VARYING MASS, NO DRAG case
solinit_mass.y = Z;
% Save the final time of the NO DRAG, CONSTANT MASS solution and use it as
% the guess for the final time for the VARYING MASS, NO DRAG case. Also
% subtract delta_tf from this guess as described before
solinit_mass.parameters(1) = tf-delta_tf;
% Run bvp4c for the VARYING MASS, NO DRAG
sol_mass = bvp4c(@ascent_odes_tf,@ascent_bcs_tf,solinit_mass);
% Evaluate the solution at all times in the nondimensional time vector tau
% and store the state variables in the matrix Z_mass.
Z_mass = deval(sol_mass,tau);
% Extract the final time from the solution with VARYING MASS, NO DRAG:
tf_mass = sol_mass.parameters(1);
% Convert back to dimensional time for plotting
time_mass = t0+tau*(tf_mass-t0);
% Extract the solution for each state variable from the matrix Z_mass and
% convert them back into dimensional units by multiplying each by their
% respective scaling constants.
x_sol_mass = Z_mass(1,:)*h/1000;
y_sol_mass = Z_mass(2,:)*h/1000;
vx_sol_mass = Z_mass(3,:)*Vc/1000;
vy_sol_mass = Z_mass(4,:)*Vc/1000;
lambda2_bar_sol_mass = Z_mass(5,:);
lambda3_bar_sol_mass = Z_mass(6,:);
lambda4_bar_sol_mass = Z_mass(7,:);

%% Parameters for the VARYING MASS AND DRAG case
CD = .5;          % Drag coefficient
eta = rhoref*CD*A/2;     % Update eta, since CD is now nonzero

%--------------------------------------------------------------------------
%% Solution for the VARYING MASS AND DRAG case
%--------------------------------------------------------------------------

% Copy initial guess for the VARYING MASS AND DRAG solution into a new
% structure named solinit_mass_drag
solinit_mass_drag = solinit_mass;
```

```
% Save the time histories of the 7 state and costate variables from the
% VARYING MASS, NO DRAG solution in the structure of the initial guess for
% the VARYING MASS AND DRAG case
solinit_mass_drag.y = Z_mass;
% Save the final time of the VARYING MASS, NO DRAG solution and use it as
% the guess for the final time for the VARYING MASS AND DRAG case
solinit_mass_drag.parameters(1) = tf_mass;
% Run bvp4c for the drag case
sol_mass_drag = bvp4c(@ascent_odes_tf,@ascent_bcs_tf,solinit_mass_drag);
% Evaluate the solution at all times in the nondimensional time vector tau
% and store the state variables in the matrix Z_mass_drag.
Z_mass_drag = deval(sol_mass_drag,tau);
% Extract the final time from the solution:
tf_mass_drag = sol_mass_drag.parameters(1);
% Convert back to dimensional time for plotting
time_mass_drag = t0+tau*(tf_mass_drag-t0);
% Extract the solution for each state variable from the matrix Z_mass_drag
% and convert them back into dimensional units by multiplying each by their
% respective scaling constants.
x_sol_mass_drag = Z_mass_drag(1,:)*h/1000;
y_sol_mass_drag = Z_mass_drag(2,:)*h/1000;
vx_sol_mass_drag = Z_mass_drag(3,:)*Vc/1000;
vy_sol_mass_drag = Z_mass_drag(4,:)*Vc/1000;
lambda2_bar_sol_mass_drag = Z_mass_drag(5,:);
lambda3_bar_sol_mass_drag = Z_mass_drag(6,:);
lambda4_bar_sol_mass_drag = Z_mass_drag(7,:);

%% Plot the solutions
figure(1)
subplot(221)
plot(time_mass_drag,x_sol_mass_drag,'k')
xlabel('Time [s]','fontsize',14)
ylabel('x [km]','fontsize',14)
xlim([t0 tf_mass_drag])

subplot(222)
plot(time_mass_drag,y_sol_mass_drag,'k')
xlabel('Time [s]','fontsize',14)
ylabel('y [km]','fontsize',14)
xlim([t0 tf_mass_drag])

subplot(223)
plot(time_mass_drag,vx_sol_mass_drag,'k')
xlabel('Time [s]','fontsize',14)
ylabel('V_x [km/s]','fontsize',14)
xlim([t0 tf_mass_drag])

subplot(224)
plot(time_mass_drag,vy_sol_mass_drag,'k')
xlabel('Time [s]','fontsize',14)
ylabel('V_y [km/s]','fontsize',14)
xlim([t0 tf_mass_drag])

figure(2)
plot(time_mass_drag,...
atand(lambda4_bar_sol_mass_drag./lambda3_bar_sol_mass_drag),'k')
xlabel('Time [s]','fontsize',14) ylabel('\theta
```

```
[deg]','fontsize',14) xlim([t0 tf_mass_drag])

figure(3)
plot(x_sol_mass_drag,y_sol_mass_drag,'k')
xlabel('Downrange Position, x [km]','fontsize',14)
ylabel('Altitude, y [km]','fontsize',14)
xlim([x_sol_mass_drag(1) x_sol_mass_drag(end)])
ylim([0 200])


function dX_dtau = ascent_odes_tf(tau,X,tf)
%
% State and Costate Differential Equation Function for the Flat-Earth
% Optimal Ascent Problem with Atmospheric Drag and linearly time-varying
% mass
%
%
% The independent variable here is the nondimensional time, tau, the state
% vector is X, and the final time, tf, is an unknown parameter that must
% also be passed to the DE function.

% Note that the state vector X has components
% X(1) = xbar, horizontal component of position
% X(2) = ybar, vertical component of position
% X(3) = Vxbar, horizontal component of velocity
% X(4) = Vybar, vertical component of velocity
% X(5) = lambda_2_bar, first costate
% X(6) = lambda_3_bar, second costate
% X(7) = lambda_4_bar, third costate

% pass in values of relevant constants as global variables
global g_accel Vc h eta beta f m0 mdot

m = m0-abs(mdot)*tau*tf;

% State and Costate differential equations in terms of d/dt:
xbardot = X(3)*Vc/h; ybardot = X(4)*Vc/h; Vxbardot =
(f/Vc*(-X(6)/sqrt(X(6)^2+X(7)^2)) ...
-eta*exp(-X(2)*beta)*X(3)*sqrt(X(3)^2+X(4)^2)*Vc)/m;
Vybardot = (f/Vc*(-X(7)/sqrt(X(6)^2+X(7)^2)) ...
-eta*exp(-X(2)*beta)*X(4)*sqrt(X(3)^2+X(4)^2)*Vc)/m-g_accel/Vc;
if sqrt(X(3)^2+X(4)^2) == 0
    lambda_2_bar = 0;
    lambda_3_bar = 0;
    lambda_4_bar = -X(5)*Vc/h;
else
  lambda_2_bar = ...
-(X(6)*X(3)+X(7)*X(4))*eta*beta*exp(-X(2)*beta)*sqrt(X(3)^2+X(4)^2)*Vc/m;
  lambda_3_bar = eta*exp(-X(2)*beta)*Vc*(X(6)*(2*X(3)^2+X(4)^2) ...
    + X(7)*X(3)*X(4))/sqrt(X(3)^2+X(4)^2)/m;
  lambda_4_bar = -X(5)*Vc/h+eta*exp(-X(2)*beta)*Vc*(X(7)*(X(3)^2 ...
                 +2*X(4)^2) ...
                 +X(6)*X(3)*X(4))/sqrt(X(3)^2+X(4)^2)/m;
end

% Nondimensionalize time (with tau = t/tf and d/dtau = tf*d/dt). We must
% multiply each differential equation by tf to convert our derivatives from
% d/dt to d/dtau.
```

```
dX_dtau = ...
tf*[xbardot; ybardot; Vxbardot; Vybardot; ...
    lambda_2_bar; lambda_3_bar; lambda_4_bar];
return

function PSI = ascent_bcs_tf(Y0,Yf,tf)
%
% Boundary Condition Function for the Flat-Earth Optimal Ascent Problem
% with Atmospheric Drag and linearly varying mass
%

% pass in values of boundary conditions and other constants as
% global variables
global xbar0 ybar0 Vxbar0 Vybar0 ybarf Vxbarf Vybarf Vc f
global eta beta g_accel m0 mdot

% the final mass is the same as the initial mass since it was assumed
% constant
mf = m0-abs(mdot)*tf;

% Create a column vector with the 7 boundary conditions on the state &
% costate variables (the 8th boundary condition, t0 = 0, is not used).
PSI =   [Y0(1) - xbar0;              % Initial Condition
        Y0(2) - ybar0;               % Initial Condition
        Y0(3) - Vxbar0;              % Initial Condition
        Y0(4) - Vybar0;              % Initial Condition
        Yf(2) - ybarf;               % Final Condition
        Yf(3) - Vxbarf;              % Final Condition
        Yf(4) - Vybarf;              % Final Condition
        % Final Value of Hamiltonian:
       (-sqrt(Yf(6)^2+Yf(7)^2)*f/mf/Vc ...
-(Yf(6)*Yf(3))*eta*exp(-beta)*sqrt(Yf(3)^2)*Vc/mf-Yf(7)*g_accel/Vc)*tf ...
+1];
return
```

C. Optimal Low-Thrust LEO to GEO Circular Orbit Transfer

Contributors: Peter J. Edelman and Kshitij Mall

We want to find the optimal transfer trajectory to get from a 300 km altitude, low-Earth orbit (LEO) to a 35,786 km altitude, geosynchronous-Earth orbit (GEO). In this problem we assume a planar transfer with a spherical, non-rotating Earth, and neglect other external forces due to the Moon, the Sun, and solar radiation pressure, etc. Additionally we assume a constant thrust magnitude and a constant mass flow rate. An illustration of the defined variables is shown in Fig. C1.

C.1. Optimization Problem

Following Bryson and Ho [1975], the optimization problem is posed as follows: Maximize:

$$J = r(t_f) \tag{C1}$$

subject to:

$$\dot{r} = u \tag{C2a}$$

$$\dot{u} = \frac{v^2}{r} - \frac{\mu}{r^2} + \frac{T \sin \alpha}{m_o - |\dot{m}|t} \tag{C2b}$$

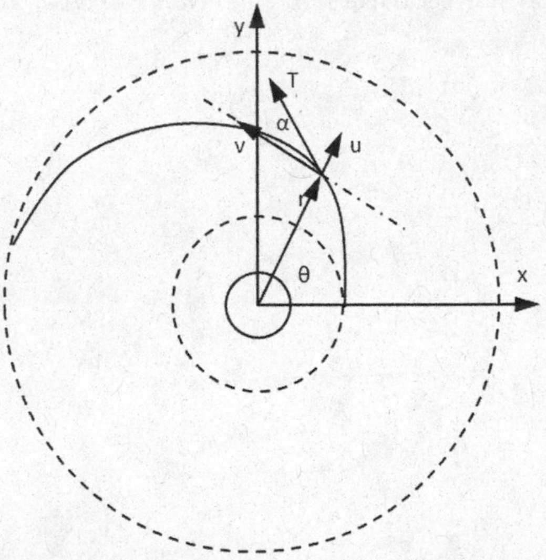

Figure C1. Schematic of a low-thrust orbit transfer from LEO to a maximum radius circular orbit in given time, t_f

$$\dot{v} = -\frac{uv}{r} + \frac{T\cos\alpha}{m_o - |\dot{m}|t} \tag{C2c}$$

$$\dot{\theta} = \frac{v}{r} \tag{C2d}$$

where the variables and constants are defined as:

r = Radius, km
u = Radial component of velocity, km/s
v = Tangential component of velocity, km/s
θ = Longitudinal angle, rad
α = Thrust angle, rad
T = Thrust magnitude, N
m_o = Initial vehicle mass, kg
$\dot{m}$ = Mass flow rate, kg/s
μ = Gravitational parameter, km^3/s^2

Note: The optimization problem is posed as a maximum radius transfer rather than a minimum time transfer with known initial and final orbit radii (due to numerical difficulties associated with the latter). We obtain the time of flight for the LEO to GEO transfer via iteration, by checking to see if the final orbit radius achieved matches the radius at GEO at the end of each iteration.

C.2. Scaling the Equations of Motion

We scale the state variables to remain close to unity to improve numerical convergence. Using the initial orbit radius and initial velocity as reference values, and normalizing the time by the final time, we define the new state variables as:

$$\bar{r} = \frac{r}{r_o}, \quad \bar{u} = \frac{u}{v_o}, \quad \bar{v} = \frac{v}{v_o}, \quad \tau = \frac{t}{t_f} \tag{C3}$$

Applying the chain rule to change variables results in the following new nondimensional equations of motion:

For $\bar{r}$:

$$\frac{d\bar{r}}{d\tau} = \frac{d(r/r_o)}{dt}\frac{dt}{d\tau} = \frac{dr}{dt}\frac{t_f}{r_o} = u\frac{t_f}{r_o} \tag{C4}$$

or,

$$\frac{d\bar{r}}{d\tau} = \bar{u}\eta \tag{C5}$$

where

$$\eta = \frac{v_o t_f}{r_o} \tag{C6}$$

For $\bar{u}$:

$$\frac{d\bar{u}}{d\tau} = \frac{d(u/v_o)}{dt}\frac{dt}{d\tau} = \frac{du}{dt}\frac{t_f}{v_o} = \left(\frac{v^2}{r} - \frac{\mu}{r^2} + \frac{T\sin\alpha}{m_o - |\dot{m}|t}\right)\frac{t_f}{v_o} \tag{C7}$$

or,

$$\frac{d\bar{u}}{d\tau} = \left(\frac{\bar{v}^2}{\bar{r}} - \frac{1}{\bar{r}^2}\right)\eta + \frac{\bar{T}\sin\alpha}{m_o - |\dot{m}|\tau\, t_f} \tag{C8}$$

where

$$\bar{T} = \frac{Tt_f}{v_o} \tag{C9}$$

For $\bar{v}$:

$$\frac{d\bar{v}}{d\tau} = \frac{d(v/v_o)}{dt}\frac{dt}{d\tau} = \frac{dv}{dt}\frac{t_f}{v_o} = \left(-\frac{uv}{r} + \frac{T\cos\alpha}{m_o - |\dot{m}|t}\right)\frac{t_f}{v_o} \tag{C10}$$

or,

$$\frac{d\bar{v}}{d\tau} = -\frac{\bar{u}\bar{v}}{\bar{r}}\eta + \frac{\bar{T}\cos\alpha}{m_o - |\dot{m}|\tau\, t_f} \tag{C11}$$

For θ:

$$\frac{d\theta}{d\tau} = \frac{d\theta}{dt}\frac{dt}{d\tau} = \frac{d\theta}{dt}t_f = \frac{v}{r}t_f \tag{C12}$$

or,

$$\frac{d\theta}{d\tau} = \frac{\bar{v}}{\bar{r}}\eta \tag{C13}$$

In summary, our newly scaled equations of motion are:

$$\frac{d\bar{r}}{d\tau} = \bar{u}\eta \tag{C14a}$$

$$\frac{d\bar{u}}{d\tau} = \left(\frac{\bar{v}^2}{\bar{r}} - \frac{1}{\bar{r}^2}\right)\eta + \frac{\bar{T}\sin\alpha}{m_o - |\dot{m}|\tau\, t_f} \tag{C14b}$$

$$\frac{d\bar{v}}{d\tau} = -\frac{\bar{u}\bar{v}}{\bar{r}}\eta + \frac{\bar{T}\cos\alpha}{m_o - |\dot{m}|\tau\, t_f} \tag{C14c}$$

$$\frac{d\theta}{d\tau} = \frac{\bar{v}}{\bar{r}}\eta \tag{C14d}$$

All of Eqs. (C14) are used to form the TPBVP from the Euler-Lagrange theorem and are used in the numerical simulation.

C.3. Applying the Euler-Lagrange Theorem

The Hamiltonian for our problem is:

$$H = \lambda_r(\bar{u}\eta) + \lambda_{\bar{u}}\left[\left(\frac{\bar{v}^2}{\bar{r}} - \frac{1}{\bar{r}^2}\right)\eta + \frac{\bar{T}\sin\alpha}{m_o - |\dot{m}|\tau\, t_f}\right] + \lambda_{\bar{v}}\left[-\frac{\bar{u}\bar{v}}{\bar{r}}\eta + \frac{\bar{T}\cos\alpha}{m_o - |\dot{m}|\tau\, t_f}\right] + \lambda_\theta\left(\frac{\bar{v}}{\bar{r}}\eta\right) \tag{C15}$$

where the λs are the costates. The Euler-Lagrange equations yield:

$$\frac{d\lambda_{\bar{r}}}{d\tau} = -\frac{\partial H}{\partial \bar{r}} = \lambda_{\bar{u}}\left(\frac{\bar{v}^2}{\bar{r}^2} - \frac{2}{\bar{r}^3}\right)\eta - \lambda_{\bar{v}}\frac{\bar{u}\bar{v}}{\bar{r}^2}\eta + \lambda_\theta\left(\frac{\bar{v}}{\bar{r}^2}\right)\eta \tag{C16a}$$

$$\frac{d\lambda_{\bar{u}}}{d\tau} = -\frac{\partial H}{\partial \bar{u}} = -\lambda_{\bar{r}}\eta + \lambda_{\bar{v}}\frac{\bar{v}}{\bar{r}} \tag{C16b}$$

$$\frac{d\lambda_{\bar{v}}}{d\tau} = -\frac{\partial H}{\partial \bar{v}} = -\lambda_{\bar{u}}\frac{2\bar{v}}{\bar{r}}\eta + \lambda_{\bar{v}}\frac{\bar{u}}{\bar{r}}\eta - \lambda_\theta\frac{1}{\bar{r}}\eta \tag{C16c}$$

$$\frac{d\lambda_\theta}{d\tau} = -\frac{\partial H}{\partial \theta} = 0 \tag{C16d}$$

We immediately note that Eq. (C16d) implies that λ_θ is constant throughout the trajectory.

The control variable, α, is an angle allowed to have the full range $[0, 2\pi]$, thus we assume that it is unbounded.

The optimal control law is then found using the procedure discussed after Eq. (4.66) by writing the thrust terms in the Hamiltonian of Eq. (C15) as $\bar{T}\boldsymbol{\mu}^T\boldsymbol{\sigma}/(m_o - |\dot{m}|\tau\, t_f)$ where

$$\boldsymbol{\mu} = \begin{bmatrix} \lambda_{\bar{u}} \\ \lambda_{\bar{v}} \end{bmatrix} \tag{C17}$$

and

$$\boldsymbol{\sigma} = \begin{bmatrix} \sin\alpha \\ \cos\alpha \end{bmatrix} \tag{C18}$$

where $\boldsymbol{\sigma}$ is a unit vector. Then H is maximized by choosing

$$\boldsymbol{\sigma} = \frac{\boldsymbol{\mu}}{\mu} \tag{C19}$$

which yields

$$\sin\alpha = \frac{\lambda_{\bar{u}}}{\sqrt{\lambda_{\bar{u}} + \lambda_{\bar{v}}}} \tag{C20a}$$

$$\cos\alpha = \frac{\lambda_{\bar{v}}}{\sqrt{\lambda_{\bar{u}} + \lambda_{\bar{v}}}} \tag{C20b}$$

C.4. Boundary Conditions and the TPBVP

We require 2n + 2 boundary conditions, where n is the number of process equations, for a well-defined TPBVP. For our problem, we require 2(4) + 2 = 10, and currently know 8. The known boundary conditions are:
Initial Conditions:

$$t_o = 0 \text{ s},\ r_o = 6{,}678 \text{ km},\ u_o = 0 \text{ km/s},\ v_o = \sqrt{\mu/r_o} = 7.726 \text{ km/s},\ \theta_o = 0 \text{ rad} \tag{C21a}$$

Final Conditions:

$$t_f = \text{known},\ u_f = 0 \text{ km/s},\ \Psi_1 = v_f - \sqrt{\mu/r_f} = 0$$

where we have used $\mu = 3.986 \times 10^5$ km^3/s^2. We apply the transversality condition to pose a well-defined TPBVP:

$$\left[Hdt - \boldsymbol{\lambda}^T d\boldsymbol{x}\right]\Big|_{t_f} + d\phi = 0 \tag{C22}$$

subject to $d\boldsymbol{\Psi} = 0$. Expanding the nonzero terms in Eq. (C22) while noting that we have a Mayer form for our optimization problem, where $\phi = r_f$:

$$(1 - \lambda_{r_f})dr_f - \lambda_{v_f} dv_{v_f} - \lambda_{\theta_f} d\theta_f = 0 \tag{C23}$$

with

$$d\Psi_1 = dv_f + \frac{1}{2}\sqrt{\frac{\mu}{r_f^3}}dr_f = 0 \tag{C24}$$

or,

$$dv_f = -\frac{1}{2}\sqrt{\frac{\mu}{r_f^3}}dr_f \tag{C25}$$

Since the differentials of the final values of the states are nonzero, the coefficients are chosen such that Eq. (C23) is satisfied. Upon substituting Eq. (C25) into Eq. (C23), we get the following new boundary conditions:

$$1 - \lambda_{r_f} + \frac{1}{2}\lambda_{v_f}\sqrt{\frac{\mu}{r_f^3}} = 0 \tag{C26a}$$

$$\lambda_{\theta_f} = 0 \tag{C27}$$

We now have enough boundary conditions for a well-defined TPBVP. First we scale all the boundary conditions so they are consistent with the differential equations. Upon doing so we obtain:

$$\tau_o = 0,\ \bar{r}_o = 1,\ \bar{u}_o = 0,\ \bar{v}_o = 1,\ \theta_o = 0 \tag{C27a}$$

$$\tau_f = 1,\ \bar{u}_f = 0,\ \bar{v}_f = \sqrt{\frac{1}{\bar{r}_f}},\ 1 - \lambda_{\bar{r}_f} + \frac{1}{2}\lambda_{\bar{v}_f}\sqrt{\frac{1}{\bar{r}_f^3}} = 0,\ \lambda_{\theta_f} = 0 \tag{C27b}$$

We note that the last boundary condition in Eq. (C27b) along with Eq. (C16d) readily implies that $\lambda_{\theta_f} = 0$ for the entire trajectory, thus we can remove it from the problem entirely. Substituting Eqs. (C20a) and (C20b) into Eqs. (C16) and reiterating all differential equations and boundary conditions gives the well-defined TPBVP:

$$\frac{d\bar{r}}{d\tau} = \bar{u}\eta \tag{C28a}$$

$$\frac{d\bar{u}}{d\tau} = \left(\frac{\bar{v}^2}{\bar{r}} - \frac{1}{\bar{r}^2}\right)\eta + \frac{\bar{T}}{m_o - |\dot{m}|\tau\ t_f}\left(\frac{\lambda_{\bar{u}}}{\sqrt{\lambda_{\bar{u}}^2 + \lambda_{\bar{v}}^2}}\right) \tag{C28b}$$

$$\frac{d\bar{v}}{d\tau} = -\frac{\bar{u}\bar{v}}{\bar{r}}\eta + \frac{\bar{T}}{m_o - |\dot{m}|\tau\ t_f}\left(\frac{\lambda_{\bar{v}}}{\sqrt{\lambda_{\bar{u}}^2 + \lambda_{\bar{v}}^2}}\right) \tag{C28c}$$

$$\frac{d\theta}{d\tau} = \frac{\bar{v}}{\bar{r}}\eta \tag{C28d}$$

$$\frac{d\lambda_{\bar{r}}}{d\tau} = \lambda_{\bar{u}}\left(\frac{\bar{v}^2}{\bar{r}^2} - \frac{2}{\bar{r}^3}\right)\eta - \lambda_{\bar{v}}\frac{\bar{u}\bar{v}}{\bar{r}^2}\eta + \lambda_\theta\left(\frac{\bar{v}}{\bar{r}^2}\right)\eta \tag{C28e}$$

$$\frac{d\lambda_{\bar{u}}}{d\tau} = -\lambda_{\bar{r}}\eta + \lambda_{\bar{v}}\frac{\bar{v}}{\bar{r}} \tag{C28f}$$

$$\frac{d\lambda_{\bar{v}}}{d\tau} = -\lambda_{\bar{u}}\frac{2\bar{v}}{\bar{r}}\eta + \lambda_{\bar{v}}\frac{\bar{u}}{\bar{r}}\eta - \lambda_\theta\frac{1}{\bar{r}}\eta \tag{C28g}$$

$$\tau_o = 0,\ \bar{r}_o = 1,\ \bar{u}_o = 0,\ \bar{v}_o = 1,\ \theta_o = 0 \tag{C29a}$$

$$\tau_f = 1,\ \bar{u}_f = 0,\ \bar{v}_f = \sqrt{\frac{1}{\bar{r}_f}},\ 1 - \lambda_{\bar{r}_f} + \frac{1}{2}\lambda_{\bar{v}_f}\sqrt{\frac{1}{\bar{r}_f^3}} = 0,\ \lambda_{\theta_f} = 0 \tag{C29b}$$

C.5. Results

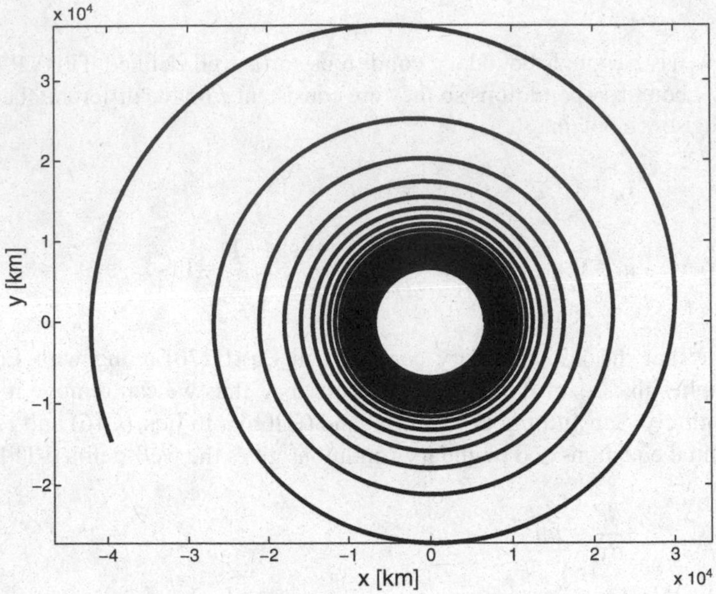

Figure C2. Low-thrust LEO to GEO spiral

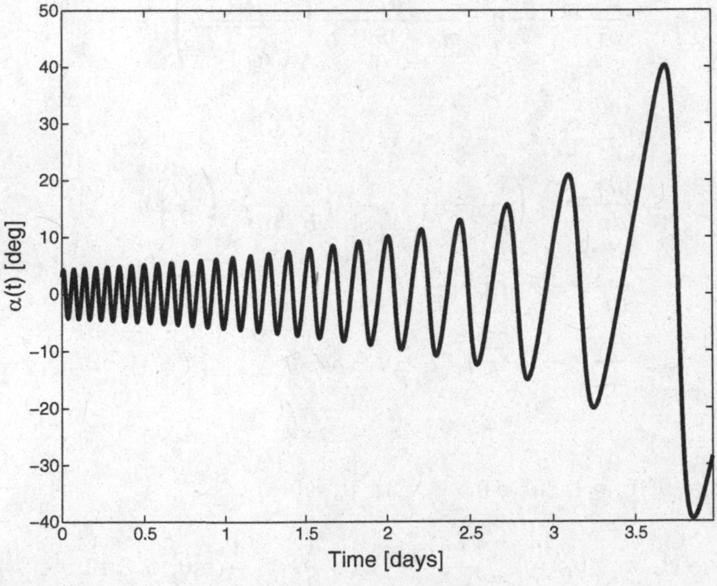

Figure C3. Control history of the thrust angle plotted as a function of time

Discussion of Results

The trajectory resembles that of a spiral with increasing width between each revolution. The increase in width can be attributed to the fact that as the vehicle gets farther from the central body, there exists a weaker gravitational force, enabling velocity changes in the radial direction to cost less propellant. The TOF from LEO to GEO with a constant thrust of 20 Newtons is approximately 4.0 days. Current technology allows low-thrust capabilities much smaller than what we used here, meaning there would be many more revolutions as well as a highly increased TOF. We performed this simulation with unrealistic parameters in order to decrease computational time as well as to give a better visualization of the trajectory.

The thrust angle (the control) oscillates about the zero position with an increasing amplitude envelope. One period of the control history corresponds to one revolution around the Earth, and the period gets larger as the vehicle gets farther away. This envelope would theoretically have an upper limit, which is when the vehicle approaches escape speed.

C.6. MATLAB Code

Executable File

```
% LEO to GEO Circular Orbit Radius Transfer Problem with MATLAB's bvp4c
%
% by Peter J. Edelman and Kshitij Mall
%
% This script uses MATLAB's bvp4c to solve the problem of finding the
% optimal transfer trajectory for launch from a 300 kilometer altitude LEO
% to a 35,786 kilometer altitude GEO orbit.
% A linearly time-varying mass is assumed.
% In addition to this script, we use two functions:
% one to provide the differential equations and another that gives the
% boundary conditions:
%
% This file: Max_Xfer.m
% State and Costate Equations: transfer_odes.m
% Boundary Conditions: transfer_bcs.m
%
% Clear figure, command and workspace histories
clear all; close all; clc;

% Define parameters of the problem

% pass these parameters to the DE and BC functions
global eta mdot Tbar tf m0
h0 = 300;                % Initial altitude [km]
rearth = 6378;           % Mean radius of the Earth [km]
mu = 3.986004418e5;      % Gravitational parameter of Earth [km^3/s^2]
T = 20;                  % Thrust of spacecraft [N]
Isp = 6000;              % Specific Impulse [sec]
mdot = T/Isp/9.80665;    % Mass flow rate [kg/s]
m0 = 1500;               % Initial spacecraft mass [kg]

%---------------------------------------------------------------------------
%% Boundary Conditions
%---------------------------------------------------------------------------

global r0bar u0bar v0bar theta0 ufbar % pass these BCs to the BC function

% Initial conditions
% Start from 300 km altitude circular orbit
% All Boundary Conditions are nondimensional
r0 = rearth+h0;          % Initial circular orbit radius [km]
u0 = 0;                  % Initial radial component of velocity [km/s]
v0 = sqrt(mu/r0);        % Initial circular orbit speed [km/s]
theta0 = 0;              % Initial longitude [rad]
t0 = 0;                  % Initial time [sec]

% Final conditions
uf = 0;                  % Final radial component of velocity [km/s]
days = .1;               % Initial amount of days to thrust. (0.1 days)
tf = 24*3600*days;       % Days in seconds

%---------------------------------------------------------------------------
%% Nondimensionalize Boundary Conditions and Important Parameters
%---------------------------------------------------------------------------
```

```
% Scaling constant eta
eta = v0*tf/r0;

% Scaled thrust Tbar
Tbar = T*tf/(v0*1000);

% Scaled Initial Conditions
r0bar = r0/r0;
u0bar = u0/v0;
v0bar = v0/v0;

% Scaled Final Conditions
ufbar = uf/v0;

%--------------------------------------------------------------------------
%% Solve the TPBVP
%--------------------------------------------------------------------------

% Nt is the number of points that the TPBVP will be discretized into. First
% 1000 mesh points will be used at each iteration to reduce computational
% time. After the very last iteration, the solution will be spline
% interpolated to 5000 mesh points and solved again, so as to ensure smooth
% plots.
Nt = 1000;
tau = linspace(0,1,Nt);     % nondimensional time vector
% list initial conditions in yinit
yinit = [r0bar u0bar v0bar 0 0 -1 0];
% Create an initial guess of the solution using the MATLAB function
% bvpinit, which requires as inputs the (nondimensional) time vector,
% initial states (or guesses, as applicable).
% The initial guess of the solution is stored in the structure solinit.
solinit = bvpinit(tau,yinit);
% Call bvp4c to solve the TPBVP. Point the solver to the functions
% containing the differential equations and the boundary conditions and
% provide it with the initial guess of the solution.
sol = bvp4c(@transfer_odes,@transfer_bcs,solinit);
% Evaluate the solution at all times in the nondimensional time vector tau
% and store the state variables in the matrix Z.
Z = deval(sol,tau);
% Implement for-loop to increment final time by 0.1 days each iteration
for days = .2:.1:2
    % update final time
    tf = 24*3600*days;
    % update Tbar and eta, which are functions of tf
    Tbar = T*tf/(v0*1000);
    eta = v0*tf/r0;
    % store the previous solution as the guess for the next iteration
    solinit.y = Z;
    solinit.x = tau;
    % solve the TPBVP
    sol = bvp4c(@transfer_odes,@transfer_bcs,solinit);
    % store the newly found solution in a matrix 'Z'
    Z = deval(sol,tau);
end
% for-loop to increment final time by 0.05 days each iteration for days 2 -
% 3.95. Due to numerical sensitivity, the step size had to be reduced.
for days = 2.05:.05:3.95
```

```
    % update final time
    tf = 24*3600*days;
    % update Tbar and eta, which are functions of tf
    Tbar = T*tf/(v0*1000);
    eta = v0*tf/r0;
    % store the previous solution as the guess for the next iteration
    solinit.y = Z;
    solinit.x = tau;
    % solve the TPBVP
    sol = bvp4c(@transfer_odes,@transfer_bcs,solinit);
    % store the newly found solution in a matrix 'Z'
    Z = deval(sol,tau);
end
% Final iteration to get to GEO (still 1000 mesh points)
days = 3.97152;
% update final time
tf = 24*3600*days;
% update Tbar and eta, which are functions of tf
Tbar = T*tf/(v0*1000);
eta = v0*tf/r0;
% store the previous solution as the guess for the next iteration
solinit.y = Z;
solinit.x = tau;
% solve the TPBVP
sol = bvp4c(@transfer_odes,@transfer_bcs,solinit);
% store the newly found solution in a matrix 'Z'
Z = deval(sol,tau);

%% Interpolate with spline to get 5000 mesh points
% Create 5000 point time vector
tau2 = linspace(0,1,5*Nt);
% Interpolate solution with spline using the new time vector
for q = 1:7
    Z2(q,:) = spline(tau,Z(q,:),tau2);
end
% Store the new solution in tau and Z
tau = tau2;
Z = Z2;

%% Solve TPBVP again with the 5000 mesh points for "smooth" answer
solinit.y = Z;
solinit.x = tau;
sol = bvp4c(@transfer_odes,@transfer_bcs,solinit);
% store the newly found solution in a matrix 'Z'
Z = deval(sol,tau);

% Convert back to dimensional time for plotting
time = t0+tau*(tf-t0);

% Extract the solution for each state variable from the matrix Z and
% convert them back into dimensional units by multiplying each by their
% respective scaling constants.
r_sol = Z(1,:)*r0;              % Radius [km]
u_sol = Z(2,:)*v0;              % Radial component of velocity [km/s]
v_sol = Z(3,:)*v0;              % Tangential component of velocity [km/s]
theta_sol = Z(4,:);             % Angle between x-axis and radius vector [rad]
lambda_rbar_sol = Z(5,:);       % 1st costate
```

```
lambda_ubar_sol = Z(6,:);    % 2nd costate
lambda_vbar_sol = Z(7,:);    % 3rd costate

% Displays final value of orbit radius
final_radius = r_sol(end)

%-----------------------------------------------------------------------------
%% Plots
%-----------------------------------------------------------------------------
figure(1)
plot(r_sol.*cos(theta_sol), r_sol.*sin(theta_sol),'k')
xlabel('x-direction [km]')
ylabel('y-direction [km]')
axis equal

% Plot the steering angle as a function of time
figure(2)
plot(time/(3600*24), atand(lambda_ubar_sol./lambda_vbar_sol),'k')
xlabel('Time [days]')
ylabel('\theta(t) [deg]')
xlim([time(1) time(end)]/(3600*24))

function dX_dtau = transfer_odes(tau,x)
%
% State and Costate Differential Equation Function for the Maximal Radius
% Low-Thrust Orbit Transfer
%
%
% The independent variable here is the nondimensional time, tau, and the
% state vector is X

% Note that the state vector X has components
% x(1) = rbar, nondimensional radius from center of Earth
% x(2) = ubar, nondimensional radial component of velocity
% x(3) = vbar, nondimensional tangential component of velocity
% x(4) = theta, angle between inertial x-axis and radius vector
% x(5) = lambda_r_bar, first costate
% x(6) = lambda_u_bar, second costate
% x(7) = lambda_v_bar, third costate

% pass in values of relevant constants as global variables
global eta mdot Tbar tf m0

m = m0-abs(mdot)*tau*tf;

drbar_dtau = x(2)*eta;
dubar_dtau = x(3)^2/x(1)*eta-eta/x(1)^2 ...
-Tbar/m*(x(6)/sqrt(x(6)^2+x(7)^2));
dvbar_dtau = -x(2)*x(3)/x(1)*eta-Tbar/m*(x(7)/sqrt(x(6)^2+x(7)^2));
dtheta_dtau = x(3)/x(1)*eta;
dlambda_r_bar_dtau = x(6)*(x(3)^2/x(1)^2*eta-2*eta/x(1)^3) ...
-x(7)*x(2)*x(3)/x(1)^2*eta;
dlambda_u_bar_dtau = -x(5)*eta+x(7)*x(3)/x(1)*eta;
dlambda_v_bar_dtau = -x(6)*2*x(3)/x(1)*eta+x(7)*x(2)/x(1)*eta;

dX_dtau = [drbar_dtau; dubar_dtau; dvbar_dtau; dtheta_dtau; ...
dlambda_r_bar_dtau; dlambda_u_bar_dtau; dlambda_v_bar_dtau];
return
```

```
function PSI = transfer_bcs(Y0,Yf)
%
% Boundary Condition Function for the Maximal Radius Low-Thrust Orbit
% Transfer
%

% pass in values of boundary conditions and other constants as global variables
global r0bar u0bar v0bar ufbar theta0

% Create a column vector with the 7 boundary conditions on the state &
% costate variables
PSI = [Y0(1)-r0bar                        % Initial Condition
    Y0(2)-u0bar                           % Initial Condition
    Y0(3)-v0bar                           % Initial Condition
    Y0(4)-theta0                          % Initial Condition
    Yf(2)-ufbar                           % Final Condition
    Yf(3)-sqrt(1/Yf(1))                   % Final Condition
   -Yf(5)+1/2*Yf(7)/Yf(1)^(3/2)-1];% Final Condition
return
```

Reference

A.E. Bryson Jr., Y.C. Ho, *Applied Optimal Control* (Hemisphere Publishing, Washington, D.C., 1975)

D. Curious Quotations

In case you don't find the concepts of the calculus of variations incipiently clear, or if after much study you now have doubts about your ideas, your intelligence, or your sanity, perhaps the following quotes from great minds and renowned authors will assuage your concerns—or at least put you in a better mood!

"A reader meeting the calculus of variations for the first time is likely to feel at this stage that a great many things are not completely clear. This is normal. We are just getting started. To feel otherwise is either a mark of genius or indication of real trouble."

George M. Ewing

"Many of the treatises of today are formulated in a half-mystical language, as though to impress the reader that he is in the permanent presence of a superman. The present book is conceived in a humble spirit and is written for humble people."

Cornelius Lanczos

"At the age of 12–16 ... I had the good fortune of hitting on books which were not too particular in their logical rigour, but which made up for this by permitting the main thoughts to stand out clearly and synoptically."

Albert Einstein

"... the application of this condition to the problems with which we shall be concerned is difficult to carry through and has, as yet (1962), received very little attention. The consequences of this additional necessary condition (first considered by Jacobi) will not therefore be explored in this book."

D.F. Lawden

"The vague, mechanical 'δ method' is avoided throughout. Thus, while no advantage is taken of a sometimes convenient shorthand tactic, there is eliminated a source of confusion which often grips the careful student when confronted with its use."

Robert Weinstock

"...a theory allowing for a larger class of admissible functions—namely, Lebesgue measurable ones—has been devised. Such a theory is beyond the scope of this book, which is devoted to what is sometimes dubbed the 'naive' theory."

George Leitmann

"Thus the main novelty arising from the discontinuities in $f(t,x)$ is that while the usual formulas of differential and integral calculus are valid, there is an occasional warning that certain equalities may hold only 'almost everywhere.' The reader who is willing to grant the correctness of this extension of the calculus can proceed with the motto, 'Damn the null sets, full differentiability ahead.' "

E. B. Lee and L. Markus

"This is a minor revision of the original 1969 publication. Some false statements have been changed."

George M. Ewing

"No attempt is made to treat problems of sufficiency or existence; no consideration is taken of the 'second variation' or of the conditions of Legendre, Jacobi, and Weierstrass. Besides being outside the scope of this book, these matters are excellently treated in the volumes of Bolza and Bliss . . ."

Robert Weinstock

"You want proof? You can't handle the proof!"

I. Michael Ross

"There is no discipline in which more correct results can be obtained by incorrect means than in the calculus of variations."

Attributed to Magnus R. Hestenes by George Leitmann

Bibliography (Books)

N.I. Akhiezer, *The Calculus of Variations* (Trans. by A. H. Fink) (Blaisdell Publishing, New York, 1962)

J.D. Anderson Jr., *Introduction to Flight*, 4th edn. (McGraw Hill, New York, 1999)

M. Athans, P.L. Falb, *Optimal Control: An Introduction to the Theory and Its Applications* (Dover, New York, 1966)

R. Barrière, *Optimal Control Theory: A Course in Automatic Control Theory* (W. B. Saunders, Philadelphia, 1967). Stresses mathematical facts and theory

R.H. Battin, *An Introduction to the Mathematics and Methods of Astrodynamics*. Rev. edn. (American Institute of Aeronautics and Astronautics, Reston, 1999). A significant work in astrodynamics. The historical notes are a delight

E.T. Bell, *Men of Mathematics* (Simon and Schuster, New York, 1965). Don't let the dated title turn you off from this inspiring and entertaining book on the lives of mathematicians from Zeno to Poincaré. A few women are mentioned including Weierstrass's favorite pupil, Sonja Kowalewski who won fame for her work on rigid body motion, and Emmy Noether who was much admired by Einstein for her theorem on symmetry and conservation laws

D.J. Bell, D.H. Jacobson, *Singular Optimal Control Problems* (Academic, New York, 1975)

R. Bellman, *Dynamic Programming* (Princeton University Press, Princeton, 1957)

R. Bellman, *Adaptive Control Processes: A Guided Tour* (Princeton University Press, Princeton, 1961)

R. Bellman, S.E. Dreyfus, *Applied Dynamic Programming* (Princeton University Press, Princeton, 1962)

R. Bellman (ed.), *Mathematical Optimization Techniques* (University of California Press, Los Angeles, 1963)

R. Bellman, *Introduction to the Mathematical Theory of Control Processes, Volume I, Linear and Quadratic Criteria* (Academic, New York, 1967)

R. Bellman, *Introduction to the Mathematical Theory of Control Processes, Volume II, Nonlinear Processes* (Academic, New York, 1971)

J.Z. Ben-Asher, *Optimal Control Theory with Aerospace Applications* (American Institute of Aeronautics and Astronautics, Reston, 2010)

L.D. Berkovitz, *Optimal Control Theory* (Springer, New York, 1974). Written for mathematicians rather than engineers; provides a mathematically precise statement and proof of the Minimum Principle

J.T. Betts, *Practical Methods for Optimal Control Using Nonlinear Programming*. SIAM Advances in Design and Control (Society for Industrial and Applied Mathematics, Philadelphia, 2001)

G.A. Bliss, *Calculus of Variations* (Open Court Publishing, Chicago, 1925)

G.A. Bliss, *Lectures on the Calculus of Variations*. Phoenix Science Series (The University of Chicago Press, Chicago, 1968). One of the classic references, the other being Bolza [1961]. These

J.M. Longuski et al., *Optimal Control with Aerospace Applications*,
Space Technology Library 32, DOI 10.1007/978-1-4614-8945-0,

references are written by mathematicians, for mathematicians and are not easy reading for the practicing engineer or the engineering student who is learning the material for the first time

V.G. Boltyanskii, *Mathematical Methods of Optimal Control* (Trans. from the Russian by K.N. Trirogoff and I. Tarnove) (Holt, Rinehart and Winston, New York, 1971)

O. Bolza, *Lectures on the Calculus of Variations* (Dover, New York, 1961). See comment on Bliss above

C.D. Brown, *Spacecraft Mission Design*, 2nd edn. (American Institute of Aeronautics and Astronautics, Reston, 1998)

A.E. Bryson Jr., *Dynamic Optimization* (Addison Wesley Longman, Menlo Park, 1999). The companion to Bryson and Ho [1975]. Contains many numerical solutions using MATLAB

A.E. Bryson Jr., Y.C. Ho, *Applied Optimal Control* (Hemisphere Publishing, Washington, D.C., 1975). The most popular and most referenced book by aerospace engineers. A must have. A bit difficult for beginners, which is why the book you are reading now was written

R. Bulirsch, A. Miele, J. Stoer, K.H. Well (eds.), *Optimal Control: Calculus of Variations, Optimal Control Theory and Numerical Methods*. International Series of Numerical Mathematics, ISNM, vol. 111 (Birkhauser, Basel, 1993)

C. Carathéodory, *Calculus of Variations and Partial Differential Equations of the First Order, Part I: Partial Differential Equations of the First Order* (Trans. by R.B. Dean and J.J. Brandstatter) (Holden-Day, San Francisco, 1965)

C. Carathéodory, *Calculus of Variations and Partial Differential Equations of the First Order, Part II: Calculus of Variations* (Trans. by R.B. Dean and J.J. Brandstatter) (Holden-Day, San Francisco, 1967)

P. Cicala, *An Engineering Approach to the Calculus of Variations* (Levrotto & Bella, Torino, 1957)

S.J. Citron, *Elements of Optimal Control* (Holt, Rinehart and Winston, New York, 1969). A nice, practical book for engineers. Emphasizes understanding from a heuristic point of view over mathematical rigor

J.C. Clegg, *Calculus of Variations* (Interscience Publishers, New York, 1968)

D.J. Clements, B.D.O. Anderson, *Singular Optimal Control: The Linear-Quadratic Problem* (Springer, New York, 1978)

B.A. Conway (ed.), *Spacecraft Trajectory Optimization* (Cambridge University Press, New York, 2010). An excellent follow up to the present text, drawing from the currently most active researchers in optimization of space trajectories

- Chapter 1, The problem of spacecraft trajectory optimization. B.A. Conway.
- Chapter 2, Primer vector theory and applications. J.E. Prussing.
- Chapter 3, Spacecraft trajectory optimization using direct transcription. B.A. Conway and S.W. Paris.
- Chapter 4, Elements of a software system for spacecraft trajectory optimization. C. Ocampo.
- Chapter 5, Low thrust trajectory optimization using orbital averaging and control parametrization. C.A. Kluever.
- Chapter 6, Analytic representation of optimal low-thrust transfer in circular orbit. J.A. Kéchichian.
- Chapter 7, Global optimization and space pruning for spacecraft trajectory design. D. Izzo.
- Chapter 8, Incremental techniques for global space trajectory design. M. Vasile and M. Ceriotti.
- Chapter 9, Optimal low-thrust trajectories using stable manifolds. C. Martin and B.A. Conway.
- Chapter 10, Swarming theory applied to space trajectory optimization. M. Pontani and B.A. Conway.

R. Courant, *Calculus of Variations and Supplementary Notes and Exercises* (New York University, New York, 1957)

H.W. Curtis, *Orbital Mechanics for Engineering Students* (Elsevier, Boston, 2007). A well written introductory text for undergraduates

J.M.A. Danby, *Fundamentals of Celestial Mechanics*, 2nd edn. (Willmann-Bell, Richmond, 1988). A must have, classic book on celestial mechanics

M.M. Denn, *Optimization by Variational Methods* (McGraw-Hill, New York, 1969)

S.E. Dreyfus, *Dynamic Programming and the Calculus of Variations* (Academic, New York, 1966)

A. Einstein, *The Meaning of Relativity* (Princeton University Press, Princeton, 1972)

L.E. Elsgolc, *Calculus of Variations* (Trans. from the Russian) (Addison-Wesley, Reading, 1962). This accessible text provides engineers with a basic understanding of calculus of variations

G.M. Ewing, *Calculus of Variations with Applications* (Dover, New York, 1985). A fundamental work requiring some "mathematical maturity." A must have reference for advanced study

C. Fox, *An Introduction to the Calculus of Variations* (Dover, New York, 1987). One of the classic introductions

I.M. Gelfand, S.V. Fomin, *Calculus of Variations* (Trans. from the Russian by R.A. Silverman) (Prentice-Hall, Englewood Cliffs, 1963). A nice balance between mathematics and applications with a bibliography that has similarly accessible books for the engineer

M.C. Gemignani, *Elementary Topology*, 2nd edn. (Dover, New York, 1990)

P.E. Gill, W. Murray, M.H. Wright, *Practical Optimization* (Academic, New York, 1981)

H.H. Goldstine, *A History of the Calculus of Variations from the 17th Through the 19th Century* (Springer, New York, 1980)

D.T. Greenwood, *Classical Dynamics* (Dover, New York, 1997). Very well written and highly accessible text on the variational principles of mechanics. Makes an excellent companion with the work by Lanczos

M.D. Griffin, J.R. French, *Space Vehicle Design*, 2nd edn. (American Institute of Aeronautics and Astronautics, Reston, 2003). An AIAA bestseller and must have for spacecraft designers

T. Hawkins, *Lebesgue's Theory of Integration: Its Origins and Development* (The University of Wisconsin Press, Madison, 1970)

M.R. Hestenes, *Calculus of Variations and Optimal Control Theory* (Wiley, New York, 1966). An important mathematical treatment of optimal control theory by one of Bliss's research assistants. Not easy reading for engineers

D.G. Hull, *Optimal Control Theory for Applications* (Springer, New York, 2003). An excellent reference for the practicing engineer and for senior and graduate students. Well written and accessible; uses uniform, modern notation

J.L. Junkins, J.D. Turner, *Optimal Spacecraft Rotational Maneuvers* (Elsevier Scientific, New York, 1986)

R.E. Kalman, The theory of optimal control and the calculus of variations, Chapter 16, in *Mathematical Optimization Techniques*, ed. by R. Bellman (University of California Press, Berkeley, 1963)

H.J. Kelley, R.E. Kopp, A.G. Moyer, Singular extremals, Chapter 3, in *Topics in Optimization*, ed. by G. Leitmann (Academic, New York, 1967)

P. Kenneth, R. McGill, Two-point boundary value problem techniques, Chapter 2, in *Advances in Control Systems: Theory and Applications*, vol. 3, ed. by C.T. Leondes (Academic, New York, 1966)

D.E. Kirk, *Optimal Control Theory: An Introduction* (Prentice Hall, Englewood Cliffs, 1970). Explains basic concepts that are often glossed over with good use of illustrations. Accessible to engineers and well written

G. Knowles, *An Introduction to Applied Optimal Control* (Academic, New York, 1981). Instead of proving the Minimum Principle, this readable text takes the practical approach of demonstrating the Minimum Principle on problems of interest to mathematics, engineering, and business students

A.V. Labunsky, O.V. Papkov, K.G. Sukhanov, *Multiple Gravity-Assist Interplanetary Trajectories* (Gordon and Breach Science, Amsterdam, 1998)

C. Lanczos, *The Variational Principles of Mechanics*, 4th edn. (Dover, New York, 1986). An enthusiastic treatment woven with historical anecdotes and philosophical insights. A must have book

D.F. Lawden, *Optimal Trajectories for Space Navigation* (Butterworths, London, 1963). This is the classic treatment of optimal space trajectories

D.F. Lawden, *Analytical Methods of Optimization* (Hafner, New York, 1975)

E.B. Lee, L. Markus, *Foundations of Optimal Control Theory* (Wiley, New York, 1967)

G. Leitmann, *The Calculus of Variations and Optimal Control* (Plenum, New York, 1981). A well written graduate-level text with care placed on mathematical details. A bit challenging for the novice

F.L. Lewis, V.L. Syrmos, *Optimal Control*, 2nd edn. (Wiley, New York, 1995). A readable book written by engineers for engineers. Focuses on practice over theory

L.A. Lyusternik, *Shortest Paths Variational Problems* (Trans. and adapted from the Russian by P. Collins and R.B. Brown) (The Macmillan, New York, 1964). An entertaining introduction to variational problems using elementary mathematics

C.R. MacCluer, *Calculus of Variations: Mechanics, Control, and Other Applications* (Pearson Prentice Hall, Upper Saddle River, 2005). An accessible book for students armed with only calculus, that includes a simple introduction to optimal control and Pontryagin's Minimum Principle. Difficulty ramps up after chapter 7 to cover advanced topics

J.P. Marec, *Optimal Space Trajectories* (Elsevier Scientific, New York, 1979)

M. Mesterton-Gibbons, *A Primer on the Calculus of Variations and Optimal Control* (American Mathematical Society, Providence, 2009)

A. Miele, *Flight Mechanics: Theory of Flight Paths*, vol. 1 (Addison-Wesley, Reading, 1962)

A. Miele (ed.), *Theory of Optimum Shapes: Extremal Problems in the Aerodynamics of Supersonic, Hypersonic, and Free-Molecular Flows* (Academic, New York, 1965)

M. Morse, *The Calculus of Variations in the Large* (American Mathematical Society, Providence, 1934)

R. Oldenburger, *Optimal Control* (Holt, Rinehart and Winston, New York, 1966)

L.A. Pars, *An Introduction to the Calculus of Variations* (Wiley, New York, 1962)

I.P. Petrov, *Variational Methods in Optimum Control Theory* (Academic, New York, 1968). This little book is a gem. Directed toward electrical engineers, it starts with fundamental concepts and builds gradually to elucidate the variational methods underlying optimum control. It includes an interesting historical survey and a glossary in the appendices. Very well written

D.A. Pierre, *Optimization Theory with Applications* (Wiley, New York, 1969)

L.S. Pontryagin, V.G. Boltyanskii, R.V. Gamkrelidze, E.F. Mishchenko, *The Mathematical Theory of Optimal Processes* (Wiley, New York, 1962). The ultimate authorities on the Minimum Principle. While the mathematical proofs are somewhat recondite, the text is clear and even conversational, the meaning of each theorem is clearly interpreted, and the examples are derived from practical problems

J.E. Prussing, B.A. Conway, *Orbital Mechanics*, 2nd edn. (Oxford University Press, New York, 2013). Fills a vacuum by supplying an excellent text for the first course on orbital mechanics

S.S. Rao, *Engineering Optimization: Theory and Practice*, 3rd edn. (Wiley-Interscience, New York, 1996)

I.M. Ross, *A Primer on Pontryagin's Principle in Optimal Control* (Collegiate, Carmel, 2009). This well written, humorous approach is a refreshing read. It drops any pretense of proving Pontryagin's Principle: "You want proof? You can't handle the proof!" and proceeds to demonstrate that the application of the Minimum Principle is easy!

H. Sagan, *Introduction to the Calculus of Variations* (McGraw-Hill, New York, 1969)

H. Schaub, J. Junkins, *Analytical Mechanics of Space Systems*, 2nd edn. (American Institute of Aeronautics and Astronautics, Reston, 2009)

D.J. Scheeres, *Orbital Motion in Strongly Perturbed Environments: Applications to Asteroid, Comet and Planetary Satellite Orbiters* (Springer, New York, 2013)

P.A. Schilpp, *Albert Einstein: Philosopher-Scientist*. The Library of Living Philosophers, vol. VII, 3rd edn. (Cambridge University Press, London, 1969)

R.F. Stengel, *Optimal Control and Estimation* (Dover, New York, 1994). A very useful and readable reference for practicing engineers

W.T. Thomson, *Introduction to Space Dynamics* (Dover, New York, 1986). A useful and friendly reference for astrodynamicists that has remained popular for half a century

P. Ulivi, D.M. Harland, *Robotic Exploration of the Solar System* (Springer, New York, 2007)

J. Vagners, Optimization techniques, in *Handbook of Applied Mathematics*, 2nd edn., ed. by C.E. Pearson (Van Nostrand Reinhold, New York, 1983), pp. 1140–1216. A wonderful summary of the fundamental elements of optimization theory. After the author admits that the literature is staggering and "unfortunately scattered throughout the various publications by engineers, mathematicians, and systems analysts," he endeavors to present sufficiently complete references and sources to guide the reader. The book is well worth purchasing just for this chapter

N.X. Vinh, A. Busemann, R.D. Culp, *Hypersonic and Planetary Flight Mechanics* (The University of Michigan Press, Ann Arbor, 1980)

N.X. Vinh, *Optimal Trajectories in Atmospheric Flight* (Elsevier Scientific, New York, 1981)

N.X. Vinh, *Flight Mechanics of High-Performance Aircraft* (Cambridge University Press, New York, 1993). A masterly treatment of aircraft performance that is accessible to advanced undergraduates

R. Weinstock, *Calculus of Variations, with Applications to Physics and Engineering* (Dover, New York, 1974)

J.R. Wertz, W.J. Larson (eds.), *Space Mission Analysis and Design*, 3rd edn. (Microcosm, El Segundo, 2008)

J.R. Wertz, with contributions by H.F. Meissinger, L.K. Newman, G.N. Smit, *Mission Geometry: Orbit and Constellation Design and Management* (Microcosm, El Segundo, 2001)

J.R. Wertz, D.F. Everett, J.J. Puschell (eds.), *Space Mission Engineering: The New SMAD* (Microcosm, Hawthorne, 2011). A must have for the practicing space mission designer

B. Wie, *Space Vehicle Dynamics and Control* (American Institute of Aeronautics and Astronautics, Reston, 1998)

L.C. Young, *Lectures on the Calculus of Variations and Optimal Control Theory* (W.B. Saunders, Philadelphia, 1969)

Bibliography (Aerospace Applications Papers and Reports)

B. Acikmese, S.R. Ploen, Convex programming approach to powered descent guidance for mars landing. J. Guid. Control Dyn. **30**(5), 1353–1366 (2007)

A.R. Archenti, N.X. Vinh, Intermediate-thrust arcs and their optimality in a central, time-invariant force field. J. Optim. Theory Appl. **11**(3), 293–304 (1973)

D.M. Azimov, R.H. Bishop, Extremal rocket motion with maximum thrust on a linear central field. J. Spacecr. Rockets **38**(5), 765–776 (2001)

X. Bai, J.L. Junkins, New results for time-optimal three-axis reorientation of a rigid spacecraft. J. Guid. Control Dyn. **32**(4), 1071–1076 (2009)

D.R. Bartz, J.L. Horsewood, Characteristics, capabilities, and costs of solar electric spacecraft for planetary missions. J. Spacecr. Rockets **7**(12), 1379–1390 (1970)

R.J. Battin, R.M. Vaughan, An elegant Lambert algorithm. J. Guid. Control Dyn. **7**, 662–670 (1984)

R.A. Beck, J.M. Longuski, Annihilation of transverse velocity bias during spinning-up maneuvers. J. Guid. Control Dyn. **20**(3), 416–421 (1997)

M. Belló-Mora, J. Rodríguez-Canabal, On the 3-d configurations of the cluster mission, in *ESA Symposium on Space Flight Dynamics*, Darmstadt, Dec 1991. SP-326 (European Space Agency, 1991), pp. 471–479

J.T. Betts, Survey of numerical methods for trajectory optimization. J. Guid. Control Dyn. **21**(2), 193–207 (1998)

J.T. Betts, *Practical Methods for Optimal Control Using Nonlinear Programming*. SIAM Advances in Design and Control (Society for Industrial and Applied Mathematics, Philadelphia, 2001)

K.D. Bilimoria, B. Wie, Time-optimal three-axis reorientation of a rigid spacecraft. J. Guid. Control Dyn. **16**(3), 446–452 (1993)

R.H. Bishop, D.M. Azimov, Analytical space trajectories for extremal motion with low-thrust exhaust-modulated propulsion. J. Spacecr. Rockets **38**(6), 897–903 (2001)

J.V. Breakwell, The optimization of trajectories. J. Soc. Ind. Appl. Math. **7**(2), 215–247 (1959)

J.V. Breakwell, H.E. Rauch, Optimal guidance for a low thrust interplanetary vehicle. AIAA J. **4**(4), 693–704 (1966)

J.V. Breakwell, H. Shoee, Minimum fuel flight paths for a given range, in *AIAA/AAS Astrodynamics Conference*, Danvers, 11–13 Aug 1980. AIAA Paper No. 80–1660

R.A. Broucke, A.F.B.A. Prado, Orbital planar maneuvers using two and three-four (through infinity) impulses. J. Guid. Control Dyn. **19**(2), 274–282 (1996)

R.G. Brusch, Constrained impulsive trajectory optimization for orbit-to-orbit transfer. J. Guid. Control **2**(3), 204–212 (1979)

R.M. Byers, Improved estimation of time-optimal control switching, in *AIAA/AAS Astrodynamics Conference*, Hilton Head Island, 10–12 Aug 1992. AIAA Paper No. 92-4590

R.M. Byers, S.R. Vadali, Quasi-closed-form solution to the time-optimal rigid spacecraft reorientation problem. J. Guid. Control Dyn. **16**(3), 453–461 (1993)

J.M. Longuski et al., *Optimal Control with Aerospace Applications*,
Space Technology Library 32, DOI 10.1007/978-1-4614-8945-0,

D.V. Byrnes, J.M. Longuski, B. Aldrin, Cycler orbit between Earth and Mars. J. Spacecr. Rockets **30**(3), 334–336 (1993)

S. Campagnola, R.P. Russell, Endgame problem part 1: V_∞-leveraging technique and the leveraging graph. J. Guid. Control Dyn. **33**(2), 463–475 (2010a)

S. Campagnola, R.P. Russell, Endgame problem part 2: multibody technique and the Tisserand-Poincaré graph. J. Guid. Control Dyn. **33**(2), 476–486 (2010b)

C.K. Carrington, J.L. Junkins, Optimal nonlinear feedback control for spacecraft attitude maneuvers. J. Guid. Control Dyn. **9**(1), 99–107 (1986)

T.E. Carter, Necessary and sufficient conditions for optimal impulsive Rendezvous with linear equations of motion. Dyn. Control **10**(3), 219–227 (2000)

L. Casalino, Singular arcs during aerocruise. J. Guid. Control Dyn. **23**(1), 118–123 (2000)

L. Casalino, G. Colasurdo, Optimization of variable-specific-impulse interplanetary trajectories. J. Guid. Control Dyn. **27**(4), 678–684 (2004)

L. Casalino, G. Colasurdo, Improved Edelbaum's approach to optimize low Earth geostationary orbits low-thrust transfers. J. Guid. Control Dyn. **30**(5), 1504–1510 (2007)

L. Casalino, G. Colasurdo, D. Pastrone, Optimization procedure for preliminary design of opposition-class Mars missions. J. Guid. Control Dyn. **21**(1), 134–140 (1998a)

L. Casalino, G. Colasurdo, D. Pastrone, Optimization of ΔV Earth-gravity-assist trajectories. J. Guid. Control Dyn. **21**(6), 991–995 (1998b)

L. Casalino, G. Colasurdo, D. Pastrone, Optimal low-thrust escape trajectories using gravity assist. J. Guid. Control Dyn. **22**(5), 637–642 (1999)

M. Ceriotti, C.R. McInnes, Generation of optimal trajectories for Earth hybrid pole sitters. J. Guid. Control Dyn. **34**(3), 991–995 (2011)

M. Ceriotti, M. Vasile, Automated multigravity assist trajectory planning with a modified ant colony algorithm. J. Aerosp. Comput. Inf. Commun. **7**, 261–293 (2010)

K.J. Chen, T.T. McConaghy, D.F. Landau, J.M. Longuski, B. Aldrin, Powered Earth-Mars cycler with three-synodic-period repeat time. J. Spacecr. Rockets **42**(5), 921–927 (2005)

D.F. Chichka, U.J. Shankar, E.M. Cliff, H.J. Kelley, Cruise-dash-climb analysis of an airbreathing missile. J. Guid. Control Dyn. **11**(4), 293–299 (1988)

C.T. Chomel, R.H. Bishop, Analytical lunar descent guidance algorithm. J. Guid. Control Dyn. **32**(3), 915–926 (2009)

R.S. Chowdhry, E.M. Cliff, F.H. Lutze, Optimal rigid-body motions. J. Guid. Control Dyn. **14**(2), 383–390 (1991)

T. Cichan, R.G. Melton, D.B. Spencer, Control laws for minimum orbital change–the satellite retrieval problem. J. Guid. Control Dyn. **24**(6), 1231–1233 (2001)

V.C. Clarke Jr., Design of lunar and interplanetary ascent trajectories. AIAA J. **1**(7), 1559–1567 (1963)

E.M. Cliff, In Memoriam—Henry J. Kelley. J. Guid. Control Dyn. **11**(4), 289–290 (1988)

G. Colasurdo, D. Pastrone, Indirect optimization method for impulsive transfers. in *AIAA/AAS Astrodynamics Conference*, Scottsdale, Aug 1994. AIAA Paper No. 94–3762

B.A. Conway, Optimal low-thrust interception of Earth-crossing asteroids. J. Guid. Control Dyn. **20**(5), 995–1002 (1997)

B.A. Conway, K.M. Larson, Collocation versus differential inclusion in direct optimization. J. Guid. Control Dyn. **21**(5), 780–785 (1998)

V. Coverstone-Carroll, J.E. Prussing, Optimal cooperative power-limited rendezvous between neighboring circular orbits. J. Guid. Control Dyn. **16**(6), 1045–1054 (1993)

B. Dachwald, B. Wie, Solar sail kinetic energy impactor trajectory optimization for an asteroid-deflection mission. J. Spacecr. Rockets **44**(4), 755–764 (2007)

R. Dai, J.E. Cochran Jr., Three-dimensional trajectory optimization in constrained airspace. J. Aircr. **46**(2), 627–634 (2009)

L.A. D'Amario, T.N. Edelbaum, Minimum impulse three-body trajectories. AIAA J. **12**(4), 455–462 (1974)

L.A. D'Amario, D.V. Byrnes, R.H. Stanford, A new method for optimizing multiple-flyby trajectories. J. Guid. Control Dyn. **4**(6), 591–596 (1981)

L.A. D'Amario, D.V. Byrnes, R.H. Stanford, Interplanetary trajectory optimization with application to Galileo. J. Guid. Control Dyn. **5**(5), 465–471 (1982)

T.J. Debban, T.T. McConaghy, J.M. Longuski, Design and optimization of low-thrust gravity-assist trajectories to selected planets, in *AIAA/AAS Astrodynamics Conference*, Monterey, 5–8 Aug 2002. AIAA Paper No. 2002-4729

P. De Pascale, M. Vasile, Preliminary design of low-thrust multiple gravity-assist trajectories. J. Spacecr. Rockets **43**(5), 1065–1076 (2006)

P.N. Desai, B.A. Conway, Six-degree-of-freedom trajectory optimization using a two-timescale collocation architecture. J. Guid. Control Dyn. **31**(5), 1308–1315 (2008)

C.N. D'Souza, An optimal guidance law for planetary landing, in *AIAA Guidance, Navigation, and Control Conference*, New Orleans, 11–13 Aug 1997. AIAA Paper No. 97-3709

D. Dunham, S. Davis, Optimization of a multiple lunar-swingby trajectory sequence, in *AIAA/AAS Astrodynamics Conference*, Seattle, 20–22 Aug 1984. AIAA Paper No. 84-1978

T.N. Edelbaum, How many impulses? Aeronaut. Astronaut. (now named Aerosp. Am.) **5**, 64–69 (1967)

T.N. Edelbaum, L.L. Sackett, H.L. Malchow, Optimal low thrust geocentric transfer, in *AIAA 10th Electric Propulsion Conference*, Lake Tahoe, 31 Oct–2 Nov 1973. AIAA Paper No. 73-1074

T. Elices, Maximum ΔV in the aerogravity assist maneuver. J. Spacecr. Rockets **32**(5), 921–922 (1995)

P.J. Enright, B.A. Conway, Optimal finite-thrust spacecraft trajectories using collocation and nonlinear programming. J. Guid. Control Dyn. **14**(5), 981–985 (1991)

P.J. Enright, B.A. Conway, Discrete approximations to optimal trajectories using direct transcription and nonlinear programming. J. Guid. Control Dyn. **15**(4), 994–1002 (1992)

R.W. Farquhar, Lunar communications with libration-point satellites. J. Spacecr. Rockets **4**(10), 1383–1384 (1967)

R.W. Farquhar, Comments on optimal controls for out-of-plane motion about the translunar libration point. J. Spacecr. Rockets **8**(7), 815–816 (1971)

R.W. Farquhar, The flight of ISEE3/ICE: origins, mission history, and a legacy. J. Astronaut. Sci. **49**(1), 23–73 (2001)

T.S. Feeley, J.L. Speyer, Techniques for developing approximate optimal advanced launch system guidance. J. Guid. Control Dyn. **17**(5), 889–896 (1994)

A. Fleming, I.M. Ross, Optimal control of spinning axisymmetric spacecraft: a pseudospectral approach, in *AIAA Guidance, Navigation, and Control Conference*, Honolulu, 18–21 Aug 2008. AIAA Paper No. 2008-7164

E.G. Gilbert, R.M. Howe, P. Lu, N.X. Vinh, Optimal aeroassisted intercept trajectories at hyperbolic speeds. J. Guid. Control Dyn. **14**(1), 123–131 (1991)

B.S. Goh, Necessary conditions for singular extremals involving multiple control variables. SIAM J. Control **4**(4), 716–722 (1966)

B.S. Goh, Compact forms of the generalized Legendre-Clebsch conditions and the computation of singular control trajectories, in *Proceedings of the American Control Conference*, Seattle, June 1995, pp. 3410–3413. Paper No. FA19-10:35

B.S. Goh, Optimal singular rocket and aircraft trajectories, in *Chinese Control and Decision Conference*, Yantai, 24 July 2008

E.D. Gustafson, D.J. Scheeres, Optimal timing of control-law updates for unstable systems with continuous control. J. Guid. Control Dyn. **32**(3), 878–887 (2009)

J.J. Guzmán, J.L. Horsewood, Mission options for rendezvous with wild 2 in 2016 using electric propulsion, in *AIAA/AAS Astrodynamics Conference*, Keystone, 21–24 Aug 2006. AIAA Paper No. 06-6175

J.J. Guzmán, L.M. Mailhe, C. Schiff, S.P. Hughes, D.C. Folta, Primer vector optimization: survey of theory, new analysis and applications, in *53rd International Astronautical Congress*, Houston, Oct 2002. Paper No. IAC-02-A.6.09

C.M. Haissig, K.D. Mease, N.X. Vinh, Canonical transformations for space trajectory optimization, in *AIAA/AAS Astrodynamics Conference*, Hilton Head Island, 10–12 Aug 1992. AIAA Paper No.1992-4509

C.D. Hall, V. Collazo-Perez, Minimum-time orbital phasing maneuvers. J. Guid. Control Dyn. **26**(6), 934–941 (2003)

M. Handelsman, Optimal free-space fixed-thrust trajectories using impulsive trajectories as starting iteratives. AIAA J. **4**(6), 1077–1082 (1966)

C.R. Hargraves, S.W. Paris, Direct optimization using nonlinear programming and collocation. J. Guid. Control Dyn. **10**(4), 338–342 (1987)

C. Hargraves, F. Johnson, S. Paris, I. Rettie, Numerical computation of optimal atmospheric trajectories. J. Guid. Control Dyn. **4**(4), 406–414 (1981)

G.A. Henning, J.M. Longuski, Optimization of aerogravity-assist trajectories for waveriders, in *AAS/AIAA Astrodynamics Conference*, Mackinac Island, 19–23 Aug 2007. AAS Paper No. 07-325

T.A. Heppenheimer, Optimal controls for out-of-plane motion about the translunar libration point. J. Spacecr. Rockets **7**(9), 1088–1092 (1970)

A.L. Herman, B.A. Conway, Direct optimization using collocation based on high-order Gauss-Lobatto quadrature rules. J. Guid. Control Dyn. **19**(3), 592–599 (1996)

A.L. Herman, B.A. Conway, Optimal, low-thrust, Earth-Moon transfer. J. Guid. Control Dyn. **21**(1), 141–147 (1998)

A.L. Herman, D.B. Spencer, Optimal, low-thrust Earth-orbit transfers using higher-order collocation methods. J. Guid. Control Dyn. **25**(1), 40–47 (2002)

L.A. Hiday, K.C. Howell, Impulsive time-free transfers between halo orbits, in *AIAA/AAS Astrodynamics Conference*, Hilton Head Island, 10–12 Aug 1992. AIAA Paper No. 92-4580

L.A. Hiday, Optimal transfers between libration-point orbits in the elliptic restricted three-body problem. PhD thesis, Purdue University, West Lafayette, Aug 1992

Y.C. Ho, J.L. Speyer, In appreciation of arthur E. Bryson, Jr. J. Guid. Control Dyn. **13**(5), 770–774 (1990)

D. Hocken, J. Schoenmaekers, Optimization of cluster constellation manoeuvres, in *16th International Symposium on Space Flight Dynamics*, Pasadena, Dec 2001

J.L. Horsewood, The optimization of low thrust interplanetary swingby trajectories, in *AAS/AIAA Astrodynamics Conference*, Santa Barbara, 19–21 Aug 1970. AIAA Paper No. 70-1041

J.L. Horsewood, F.I. Mann, Optimization of low-thrust heliocentric trajectories with large launch asymptote declinations. AIAA J. **13**(10), 1304–1310 (1975)

J.L. Horsewood, M.A. Suskin, The effect of multi-periapse burns on planetary escape and capture, in *AIAA/NASA/OAI Conference on Advanced SEI Technologies*, Cleveland, 4–6 Sept 1991. AIAA Paper No. 91-3405

J.L. Horsewood, J.D. Vickery, Mission window definition for Jupiter swingbys to the outer planets. J. Spacecr. Rockets **6**(5), 525–531 (1969)

K.C. Howell, L.A. Hiday-Johnston, Transfers between libration points in the elliptic restricted three body problem. Celest. Mech. Dyn. Astron. **58**(4), 317–337 (1994)

K.C. Howell, B.T. Barden, M.W. Lo, Application of dynamical systems theory to trajectory design for a libration point mission. J. Astronaut. Sci. **45**(2), 161–178 (1997)

F.-K. Hsu, T.-S. Kuo, J.-S. Chern, C.-C. Lin, Optimal aeroassisted orbital plane change with heating rate constraint, in *AIAA 26th Aerospace Sciences Meeting*, Reno, 11–14 Jan 1988. AIAA Paper No. 88-0301

S.P. Hughes, L.M. Mailhe, J.J. Guzmán, A comparison of trajectory optimization methods for the impulsive minimum fuel rendezvous problem, in *26th Annual AAS Guidance and Control Conference*, Breckenridge, 5–9 Feb 2003. AAS Paper No. 03-006

D. Izzo, Optimization of interplanetary trajectories for impulsive and continuous asteroid deflection. J. Guid. Control Dyn. **30**(2), 401–408 (2007)

D. Izzo, V.M. Becerra, D.R. Myatt, S.J. Nasuto, J.M. Bishop, Search space pruning and global optimisation of multiple gravity assist spacecraft trajectories. J. Glob. Optim. **38**, 283–296 (2007)

K. Jackson, V.L. Coverstone, Optimal lunar launch trajectories to the Sun-Earth L1 vicinity. J. Guid. Control Dyn. **31**(3), 712–719 (2008)

S. Jain, P. Tsiotras, Trajectory optimization using multiresolution techniques. J. Guid. Control Dyn. **31**(5), 1424–1436 (2008)

B.R. Jamison, V. Coverstone, Analytical study of the primer vector and orbit transfer switching function. J. Guid. Control Dyn. **33**(1), 235–245 (2010)

M. Jesick, C. Ocampo, Optimal lunar orbit insertion from a variable free return with analytical gradients, in *AIAA/AAS Astrodynamics Conference*, Toronto, 2–5 Aug 2010. AIAA Paper No. 2010-8388

M. Jesick, C. Ocampo, Automated generation of symmetric lunar free-return trajectories. J. Guid. Control Dyn. **34**(1), 98–106 (2011)

D.J. Jezewski, Primer vector theory and applications. Technical report, NASA TR R-454. NASA Johnson Space Center, Houston, Nov 1975

D.J. Jezewski, N.L. Faust, Inequality constraints in primer-optimal, N-impulse solutions. AIAA J. **9**(4), 760–763 (1971)

D.J. Jezewski, H.L. Rozendaal, An efficient method for calculating optimal free-space N-impulse trajectories. AIAA J. **6**(11), 2160–2165 (1968)

D.J. Jezewski, J.P. Brazzel, E.E. Prust, B.G. Brown, T.A. Mulder, D.B. Wissinger, A survey of rendezvous trajectory planning, in *AAS/AIAA Astrodynamics Conference*, Durango, Aug 1991. AAS Paper No. 91-505

J.-W. Jo, J.E. Prussing, Procedure for applying second-order conditions in optimal control problems. J. Guid. Control Dyn. **23**(2), 241–250 (2000)

J.R. Johannesen, N.X. Vinh, K.D. Mease, Effect of maximum lift to drag ratio on optimal aeroassisted plane change, in *Atmospheric Flight Mechanics Conference*, Snowmass, 19–21 Aug 1985. AIAA Paper No. 85-1817

W.R. Johnson, J.M. Longuski, Design of aerogravity-assist trajectories. J. Spacecr. Rockets **39**(1), 23–30 (2002)

W.R. Johnson, J.M. Longuski, D.T. Lyons, Nondimensional analysis of reaction-wheel control for aerobraking. J. Guid. Control Dyn. **26**(6), 861–868 (2003)

M.D. Jokic, J.M. Longuski, Design of tether sling for human transportation systems between Earth and Mars. J. Spacecr. Rockets **41**(6), 1010–1015 (2004)

M.D. Jokic, J.M. Longuski, Artificial gravity and abort scenarios via tethers for human missions to Mars. J. Spacecr. Rockets **42**(5), 883–889 (2005)

J.B. Jones, A solution of the variational equations for elliptic orbits in rotating coordinates, in *AIAA/AAS Astrodynamics Conference*, Danvers, 11–13 Aug 1980. AIAA Paper No. 80-1690

J.L. Junkins, R.C. Thompson, An asymptotic perturbation method for nonlinear optimal control problems. J. Guid. Control Dyn. **9**(5), 391–396 (1986)

J.L. Junkins, J.D. Turner, Optimal continuous torque attitude maneuvers. J. Guid. Control Dyn. **3**(3), 210–217 (1980)

J.L. Junkins, Z.H. Rahman, H. Bang, Near-minimum-time control of distributed parameter systems: analytical and experimental results. J. Guid. Control Dyn. **14**(2), 406–415 (1991)

J.A. Kechichian, Optimal steering for North-South stationkeeping of geostationary spacecraft. J. Guid. Control Dyn. **20**(3), 435–444 (1997a)

J.A. Kechichian, Optimal low-Earth-orbit–geostationary-Earth-orbit intermediate acceleration orbit transfer. J. Guid. Control Dyn. **20**(4), 803–811 (1997b)

J.A. Kechichian, Trajectory optimization using eccentric longitude formulation. J. Spacecr. Rockets **35**(3), 317–326 (1998)

J.A. Kechichian, Minimum-time low-thrust rendezvous and transfer using epoch mean longitude formulation. J. Guid. Control Dyn. **22**(3), 421–432 (1999)

J.A. Kechichian, Low-thrust trajectory optimization based on epoch eccentric longitude formulation. J. Spacecr. Rockets **36**(4), 543–553 (1999)

J.A. Kechichian, Minimum-time constant acceleration orbit transfer with first-order oblateness effect. J. Guid. Control Dyn. **23**(4), 595–603 (2000)

J.A. Kechichian, M.I. Cruz, E.A. Rinderle, N.X. Vinh, Optimization and closed-loop guidance of drag-modulated aeroassisted orbital transfer, in *AIAA Atmospheric Flight Mechanics Conference*, Gatlinburg, 15–17 Aug 1983. AIAA Paper No. 83-2093

H.J. Kelley, A second variation test for singular extremals. AIAA J. **2**, 1380–1382 (1964)

H.J. Kelley, E.M. Cliff, F.H. Lutze, Boost-glide range-optimal guidance, in *AIAA Guidance and Control Conference*, Albuquerque, 19–21 Aug 1981. AIAA Paper No. 81-1781

E.A. Kern, D.T. Greenwood, Minimum-fuel thrust-limited transfer trajectories between coplanar elliptic orbits. AIAA J. **8**(10), 1772–1779 (1970)

M. Kim, C.D. Hall, Symmetries in optimal control of solar sail spacecraft. Celest. Mech. Dyn. Astron. **92**, 273–293 (2005)

Y.H. Kim, D.B. Spencer, Optimal spacecraft rendezvous using genetic algorithms. J. Spacecr. Rockets **39**(6), 859–865 (2002)

C.A. Kluever, Optimal Earth-Moon trajectories using combined chemical-electric propulsion. J. Guid. Control Dyn. **20**(2), 253–258 (1997)

C.A. Kluever, K-R. Chang, Electric-propulsion spacecraft optimization for lunar missions. J. Spacecr. Rockets **33**(2), 235–245 (1996)

C.A. Kluever, B.L. Pierson, Optimal low-thrust three-dimensional Earth-Moon trajectories. J. Guid. Control Dyn. **18**(4), 830–837 (1995)

C.A. Kluever, B.L. Pierson, Optimal Earth-Moon trajectories using nuclear propulsion. J. Guid. Control Dyn. **20**(2), 239–245 (1997)

R.E. Kopp, A.G. Moyer, Necessary conditions for singular extremals. AIAA J. **3**(8), 1439–1444 (1965)

J.-P. Kremer, K.D. Mease, Near-optimal control of altitude and path angle during aerospace plane ascent. J. Guid. Control Dyn. **20**(4), 789–796 (1997)

R. Kumar, H.J. Kelley, Singular optimal atmospheric rocket trajectories. J. Guid. Control Dyn. **11**(4), 305–312 (1988)

D.F. Landau, J.M. Longuski, Trajectories for human missions to Mars, part 1: impulsive transfers. J. Spacecr. Rockets **43**(5), 1035–1042 (2006a)

D.F. Landau, J.M. Longuski, Trajectories for human missions to Mars, part 2: low-thrust transfers. J. Spacecr. Rockets **43**(5), 1043–1047 (2006b)

D.F. Landau, T. Lam, N. Strange, Broad search and optimization of solar electric propulsion trajectories to Uranus and Neptune, in *AAS/AIAA Astrodynamics Conference*, Pittsburgh, 9–13 Aug 2009. AAS Paper No. 09-428

D.F. Lawden, Impulsive transfer between elliptic orbits, in *Optimization Techniques*, ed. by G. Leitmann (Academy, New York, 1962), pp. 323–351

D.F. Lawden, Rocket trajectory optimization: 1950–1963. J. Guid. Control Dyn. **14**(4), 705–711 (1991). Lawden gives a brief history of key technologies and some personal reflections

M.A. LeCompte, T.R. Meyer, J.L. Horsewood, C.P. McKay, D.D. Durda, Early, short-duration, near-Earth asteroid rendezvous missions, J. Spacecr. Rockets **49**(4), 731–741 (2012)

D. Lee, J.E. Cochran Jr., J.H. Jo, Solutions to the variational equations for relative motion of satellites. J. Guid. Control Dyn. **30**(3), 669–678 (2007)

G. Leitmann, On a class of variational problems in rocket flight. J. Aerosp. Sci. **26**(9), 586–591 (1959)

C.L. Leonard, W.M. Hollister, E.V. Bergmann, Orbital formationkeeping with differential drag. J. Guid. Control Dyn. **12**(1), 108–113 (1989)

J.M. Lewallen, B.D. Tapley, S.D. Williams, Iteration procedures for indirect trajectory optimization methods. J. Spacecr. Rockets **5**(3), 321–327 (1968)

J.M. Lewallen, O.A. Schwausch, B.D. Tapley, Coordinate system influence on the regularized trajectory optimization problem. J. Spacecr. Rockets **8**(1), 15–20 (1971)

F. Li, P.M. Bainum, Numerical approach for solving rigid spacecraft minimum time attitude maneuvers. J. Guid. Control Dyn. **13**(1), 38–45 (1990)

F. Li, P.M. Bainum, Analytic time-optimal control synthesis of fourth-order system and maneuvers of flexible structures. J. Guid. Control Dyn. **17**(6), 1171–1178 (1994)

P.M. Lion, M. Handelsman, Primer vector on fixed-time impulsive trajectories. AIAA J. **6**(1), 127–132 (1968)

M.W. Lo, M.-K.J. Chung, Lunar sample return via the interplanetary superhighway, in *AIAA/AAS Astrodynamics Conference*, Monterey, 5–8 Aug 2002. AIAA Paper No. 2002-4718

M.W. Lo, B.G. Williams, W.E. Bollman, D. Han, Y. Hahn, J.L. Bell, E. A. Hirst, R.A. Corwin, P.E. Hong, K.C. Howell, B. Barden, R. Wilson, Genesis mission design. J. Astronaut. Sci. **49**(1), 169–184 (2001)

F.A. Lohar, A.K. Misra, D. Mateescu, Optimal atmospheric trajectory for aerogravity assist with heat constraint. J. Guid. Control Dyn. **18**(4), 723–730 (1995)

F.A. Lohar, A.K. Sherwani, A.K. Misra, Optimal transfer between coplanar elliptical orbits using aerocruise, in *AIAA/AAS Astrodynamics Conference*, San Diego, 29–31 July 1996. AIAA Paper No. 96-3594

P. Lu, B. Pan, Highly constrained optimal launch ascent guidance. J. Guid. Control Dyn. **33**(2), 404–411 (2010)

P. Lu, B.L. Pierson, Optimal aircraft terrain-following analysis and trajectory generation. J. Guid. Control Dyn. **18**(3), 555–560 (1995)

P. Lu, N.X. Vinh, Optimal control problems with maximum functional. J. Guid. Control Dyn. **14**(6), 1215–1223 (1991)

P. Lu, B.J. Griffin, G.A. Dukeman, F.R. Chavez, Rapid optimal multiburn ascent planning and guidance. J. Guid. Control Dyn. **31**(6), 1656–1664 (2008)

D.T. Lyons, E. Sklyanskiy, J. Casoliva, A.A. Wolf, Parametric optimization and guidance for an aerogravity assisted atmospheric sample return from Mars and Venus, in *AIAA/AAS Astrodynamics Conference*, Honolulu, 18–21 Aug 2008. AIAA Paper No. 2008–7353

M. Mangad, M.D. Schwartz, Guidance, flight mechanics and trajectory optimization. Volume IV: the calculus of variations and modern applications. NASA contractor report, NASA CR-1003, Washington, D.C., Jan 1968

J.E. Marsden, S.D. Ross, New methods in celestial mechanics and mission design. Bull. Am. Math. Soc. **43**, 43–73 (2005)

J.V. McAdams, J.L. Horsewood, C.L. Yen, Discovery-class Mercury orbiter trajectory design for the 2005 launch opportunity, in *Astrodynamics Conference*, Boston, 10–12 Aug 1998. AIAA Paper No. 98-4283

J.V. McAdams, D.W. Dunham, R.W. Farquhar, A.H. Taylor, B.G. Williams, Trajectory design and maneuver strategy for the MESSENGER mission to Mercury. J. Spacecr. Rockets **43**(5), 1054–1064 (2006)

T.T. McConaghy, J.M. Longuski, Parameterization effects on convergence when optimizing a low-thrust trajectory with gravity assists, in *AIAA/AAS Astrodynamics Conference*, Providence, 16–19 Aug 2004. AIAA Paper No. 2004-5403

T.T. McConaghy, T.J. Debban, A.E. Petropoulos, J.M. Longuski, Design and optimization of low-thrust trajectories with gravity assists. J. Spacecr. Rockets **40**(3), 380–387 (2003)

T.T. McConaghy, J.M. Longuski, D.V. Byrnes, Analysis of a class of Earth-Mars cycler trajectories. J. Spacecr. Rockets **41**(4), 622–628 (2004)

J.P. McDanell, W.F. Powers, Necessary conditions for joining optimal singular and nonsingular subarcs. SIAM J. Control **9**(2), 161–173 (1971)

J.E. McIntyre, Guidance, flight mechanics and trajectory optimization. Volume VII: the Pontryagin maximum principle. NASA contractor report, NASA CR-1006, Washington, D.C., Mar 1968

E.J. McShane, On multipliers for lagrange problems. Am. J. Math. **61**(4), 809–819 (1939)

K.D. Mease, N.X. Vinh, Minimum-fuel aeroassisted coplanar orbit transfer using lift modulation. J. Guid. Control Dyn. **8**(1), 134–141 (1985)

K.D. Mease, N.X. Vinh, S.H. Kuo, Optimal plane change during constant altitude hypersonic flight. J. Guid. Control Dyn. **14**(4), 797–806 (1991)

S. Medepalli, N.X. Vinh, A Lie bracket solution of the optimal thrust magnitude on a singular arc in atmospheric flight, in *AIAA Atmospheric Flight Mechanics Conference*, Hilton, 10–12 Aug 1992. AIAA Paper No. 1992-4345

W.G. Melbourne, C.G. Sauer Jr., Optimum interplanetary rendezvous with power-limited vehicles. AIAA J. **1**(1), 54–60 (1963)

W.G. Melbourne, C.G. Sauer Jr., Constant-attitude thrust program optimization. AIAA J. **3**(8), 1428–1431 (1965)

R.G. Melton, Comparison of direct optimization methods applied to solar sail problems, in *AIAA/AAS Astrodynamics Conference*, Monterey, 5–8 Aug 2002. AIAA Paper No. 2002-4728

R.G. Melton, D.S. Rubenstein, H.L. Fisher, Optimum detumbling of space platforms via a dynamic programming algorithm, in *AIAA Guidance, Navigation and Control Conference*, Williamsburg, 18–20 Aug 1986. AIAA Paper No. 86-2154

R.G. Melton, K.M. Lajoie, J.W. Woodburn, Optimum burn scheduling for low-thrust orbital transfers. J. Guid. Control Dyn. **12**(1), 13–18 (1989)

A. Miele, Flight mechanics and variational problems of a linear type. J. Aerosp. Sci. **25**(9), 581–590 (1958)

A. Miele, M.W. Weeks, M. Ciarcià, Optimal trajectories for spacecraft rendezvous. J. Optim. Theory Appl. **132**, 353–376 (2007)

W.E. Miner, J.F. Andrus, Necessary conditions for optimal lunar trajectories with discontinuous state variables and intermediate point constraints. NASA technical memorandum, TM X-1353, Washington, D.C., Apr 1967

R.S. Nah, S.R. Vadali, E. Braden, Fuel-optimal, low-thrust, three-dimensional Earth-Mars trajectories. J. Guid. Control Dyn. **24**(6), 1100–1107 (2001)

M. Okutsu, C.H. Yam, J.M. Longuski, Low-thrust trajectories to Jupiter via gravity assists from Venus, Earth, and Mars, in *AIAA/AAS Astrodynamics Conference*, Keystone, 21–24 Aug 2006. AIAA Paper No. 2006-6745

B. Paiewonsky, Optimal control: a review of theory and practice. AIAA J. **3**(11), 1985–2006 (1965)

B. Paiewonsky, P.J. Woodrow, Three-dimensional time-optimal rendezvous. J. Spacecr. Rockets **3**(11), 1577–1584 (1966)

S.W. Paris, J.P. Riehl, W.K. Sjauw, Enhanced procedures for direct trajectory optimization using nonlinear programming and implicit integration, in *AIAA/AAS Astrodynamics Conference*, Keystone, 21–24 Aug 2006. AIAA Paper No. 2006-6309

S.-Y. Park, S.R. Vadali, Touch points in optimal ascent trajectories with first-order state inequality constraints. J. Guid. Control Dyn. **21**(4), 603–610 (1998)

C. Park, V. Guibout, D.J. Scheeres, Solving optimal continuous thrust rendezvous problems with generating functions. J. Guid. Control Dyn. **29**(2), 321–331 (2006)

H.J. Pernicka, D.P. Scarberry, S.M. Marsh, T.H. Sweetser, A search for low ΔV Earth-to-Moon trajectories, in *AIAA/AAS Astrodynamics Conference*, Scottsdale, 1–3 Aug 1994. AIAA Paper No. 94-3772

A.E. Petropoulos, J.M. Longuski, Shape-based algorithm for automated design of low-thrust, gravity-assist trajectories. J. Spacecr. Rockets **41**(5), 787–796 (2004)

A.E. Petropoulos, R.P. Russell, Low-thrust transfers using primer vector theory and a second-order penalty method, in *AIAA/AAS Astrodynamics Conference*, Honolulu, 18–21 Aug 2008. AIAA Paper No. 2008-6955

A.E. Petropoulos, G.J. Whiffen, J.A. Sims, Simple control laws for continuous-thrust escape or capture and their use in optimisation, in *AIAA/AAS Astrodynamics Conference*, Monterey, 5–8 Aug 2002. AIAA Paper No. 2002-4900

B.L. Pierson, C.A. Kluever, Three-stage approach to optimal low-thrust Earth-Moon trajectories. J. Guid. Control Dyn. **17**(6), 1275–1282 (1994)

M. Pontani, Simple methods to determine globally optimal orbit transfers. J. Guid. Control Dyn. **32**(3), 899–914 (2009)

M. Pontani, B.A. Conway, Optimal interception of evasive missile warheads: numerical solution of the differential game. J. Guid. Control Dyn. **31**(4), 1111–1122 (2008)

W.F. Powers, Hamiltonian perturbation theory for optimal trajectory analysis. M.S. thesis, Department of Aerospace Engineering and Engineering Mechanics, The University of Texas, Austin, 1966

W.F. Powers, On the order of singular optimal control problems. J. Optim. Theory Appl. **32**(4), 479–489 (1980)

W.F. Powers, J.P. McDanell, Switching conditions and a synthesis technique for the singular saturn guidance problem. J. Spacecr. Rockets **8**(10), 1027–1032 (1971)

W.F. Powers, B.D. Tapley, Canonical transformation applications to optimal trajectory analysis. AIAA J. **7**(3), 394–399 (1969)

W.F. Powers, B.-D. Cheng, E.R. Edge, Singular optimal control computation. J. Guid. Control, **1**(1), 83–89 (1978)

J.E. Prussing, Optimal four-impulse fixed-time rendezvous in the vicinity of a circular orbit. AIAA J. **7**(5), 928–935 (1969)

J.E. Prussing, Optimal two- and three-impulse fixed-time rendezvous in the vicinity of a circular orbit. AIAA J. **8**(7), 1221–1228 (1970)

J.E. Prussing, Optimal impulsive linear systems: sufficient conditions and maximum number of impulses. J. Astronaut. Sci. **43**(2), 195–206 (1995)

J.E. Prussing, A class of optimal two-impulse rendezvous using multiple revolution Lambert solutions. J. Astronaut. Sci. **48**(2–3), 131–148 (2000)

J.E. Prussing, J.-H. Chiu, Optimal multiple-impulse time-fixed rendezvous between circular orbits. J. Guid. Control **9**(1), 17–22 (1986)

J.E. Prussing, S.L. Sandrik, Second-order necessary conditions and sufficient conditions applied to continuous-thrust trajectories. J. Guid. Control Dyn. (Engineering Note) **28**(4), 812–816 (2005)

J. Puig-Suari, Optimal mass flexible tethers for aerobraking maneuvers. J. Guid. Control Dyn. **20**(5), 1018–1024 (1997)

U.P. Rajeev, M. Seetharama Bhat, S. Dasgupta, Predicted target flat Earth guidance algorithm for injection in to a fixed ellipse using iteration unfolding, in *AIAA Guidance, Navigation, and Control Conference*, Providence, 16–19 Aug 2004. AIAA Paper No. 2004-4902

C.L. Ranieri, C.A. Ocampo, Optimization of roundtrip, time-constrained, finite burn trajectories via an indirect method. J. Guid. Control Dyn. **28**(2), 306–314 (2005)

C.L. Ranieri, C.A. Ocampo, Indirect optimization of spiral trajectories. J. Guid. Control Dyn. **29**(6), 1360–1366 (2006)

C.L. Ranieri, C.A. Ocampo, Indirect optimization of two-dimensional finite burning interplanetary transfers including spiral dynamics. J. Guid. Control Dyn. **31**(3), 720–728 (2008)

C.A. Renault, D.J. Scheeres, Optimal placement of statistical maneuvers in an unstable orbital environment, in *AIAA/AAS Astrodynamics Conference*, Monterey, 5–8 Aug 2002. AIAA Paper No. 2002-4725

H.M. Robbins, A generalized Legendre-Clebsch condition for the singular cases of optimal control. IBM J. Res. Dev. **11**(4), 361–372 (1967)

R.D. Robinett, G.G. Parker, H. Schaub, J.L. Junkins, Lyapunov optimal saturated control for nonlinear systems. J. Guid. Control Dyn. **20**(6), 1083–1088 (1997)

J. Rodríguez-Canabal, M. Belló-Mora, Cluster: consolidated report on mission analysis. Technical report, ESOC CL-ESC-RP-0001, European Space Operation Centre, Darmstadt, July 1990

I.M. Ross, Extremal angle of attack over a singular thrust arc in rocket flight. J. Guid. Control Dyn. (Engineering Note) **20**(2), 391–393 (1997)

I.M. Ross, S.-Y. Park, S.D.V. Porter, Gravitational effects of Earth in optimizing ΔV for deflecting Earth-crossing asteroids. J. Spacecr. Rockets **38**(5), 759–764 (2001)

I.M. Ross, C. D'Souza, F. Fahroo, J.B. Ross, A fast approach to multi-stage launch vehicle trajectory optimization, in *AIAA Guidance, Navigation, and Control Conference*, Austin, 11–14 Aug 2003. AIAA Paper No. 2003-5639

R.P. Russell, Primer vector theory applied to global low-thrust trade studies. J. Guid. Control Dyn. **30**(2), 460–472 (2007)

R.P. Russell, C.A. Ocampo, Global search for idealized free-return Earth-Mars cyclers. J. Guid. Control Dyn. **28**(2), 194–208 (2005)

R.P. Russell, C.A. Ocampo, Optimization of a broad class of ephemeris model Earth-Mars Cyclers. J. Guid. Control Dyn. **29**(2), 354–367 (2006)

R.P. Russell, N.J. Strange, Cycler trajectories in planetary moon systems. J. Guid. Control Dyn. **32**(1), 143–157 (2009)

L.L. Sackett, T.N. Edelbaum, Optimal high- and low-thrust geocentric transfer, in *AIAA Mechanics and Control of Flight Conference*, Anaheim, 5–9 Aug 1974. AIAA Paper No. 74-801

C.G. Sauer Jr., Optimization of multiple target electric propulsion trajectories, in *AIAA 11th Aerospace Sciences Meeting*, Washington, D.C., 10–12 Jan 1973. AIAA Paper No. 73-205

C.G. Sauer Jr., Optimum solar-sail interplanetary trajectories, in *AIAA/AAS Astrodynamics Conference*, San Diego, 18–20 Aug 1976. AIAA Paper No. 76-792

H. Schaub, J.L. Junkins, R.D. Robinett, New penalty functions and optimal control formulation for spacecraft attitude control problems. J. Guid. Control Dyn. **20**(3), 428–434 (1997)

W.A. Scheel, B.A. Conway, Optimization of very-low thrust, many-revolution spacecraft trajectories. J. Guid. Control Dyn. **17**(6), 1185–1192 (1994)

J. Schoenmaekers, Cluster: fuel optimum spacecraft formation control, in *ESA Symposium on Space Flight Dynamics*, Darmstadt, Dec 1991, ESA SP-326 (European Space Agency, 1991), pp. 419–425

J. Schoenmaekers, Post-launch optimisation of the SMART-1 low-thrust trajectory to the Moon, in *18th International Symposium on Space Flight Dynamics*, Munich, 11–15 Oct 2004, ESA SP-548, pp. 505–510

J. Schoenmaekers, D. Horas, J.A. Pulido, SMART-1: with solar electric propulsion to the Moon, in *16th International Symposium on Space Flight Dynamics*, Pasadena, 3–7 Dec 2001

C.J. Scott, D.B. Spencer, Optimal reconfiguration of satellites in formation. J. Spacecr. Rockets **44**(1), 230–239 (2007)

S.L. Scrivener, R.C. Thompson, Survey of time-optimal attitude maneuvers. J. Guid. Control Dyn. **17**(2), 225–233 (1994)

M.R. Sentinella, L. Casalino, Hybrid evolutionary algorithm for the optimization of interplanetary trajectories. J. Spacecr. Rockets **46**(2), 365–372 (2009)

R. Serban, W.S. Koon, M. Lo, J.E. Marsden, L.R. Petzold, S.D. Ross, R.S. Wilson, Optimal control for halo orbit missions. in *IFAC Workshop on Lagrangian and Hamiltonian Methods for Nonlinear Control*, Princeton University, Princeton, 16–18 Mar 2000

H. Seywald, E.M. Cliff, Goddard problem in the presence of a dynamic pressure limit. J. Guid. Control Dyn. **16**(4), 776–781 (1993a)

H. Seywald, E.M. Cliff, The generalized Legendre-Clebsch condition on state/control constrained Arcs, in *AIAA Guidance, Navigation, and Control Conference*, Monterey, 11–13 Aug 1993b. AIAA Paper No. 93–3746

H. Seywald, E.M. Cliff, Neighboring optimal control based feedback law for the advanced launch system. J. Guid. Control Dyn. **17**(6), 1154–1162 (1994)

H. Seywald, R.R. Kumar, Singular control in minimum time spacecraft reorientation. J. Guid. Control Dyn. **16**(4), 686–694 (1993)

H. Seywald, R.R. Kumar, E.M. Cliff, New proof of the Jacobi necessary condition. J. Guid. Control Dyn. (Engineering Note) **16**(6), 1178–1181 (1993)

U. Shankar, E. Cliff, H. Kelley, Singular perturbations analysis of optimal climb-cruise-dash, in *AIAA Guidance, Navigation, and Control Conference*, Monterey, 17–19 Aug 1987. AIAA Paper No. 87-2511

U. Shankar, E. Cliff, H. Kelley, Relaxation oscillations in aircraft cruise-dash optimization, in *AIAA Guidance, Navigation, and Control Conference*, Minneapolis, 15–17 Aug 1988a. AIAA Paper No. 88-4161

U. Shankar, E. Cliff, H. Kelley, Aircraft cruise-dash optimization: periodic versus steady-state solutions, in *AIAA Guidance, Navigation, and Control Conference*, Minneapolis, 15–17 Aug 1988b. AIAA Paper No. 88-4162

H. Shen, P. Tsiotras, Time-optimal control of axisymmetric rigid spacecraft using two controls. J. Guid. Control Dyn. **22**(5), 682–694 (1999)

H. Shen, P. Tsiotras, Optimal two-impulse rendezvous using multiple-revolution Lambert solutions. J. Guid. Control Dyn. **26**(1), 50–61 (2003)

J.A. Sims, A.J. Staugler, J.M. Longuski, Trajectory options to Pluto via gravity assists from Venus, Mars, and Jupiter. J. Spacecr. Rockets **34**(3), 347–353 (1997)

J.A. Sims, P.A. Finlayson, E.A. Rinderle, M.A. Vavrina, T.D. Kowalkowski, Implementation of a low-thrust trajectory optimization algorithm for preliminary design, in *AIAA/AAS Astrodynamics Conference*, Keystone, 21–24 Aug 2006. AIAA Paper No. 2006-6746

T. Singh, S.R. Vadali, Robust time-optimal control: frequency domain approach. J. Guid. Control Dyn. **17**(2), 346–353 (1994)

J.L. Speyer, A.E. Bryson Jr., Optimal programming problems with a bounded state space. AIAA J. **6**(8), 1488–1491, (1968)

S.A. Stanton, B.G. Marchand, Finite set control transcription for optimal control applications. J. Spacecr. Rockets **47**(3), 457–471 (2010)

R.G. Stern, Singularities in the analytic solution of the linearized variational equations of elliptical motion. Technical report, Report RE-8, Experimental Astronomy Laboratory, Massachusetts Institute of Technology, Cambridge, May 1964

R. Stevens, I.M. Ross, Preliminary design of Earth-Mars cyclers using solar sails. J. Spacecr. Rockets **42**(1), 132–137 (2005)

K.S. Tait, Singular problems in optimal control. PhD thesis, Harvard University, Cambridge, 1965

Z. Tan, P.M. Bainum, Optimal linear quadratic Gaussian digital control of an orbiting tethered antenna/reflector system. J. Guid. Control Dyn. **17**(2), 234–241 (1994)

B.D. Tapley, V. Szebehely, J.M. Lewallen, Trajectory optimization using regularized variables. AIAA J. **7**(6), 1010–1017 (1969)

D.-R. Taur, V. Coverstone-Carroll, J.E. Prussing, Optimal impulsive time-fixed orbital rendezvous and interception with path constraints. J. Guid. Control Dyn. **18**(1), 54–60 (1995)

J.D. Thorne, C.D. Hall, Minimum-time continuous-thrust orbit transfers using the Kustaanheimo-Stiefel transformation. J. Guid. Control Dyn. (Engineering Note) **20**(4), 836–838 (1997)

S.G. Tragesser, A. Tuncay, Orbital design of Earth-oriented tethered satellite formations, in *AIAA/AAS Astrodynamics Conference*, Monterey, 5–8 Aug 2002. AIAA Paper No. 2002-4641

S.G. Tragesser, J.M. Longuski, J. Puig-Suari, J.P. Mechalas, Analysis of the optimal mass problem for aerobraking tethers, in *AIAA/AAS Astrodynamics Conference*, Scottsdale, 1–3 Aug 1994. AIAA Paper No. 94-3747

S.G. Tragesser, J.M. Longuski, J. Puig-Suari, Global minimum mass for aerobraking tethers. J. Guid. Control Dyn. (Engineering Note) **20**(6), 1260–1262 (1997)

E. Trélat, Optimal control and applications to aerospace: some results and challenges. J. Optim. Theory Appl. **154**(3), 713–758 (2012)

P. Tsiotras, Optimal regulation and passivity results for axisymmetric rigid bodies using two controls. J. Guid. Control Dyn. **20**(3), 457–463 (1997)

P. Tsiotras, H.J. Kelley, Goddard problem with constrained time of flight. J. Guid. Control Dyn. **15**(2), 289–296 (1992)

S.R. Vadali, J.L. Junkins, Spacecraft large angle rotational maneuvers with optimal momentum transfer, in *AIAA/AAS Astrodynamics Conference*, San Diego, 9–11 Aug 1982. AIAA Paper No. 82-1469

S.R. Vadali, R. Sharma, Optimal finite-time feedback controllers for nonlinear systems with terminal constraints. J. Guid. Control Dyn. **29**(4), 921–928 (2006)

S.R. Vadali, R. Nah, E. Braden, I.L. Johnson Jr., Fuel-optimal planar Earth-Mars trajectories using low-thrust exhaust-modulated propulsion. J. Guid. Control Dyn. **23**(3), 476–482 (2000)

M. Vasile, E. Minisci, M. Locatelli, On testing global optimization algorithms for space trajectory design, in *AIAA/AAS Astrodynamics Conference*, Honolulu, 18–21 Aug 2008. AIAA Paper No. 2008-6277

M.A. Vavrina, K.C. Howell, Global low-thrust trajectory optimization through hybridization of a genetic algorithm and a direct method, in *AIAA/AAS Astrodynamics Conference*, Honolulu, 18–21 Aug 2008. AIAA Paper No. 2008-6614

N.X. Vinh, Cuspidal point on the primer locus. AIAA J. **9**(11), 2239–2244 (1971)

N.X. Vinh, Minimum fuel rocket maneuvers in horizontal flight. AIAA J. **11**(2), 165–169 (1973a)

N.X. Vinh, Integrals of the motion for optimal trajectories in atmospheric flight. AIAA J. **11**(5), 700–703 (1973b)

N.X. Vinh, Optimal control of orbital transfer vehicles, in *AIAA Atmospheric Flight Mechanics Conference*, Gatlinburg, 15–17 Aug 1983. AIAA Paper No. 83-2092

N.X. Vinh, D.-M. Ma, Optimal plane change by low aerodynamic forces, in *AIAA Atmospheric Flight Mechanics Conference*, Portland, 20–22 Aug 1990. AIAA Paper No. 90-2831

N.X. Vinh, C.-Y. Yang, J.-S. Chern, Optimal trajectories for maximum endurance gliding in a horizontal plane. J. Guid. Control Dyn. **7**(2), 246–248 (1984)

N.X. Vinh, J.R. Johannesen, K.D. Mease, J.M. Hanson, Explicit guidance of drag-modulated aeroassisted transfer between elliptical orbits. J. Guid. Control Dyn. **9**(3), 274–280 (1986)

N.X. Vinh, P.T. Kabamba, T. Takehira, Optimal interception of a maneuvering long range missile, in *AIAA/AAS Astrodynamics Conference*, Boston, 10–12 Aug 1998. AIAA Paper No. 98-4547

B.J. Wall, B.A. Conway, Shape based approach to low-thrust rendezvous trajectory design. J. Guid. Control Dyn. **32**(1), 95–101 (2009)

P.K.C. Wang, F.Y. Hadaegh, Optimal formation reconfiguration for multiple spacecraft, in *AIAA Guidance, Navigation, and Control Conference*, Boston, 10–12 Aug 1998. AIAA Paper No. 98-4226

L.J. Wellnitz, J.E. Prussing, Optimal trajectories for time-constrained rendezvous between arbitrary conic orbits, in *AAS/AIAA Astrodynamics Conference*, Kalispell, 10–13 Aug 1987. AAS Paper No. 87-539

J.R. Wertz, Rapid interplanetary round trips at moderate energy, in *International Astronautics Federation Congress*, Vancouver, 4–8 Oct 2004. Paper No. IAC-04-Q.2.A.11

J.R. Wertz, T.L. Mullikin, R.F. Brodsky, Reducing the cost and risk of orbit transfer. J. Spacecr. Rockets **25**(1), 75–80 (1988)

G. Whiffen, Mystic: implementation of the static dynamic optimal control algorithm for high-fidelity, low-thrust trajectory design, in *AIAA/AAS Astrodynamics Conference*, Keystone, 21–24 Aug 2006. Paper No. 06-2356

B. Wie, C.-H. Chuang, J. Sunkel, Minimum-time pointing control of a two link manipulator. J. Guid. Control Dyn. **13**(5), 867–873 (1990)

B. Wie, R. Sinha, J. Sunkel, K. Cox, Robust fuel- and time-optimal control of uncertain flexible space structures, in *AIAA Guidance, Navigation, and Control Conference*, Monterey, 11–13 Aug 1993. AIAA Paper No. 93-3804

P. Williams, Optimal control of a tethered payload capture maneuver, in *41st AIAA/ASME/SAE/ASEE Joint Propulsion Conference*, Tucson, 10–13 July 2005. AIAA Paper No. 2005-4114

P. Williams, Optimal deployment/retrieval of a tethered formation spinning in the orbital plane. J. Spacecr. Rockets **43**(3), 638–650 (2006a)

P. Williams, Optimal deployment and offset control for a spinning flexible tethered formation, in *AIAA Guidance, Navigation, and Control Conference*, Keystone, 21–24 Aug 2006b. AIAA Paper No. 2006-6041

S.N. Williams, V. Coverstone-Carroll, Mars missions using solar electric propulsion. J. Spacecr. Rockets **37**(1), 71–77 (2000)

E.A. Williams, W.A. Crossley, Empirically-derived population size and mutation rate guidelines for a genetic algorithm with uniform crossover. in *Soft Computing in Engineering Design and Manufacturing*, ed. by P.K. Chawdhry, R. Roy, R.K. Pant (Springer, London/New York, 1998), pp. 163–172

P. Williams, P. Trivailo, On the optimal deployment and retrieval of tethered satellites, in *41st AIAA/ASME/SAE/ASEE Joint Propulsion Conference*, Tucson, 10–13 July 2005. AIAA Paper No. 2005-4291

P. Williams, C. Blanksby, P. Trivailo, Tethered planetary capture maneuvers. J. Spacecr. Rockets **41**(4), 603–613 (2004)

R.S. Wilson, K.C. Howell, M. Lo, Optimization of insertion cost for transfer trajectories to libration point orbits, in *AIAA/AAS Astrodynamics Conference*, Girdwood, 16–19 Aug 1999. AAS Paper No. 99-041

B. Woo, V.L. Coverstone, M. Cupples, Low-thrust trajectory optimization procedure for gravity-assist, outer-planet missions. J. Spacecr. Rockets **43**(1), 121–129 (2006)

L.J. Wood, Perturbation guidance for minimum time flight paths of spacecraft, in *AIAA/AAS Astrodynamics Conference*, Palo Alto, 11–12 Sept 1972. AIAA Paper No. 72-915

L.J. Wood, Second-order optimality conditions for the Bolza problem with both endpoints variable. J. Aircr. **11**(4), 212–221 (1974)

L.J. Wood, A.E. Bryson Jr., Second-order optimality conditions for variable end time terminal control problems. AIAA J. **11**(9), 1241–1246 (1973)

C.H. Yam, J.M. Longuski, Reduced parameterization for optimization of low-thrust gravity-assist trajectories: case studies, in *AIAA/AAS Astrodynamics Conference*, Keystone, 21–24 Aug 2006. AIAA Paper No. 2006-6744

H. Yan, K.T. Alfriend, S.R. Vadali, P. Sengupta, Optimal design of satellite formation relative motion orbits using least-squares methods. J. Guid. Control Dyn. **32**(2), 599–604 (2009)

S. Zimmer, C. Ocampo, Use of analytical gradients to calculate optimal gravity-assist trajectories. J. Guid. Control Dyn. **28**(2), 324–332 (2005a)

S. Zimmer, C. Ocampo, Analytical gradients for gravity-assist trajectories using constant specific impulse engines. J. Guid. Control Dyn. **28**(4), 753–760 (2005b)

Index

J.M. Longuski et al., *Optimal Control with Aerospace Applications*,
Space Technology Library 32, DOI 10.1007/978-1-4614-8945-0,

Printed in the United States
By Bookmasters